Industrial Applications
for Microprocessors

Industrial Applications for Microprocessors

A. D. Steckhahn

J. Den Otter

Reston Publishing Company, Inc., Reston, Virginia
A Prentice-Hall Company

Library of Congress Cataloging in Publication Data

Steckhahn, A. D.
  Industrial applications for microprocessors.

  Bibliography: p.
  Includes index.
  1. Electric motors—Automatic control.   2. Microprocessors.
I. Den Otter, J.   II. Title.
TK2851.S858   629.8′043   82-5407
ISBN 0-8359-3067-X   AACR2

© 1982 by
Reston Publishing Company, Inc.
A Prentice-Hall Company
Reston, Virginia

10 9 8 7 6 5 4 3 2 1

Printed in the United States of America

Contents

Preface

This text has been written to assist technology students, technologists, and engineers in the art of microprocessor-controlled control systems. The various aspects of building, programming, and interfacing of a processor for motor control are discussed. No text, however, can cover every conceivable problem that exists, but if the techniques that have been illustrated are studied and applied, it is felt that the difficult road that lies ahead of those entering the field of processor control will be made much easier to travel.

Most of the material presented in this text has been developed and taught over the past three years by the authors at the Northern Alberta Institute of Technology in Edmonton, both during the full-time day classes of Electrical Engineering Technology and in the evening programs, to people from industry. The area surrounding Edmonton has a large petrochemical industry, which makes use of the most modern technological developments in the area of microprocessors and microprocessor-based programmable controllers.

The text concentrates on the MC 6800 and MC 6805 E2 microprocessors and the Modicon 484 programmable controller. All the examples presented are industrial problems based upon our experience with students from industry or our experience directly with industry. The text has many worked out tutorial examples as it is felt by the authors that this is the way to learn this technology. The text starts with the basic numbering systems, that is, decimal, binary, octal, binary-coded decimal, and hexadecimal, and the conversion from one number system to another.

A topic is not necessarily completely developed at its first introduction, since this can be too complex and may lead to confusion; however, enough information is presented at the time that the topic is introduced to give the reader sufficient understanding to proceed to the next section.

Chapter 2 is an introduction to the design of motor-control systems via logic gates and *JK* flip-flops. It commences with simple series and/or parallel control systems, covers the design of stop-start motor-control circuits, and ends with state diagrams and state reduction techniques. This chapter is essential for the design of the logic interfaces required for decoding addresses and routing control signals in microprocessor applications.

The importance of fail-safe is introduced, and if reference to this topic becomes boring in this and later chapters, the authors apologize; but safety in industrial applications cannot be overemphasized. The older industrial applications used magnet-actuated devices, which were inherently fail-safe, but solid-state devices can and do fail; thus every effort must be made to incorporate fail-safe techniques and methods of stopping the process must be considered. Control of the plant must always be maintained.

Chapter 3 introduces the reader to the MC 6800 microprocessor and its peripherals, such as RAM and ROM. A timing diagram illustrating the method of operation of a microprocessor is introduced; also details of the PIA, the parallel data interface, the ACIA, the serial data interface, and the PTM, a programmable timer module, are introduced, as well as the processor's addressing modes. Details of how to read the instruction set are given, as well as simple programming examples. More complete details of the index register and its applications will be found in later chapters.

Chapter 4 introduces the reader to structured programming, with many examples of its use. This technology is essential for the development of workable, error-free programs. Once again, numerous worked out examples are presented.

Chapter 5 describes the MC 6800 in more detail and also describes how to build a microprocessor-controlled system. The chapter is basic from the point of view that the system developed is a minimum system and that the program executed is very basic; it merely flashes an LED. But once the reader understands this, more complex tasks are just around the corner. Many students believe that building a microprocessor-controlled control system is out of their reach; this is not so, as this chapter illustrates.

Chapter 6 covers the MC 6805 E2 microprocessor chip; this device was selected since it is virtually a stand-alone chip. The circuit to interface the processor to a CRT is illustrated along with the start-up program. This chapter illustrates the ease with which plant maintenance personnel can build a microprocessor-based process control system for their particular application.

Chapter 7 is a chapter of worked-out examples; they are more complex than those presented in the preceding chapters and are applied to both processors, the 6800 and the 6805. The selection is by no means exhaustive, and the examples are selected to illustrate to the reader that complex programs can be resolved into relatively simple problems. Failsafe is always a factor incorporated in all designs. Examples to interface a CRT for communication with the processor are included.

Chapter 8 introduces the reader to the programmable controller, a microprocessor-based system, which includes circuitry to permit control logic to be entered in a form similar to a relay ladder diagram. These controllers were introduced in the late 1960s in the automobile industry to avoid the costly and time consuming rewiring of control systems at model changeover. They have evolved rapidly with the advent of more sophisticated solid-state devices and are now found in every industry.

Chapter 9 is a further development of the programmable controller; it covers the actual interface with the MC 6800 microprocessor. This processor or a computer can be used for master control of several controllers and for storage of important data.

Chapter 10 deals with the very important concept of interfacing controllers and microprocessors. Without a proper understanding of this topic, no reliable process control is possible. This chapter therefore is pertinent to all other chapters of this text, although it can be studied in isolation. Information about solid state input and output circuits and noise considerations has been supplied by Allen Bradley; see Bulletin 1772-820-1.

Although this text concentrates on the MC 6800 and MC6805 processors and the 484 controller, the material in this book can be applied to MC 6802D5, as well, and to the Allen-Bradley PLC.

The authors wish to extend their thanks to Motorola, Inc. and Gould-Modicon Division for their cooperation and courtesy in the preparation of this text, and also to one another for the cooperation that made the task very pleasant. We thank our wives Mary and Sina for their forbearance, understanding, and help in the preparation of this text.

A. D. STECKHAHN
J. DEN OTTER

Chapter 1

Numbering Systems

The decimal numbering system is an integral part of our everyday life; our currency is decimal and decimalization or metrication is rapidly becoming a household word. It is therefore difficult to imagine counting and calculations being performed by some other number system or systems; but actually this is what is happening in all digital computers, microprocessors and the like. These devices all perform their daily tasks by a numbering system called the *binary system*. Therefore, this system and its related forms, that is, the binary coded decimal (BCD), the octal, and the hexadecimal systems, will all be examined in this chapter, as well as binary multiplication and addition. Only integers will be considered throughout this text.

1.1 DECIMAL SYSTEM

The decimal system is based on 10 symbols or digits: 0, 1, 2, 3, 4, 5, 6, 7, 8, 9. Any number can be represented by an ordered set of these, for example, 5053_{10} (the subscript indicates the base or radix of the system). This number is interpreted as follows:

$$
\begin{array}{llll}
10^3 & 10^2 & 10^1 & 10^0 \\
1000 & 100 & 10 & 1 \\
5 & 0 & 5 & 3 = 5 \times 1000 + 0 \times 100 + 5 \times 10 + 3 \times 1
\end{array}
\tag{1.1}
$$

Notice that the numerical *weight* of each digit increases by a constant factor (10) as the set of digits is traversed from right to left. All the number systems considered are based on this *position-weighting* system.

1.2 BINARY NUMBER SYSTEM

The binary system is based on two digits, 0 and 1. Any number can be represented by these digits, for example, 1101_2; this number may be interpreted as

$$
\begin{array}{llll}
2^3 & 2^2 & 2^1 & 2^0 \\
8 & 4 & 2 & 1 \\
1 & 1 & 0 & 1 & = 1 \times 8 + 1 \times 4 + 0 \times 2 + 1 \times 1 = 13_{10}
\end{array}
\tag{1.2}
$$

Notice that the numerical weight of each digit increases by a constant factor of 2 as the number is traversed from right to left. Therefore, the number 13_{10} is the decimal equivalent of 1101_2.

1.3 OCTAL NUMBER SYSTEM

The octal number system is based on 8 digits, 0, 1, 2, 3, 4, 5, 6, 7. Any number can be represented by these 8 digits, for example, 675_8; this number may be considered as

$$
\begin{array}{lll}
8^2 & 8^1 & 8^0 \\
64 & 8 & 1 \\
6 & 7 & 5 & = 6 \times 64 + 7 \times 8 + 5 \times 1 = 445_{10}
\end{array}
\tag{1.3}
$$

Notice that the numerical weight increases in a manner similar to the previous examples; furthermore, the conversion to decimal form was easily achieved and illustrated that $445_{10} = 675_8$.

The octal form of the binary number is realized as follows:

4 2 1	4 2 1	4 2 1	Binary weighting	(1.4)
1 1 0	1 1 1	1 0 1	Binary representation	
6	7	5	Octal representation	

Thus $675_8 = 445_{10} = 110111101_2$.

1.4   BINARY CODED DECIMAL (BCD) SYSTEM

The BCD system is a binary system based on a 8421 weighting and any decimal digit can be represented by these four digits; for example,

8421	8421	8421	Weighting	(1.5)
1001	0111	0000	BCD presentation	
9	7	0	Decimal presentation	

Thus, $970_{10} = 1001\ 0111\ 0000$ in BCD notation.

The disadvantage with the BCD system is that its largest digit is 1001 and thus the following codes are wasted:

$$
\begin{array}{ccc}
Binary & & Decimal \\
1010 & = & 10 \\
1011 & = & 11 \\
1100 & = & 12 \\
1101 & = & 13 \\
1110 & = & 14 \\
1111 & = & 15
\end{array}
\qquad (1.6)
$$

However, this system is used extensively with digital meters.

1.5   HEXADECIMAL NUMBER SYSTEM

The hexadecimal system is a system based on 16 digits and/or symbols, 0, 1, 2, 3, 4, 5, 6, 7, 8, 9, A, B, C, D, E, F. The system utilizes all the combinations of the 8421 weighted system in the following way. Digits 0 to 9 assume the binary numbers 0000 to 1001, respectively; the remaining binary combinations now take on alphabetic characters.

$$
\begin{array}{ccc}
Decimal & 8421\ Code & Hexadecimal\ Code \\
10 & 1010 & A \\
11 & 1011 & B \\
12 & 1100 & C \\
13 & 1101 & D \\
14 & 1110 & E \\
15 & 1111 & F
\end{array}
\qquad (1.7)
$$

A number represented in hexadecimal notation could appear in the following form: $30F6_{16}$. This number can be converted into binary and decimal notation in a manner similar to the previous examples.

Binary Conversion

8421	8421	8421	8421	Weighting
0011	0000	1111	0110	Binary representation
3	0	F	6	Hex representation

$$(1.8)$$

Thus $30F6_{16} = 0011\ 0000\ \ 1111\ 0110_2$.

Decimal Conversion

$$(1.9)$$

$16^3 \quad 16^2 \quad 16^1 \quad 16^0$
$4096 \quad 256 \quad 16 \quad 1$
$\ \ 3 \quad \ \ 0 \quad \ \ F \quad \ \ 6 = 3 \times 4096 + 0 \times 256 + F \times 16 + 6 \times 1 = 12534_{10}$
$\qquad\qquad\qquad$ (remember, $F = 15$)

Thus, $12534_{10} = 30F6_{16} = 0011\ 0000\ 1111\ 0110_2$.

The binary number systems just presented form the basis for all calculations performed in computers and microprocessors, and it is important that the reader can relate to all the systems and be able to convert from one to another.

The method of converting to decimal notation has been demonstrated for each number system; converting from decimal to binary and so on is just as straightforward and will now be presented.

1.6 DECIMAL CONVERSION

The conversion of a decimal number to a binary number or a binary-related number system can be performed as shown in Equation (1.10).

Example 1. Convert 25_{10} to binary form. Divide the decimal number by 2 and record the remainders as shown:

$$(1.10)$$

$\frac{25}{2} = 12 +$ remainder of 1

$\frac{12}{2} = \ \ 6 +$ remainder of 0

$\frac{6}{2} = \ \ 3 +$ remainder of 0

$\frac{3}{2} = \ \ 1 +$ remainder of 1

$\frac{1}{2} = \ \ 0 +$ remainder of 1 $\qquad\qquad$ least significant digit

$\qquad\qquad\qquad\qquad\qquad 1\ \ 1\ \ 0\ \ 0\ \ 1$

which can be written in 8-bit form as 00011001_2. Therefore, $25_{10} = 00011001_2$.

The reader should now convert this binary number back to the original decimal number as an exercise. Similarly, we convert from decimal to octal code as follows.

Example 2. Convert 673_{10} to octal form.

Divide the decimal number by 8 and record the remainders as follows:

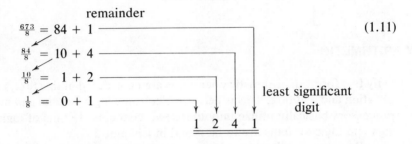

$$(1.11)$$

The resulting octal number is 1241.

The method of converting a decimal number to a hexadecimal number is the same as that shown in Examples 1 and 2, except that the divisor is 16.

Example 3. Convert 227_{10} to a hexadecimal number.

remainder

$$\frac{227}{16} = 14 + 3 \qquad\qquad\qquad (1.12)$$

$$\frac{14}{16} = 0 + 14$$

$$\underline{E \quad 3} \quad \text{least significant digit}$$

The resulting hex number is $E3_{16}$, and it is suggested that the reader convert this result and that of Example 2 back to the decimal form as an exercise.

Quite naturally the reader will ask, why do microprocessors work in the binary notation? It is because the binary system can represent any system that has only two states, such as an electrical signal, which is either present, say $+5$ volts (V) with respect to ground or common rail, or it is absent, say 0 volts. Therefore, if the voltage on a line is high, $+5$ V to ground, the line is said to be at binary 1; on the other hand, if the line is 0 V to ground, the line is at binary 0. Five volts was chosen as the high-level voltage because microprocessors usually operate on a 5 V supply;

thus the two levels of an electrical signal are relatively easy to detect. But at what point between 0 and 5 V does the 0 become 1? The following is the generally agreed standard:

$$0 < \text{binary } 0 < 1.5 \text{ V} \qquad (1.13)$$
$$3.5 \text{ V} < \text{binary } 1 \le 5 \text{ V}$$

The voltage between 1.5 and 3.5 is a gray area, and the response of the electronic devices that make up a microprocessor are unpredictable in this range and misoperation could result.

1.7 BINARY ARITHMETIC

Only two forms of arithmetic operations are considered in this text, multiplication and addition. This is because electronic logic circuits and microprocessors basically utilize only these two concepts. Details of logic design and logic systems are to be found in Chapter 2.

Binary multiplication is identical to that of ordinary arithmetic:

$$1.0 = 1 \times 0 = 0 \qquad (1.14)$$
$$0.1 = 0 \times 1 = 0$$
$$0.0 = 0 \times 0 = 0$$
$$1.1 = 1 \times 1 = 1$$

Binary addition, as used in binary arithmetic differs from that of ordinary arithmetic in just one case, as shown:

$$1 + 1 = 10 = 0 \quad \text{(with a carry of 1 into the next position)} \quad (1.15)$$
$$1 + 0 = 1$$
$$0 + 1 = 1$$
$$0 + 0 = 0$$

It was previously stated that only multiplication and addition are utilized by microprocessors. How then can two numbers be subtracted? The answer is by a mathematical process known as the 2's complement method.

A binary number is converted into a 2's complement number as follows:

Example 4. Convert 00001011 into its 2's complement form.

Step 1. Form a 1's complement of the original number; that is, change the 1's to 0's and 0's to 1's, as follows:

Original No.	1's Complement	
00001011	11110100	(1.16)

Step 2. Add 1 at the least significant position of the 1's complement form; that is,

Original No.	1's Complement	2's Complement
00001011	11110100	11110100 + 1 = 11110101

Thus, 11110101 is the 2's complement of the number 00001011.

Binary subtraction by the 2's complement method is illustrated in Example 5.

It must be noted that when two numbers are being subtracted by the 2's complement form the subtrahend and the minuend *must* have the same number of binary digits. If one number has less than the other, the smaller number must be made up to the length of the larger by adding zero's at the left *before 2's complement conversion takes place;* otherwise, the result may be in error.

Example 5. Subtract 7 from 11.

Decimal	Binary		
11	00001011	(minuend)	(1.17)
− 7	00000111	(subtrahend)	
4		result	

Step 1. Convert the subtrahend into 2's complement form.

Original No.	1's Complement	2's Complement
00000111	11111000	11111000 + 1 = 11111001

Step 2. Add the 2's complement form of the subtrahend to the minuend by the method of addition previously shown.

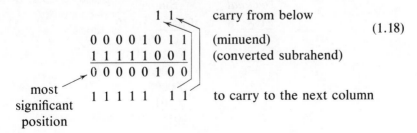

The 0 in the most significant position indicates that the result is positive; therefore $+00000100_2 = 4_{10}$.

Negative numbers obey a similar set of rules, and Example 6 will be used to illustrate this.

Example 6. Subtract 12 from 5.

Step 1. Convert the binary form of 12 into 2's complement form.

Original No.	*1's Complement*	*2's Complement*
00001100	11110011	$11110011 + 1 = 11110100 = -12$

Step 2. Add the 2's complement of the subtrahend to the minuend.

$$
\begin{array}{cccccccc}
 & & & & & 1 & & & \text{carry from below} \qquad (1.19)\\
0 & 0 & 0 & 0 & 0 & 1 & 0 & 1 & \text{(minuend)}\\
1 & 1 & 1 & 1 & 0 & 1 & 0 & 0 & \text{(converted subtrahend)}\\
\hline
1 & 1 & 1 & 1 & 1 & 0 & 0 & 1 & \\
 & & & & & 1 & & & \text{1 to carry to the next column}
\end{array}
$$

The 1 in the most significant position indicates that the result is negative and is in 2's complement form. Therefore, to obtain the correct answer, the 2's complement of the result must be taken.

Original No.	*1's Complement*	*2's Complement*
11111001	00000110	$00000110 + 1 = 00000111 = 7_{10}$

Therefore, the answer is 7_{10} and it is negative. Thus, the final answer is -7_{10}.

Multiplication is performed in the same manner as in ordinary arithmetic, that is, multiply, shift and add.

Example 7. Multiply 21 by 18.

$$
\begin{array}{ll}
\textit{Decimal} & \textit{Binary}\\
\underline{21 \times 18} & \underline{10101 \times 10010} \qquad (1.20)\\
168 & 00000\\
\underline{21} & 10101\\
378 & 00000\\
 & 00000\\
 & \underline{10101}\\
 & 101111010
\end{array}
$$

The reader should convert the binary number back to decimal form to confirm the result.

1.8 SIGNED AND UNSIGNED BINARY NUMBERS

Binary arithmetic is performed in the arithmetic logic unit (ALU) of the microprocessor and utilizes only 8-digit binary numbers (commonly called a byte). It can perform arithmetic operations on what are known as signed and unsigned numbers and 2's complement representation. An 8-digit number can be represented as

$$
\begin{array}{cccccccc}
2^7 & 2^6 & 2^5 & 2^4 & 2^3 & 2^2 & 2^1 & 2^0 \\
\hline
128 & 64 & 32 & 16 & 8 & 4 & 2 & 1 \quad \text{Binary weighting} \\
\hline
b7 & b6 & b5 & b4 & b3 & b2 & b1 & b0 \quad \text{Bit positions}
\end{array}
$$

(1.21)

where b = bit, which means binary digit.

If the number is a signed number or a 2's complement representation, then b7 is used as a sign bit, and not a weighting bit, and will illustrate whether the number represented by the remaining digits b0 to b6 is positive or negative. For example, Table 1.1 illustrates some typical values of 2's complement representation.

Table 1.1 Binary Numbers in 2's Complement Representation

b7	b6	b5	b4	b3	b2	b1	b2	Decimal Number
1	0	0	0	0	0	0	0	−128
1	0	0	0	0	0	0	1	−127
1	1	1	1	1	1	1	1	−1
0	0	0	0	0	0	0	0	0
0	0	0	0	0	0	0	1	+1
0	1	1	1	1	1	1	1	+127

Note. If b7 = 1, the number is negative, and if b7 = 0, the number is positive; furthermore, the number 0 is treated as positive. Therefore, a byte can represent any number in the range − 128 to + 127.

Example 8. Determine the 8-bit 2's complement representation of -63_{10}.

Step 1. Convert 63_{10} to binary format.

$$63_{10} = 00111111_2 \quad \text{(in 8-bit form)}$$

Step 2. Convert 00111111 to its 2's complement of 00111111_2.

$$11000001_2 = 2\text{'s complement of } 00111111_2 \qquad (1.22)$$

$$\therefore \quad -63_{10} = 11000001_2$$

The reader should now verify that -128_{10}, -127_{10}, and -1_{10} are as illustrated in Table 1.1.

If the number is an unsigned number, b7 is considered as a weighted bit, that is, $2^7 = 128$, and then the binary representation has a range from 0 to $+255$. Table 1.2 illustrates some typical values in the range.

Table 1.2 Binary Numbers in Unsigned Form

128	64	32	16	8	4	2	1	*Binary Weighting*
b7	*b6*	*b5*	*b4*	*b3*	*b2*	*b1*	*b0*	*Decimal Number*
0	0	0	0	0	0	0	0	0
0	0	0	0	0	0	0	1	1
0	1	0	0	0	0	0	0	64
1	0	0	0	0	0	0	0	128
1	1	1	1	1	1	1	1	255

Eight-bit signed numbers will be considered in Appendix A and can be studied as and when they occur and are required. The ALU of the microprocessor utilizes the concept of 2's complement numbers in the operation of subtraction, similar to worked Examples 5 and 6; but the capacity of this device is limited to 8 bits, and if bit 7 is the sign bit, then it is possible that this bit is in error (i.e., it indicates the incorrect sign). The sign bit will be correct if

$$-128 \leq \text{true result} \leq +127$$

Thus the examples offered in this chapter will be restricted to this range; examples outside of the range will be considered in Chapter 3.

1.9 BINARY-CODED DECIMAL

An 8-bit number can also be used to represent one or two BCD numbers; for example,

8	4	2	1	8	4	2	1	Weighting
b7	b6	b5	b4	b3	b2	b1	b0	Digit number
0	0	0	0	1	0	0	0	Binary representation
		0				8		Decimal numbers

A single BCD number can be represented by the lower 4 digits, as shown. Similarly, a 2-digit BCD number can be represented by utilizing the upper and lower 4 binary digits. To assist the reader in relating to the binary system and its related forms, a series of worked examples will now be offered.

1.10 TUTORIAL EXAMPLES

Example 1.   Convert FF_{16} to octal code.

Step 1.   Write out FF_{16} in its binary form.

$$FF_{16} = 1111\ 1111$$

Step 2.   Partition the binary number into its octal format.

$$1\ 1 \quad 1\ 1\ 1 \quad 1\ 1\ 1$$

Step 3.   Convert the binary octal format to octal numbers.

$$\begin{array}{ccc} 1\ 1 & 1\ 1\ 1 & 1\ 1\ 1 \\ 3 & 7 & 7 \end{array} \qquad \therefore\ FF_{16} = 377_8$$

Example 2.   Subtract 65 from 37 by 2's complement addition (the answer is in the range from -128 to $+127$).

Step 1.   Convert 65 and 37 into 8-bit binary form.

$$+37 = 00100101 \quad \text{(minuend)}$$
$$+65 = 01000001 \quad \text{(subtrahend)}$$

Step 2.   Change the subtrahend into 2's complement form.

Decimal	*Binary*	*2's Complement*
$+65$	01000001	10111111

Step 3. Add the 2's complement of the subtrahend to the minuend.

Binary

```
0  0  1  0  0  1  0  1
1  0  1  1  1  1  1  1      2's complement
1  1  1  0  0  1  0  0
```

Step 4. Since b7 = 1, the result is in 2's complement form and negative. Take the 2's complement of the result and precede it by a negative sign.

Result	*2's Complement*
11100100	00011100 = 28

Answer = -28_{10}

Example 3. Multiply 1001 × 0110; express the result in 8-bit form.

```
 1001 × 0110
 0000
 1001
 1001
0000
0110110 = 00110110    (include a 0 on the left-hand end of the number
                        to make it 8 bits long)
```

Answer = 00110110

Example 4. Convert the BCD number 001101111001 to its decimal equivalent.

0011	0111	1001
3	7	9

Example 5. Convert the hexadecimal number 8C4 to its binary equivalent.

8	C	4
1000	1100	0100

Problems

1-1. Convert the following decimal numbers to binary: 531, 29, 87, 1063, 35.

1-2. Convert the following binary numbers to decimal: 00001111, 00101101, 10101010, 11100011, 00110011.

1-3. Convert the following binary numbers to octal form: 11001100, 11100011, 00011100, 10101010, 01010101.

1-4. Subtract -35_{10} from $+83_{10}$ by the 2's complement technique.

1-5. Subtract 37_{10} from 96_{10} by the 2's complement technique.

1-6. Complete the following table:

Decimal	Binary	Octal	Hexadecimal
63			
	10110110		
		29	
			AD

Chapter 2

Logic Design

Logic design in its present state is basically the result of the efforts of many individuals: E. C. Shannon, S. H. Caldwell, W. V. Quine, E. J. McCluskey, D. Zissos, and many others too numerous to mention. Design techniques have advanced from the methods of cut and try in the pre-Shannon era, to the present-day advanced methods of Boolean equations and the reduction techniques of Zissos. Thus the students of today can use a set of very simple and clear instructions to design complex circuits, which greatly increases the knowledge input per unit of instruction time.

This chapter will briefly explain the Zissos method and, by a series of examples, will demonstrate the technique.[1]

2.1 INTRODUCTION

Present-day techniques are based on the premise that everything is either this or that; that is, only two states exist. A contact is either opened or closed, an electric motor is either energized or deenergized, a lamp is either on or off, everything is either black or white. There are no shades of gray.

To put these concepts into mathematical terms, binary numbers have been used. A 0 (zero) is used to mathematically represent an open

[1] For more complete details, see D. Zissos, *Logic Design Algorithms*, Oxford University Press, New York, 1972.

contact, a deenergized motor or relay, an extinguished lamp. On the other hand, a 1 (one) is used to mathematically represent a closed contact, an energized motor or relay, a lamp that is lit. Each contact, relay, motor, and so on, is given an identity, usually a capital letter, and since a contact is either normally open (N.O.) or normally closed (N.C.), the method of identification of Figure 2.1 is used. The bar over the identification letter signifies that the contact is *not* open, and since only two states exist, it follows that if the contact is not open, it must be closed. The identification letters, A, B, C, and so on, are called true literals, or variables, whereas $\overline{A}, \overline{B}, \overline{C}$, and so on, are called inverted literals or the complemented form of a true literal. Since the complement of open is closed and the complement of closed is open, then the complement of 1 is 0, and vice versa.

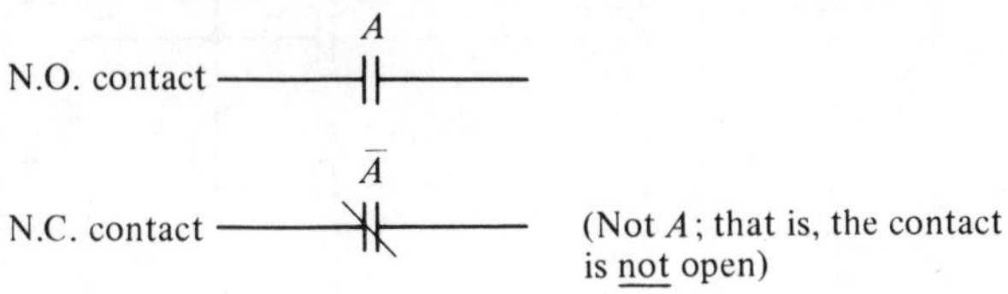

Figure 2.1

N.O. contact ——————A———————

N.C. contact ——————$\overline{A}$——————— (Not A; that is, the contact is <u>not</u> open)

2.2 AND, OR, NOT

When George Boole wrote his classic work, *The Laws of Thought,* he proposed three basic logical concepts, AND, OR, and NOT:

$$A \text{ AND } B \quad (A \cdot B = AB) \qquad (2.1)$$

Mathematically, the dot indicates multiplication and it can be omitted without loss of clarity.

$$A \text{ OR } B \quad (A + B) \qquad (2.2)$$

Mathematically the plus sign indicates addition.

$$\text{NOT } A \quad (\overline{A}) \qquad (2.3)$$

If a motor is *not* energized, it follows that it is deenergized.

In electrical systems, there are only two types of electrical circuits, series and parallel. Series circuits obey the logical concept AND (see Figure 2.2). The lamp will only be lit when contacts A and B and C are closed; written as a Boolean expression,

$$L = ABC \qquad (2.4)$$

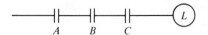

Figure 2.2

Parallel circuits obey the logical concept OR (see Figure 2.3). The lamp will be lit when contacts A or B or C are closed; written as a Boolean expression,

$$L = A + B + C \tag{2.5}$$

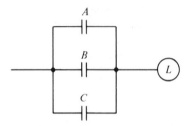

Figure 2.3

Combinations of both can also occur as shown for the simple motor start/stop circuit of Figure 2.4. Written as a Boolean expression,

$$M = \overline{A}(B + M)\,\overline{0}_1 \cdot \overline{0}_2 \cdot \overline{0}_3 \tag{2.6}$$

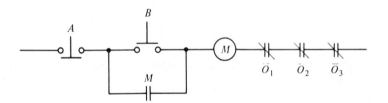

Figure 2.4

The motor magnetic starter will only be energized if the stop button is *not* depressed *and* the overloads are *not* opened *and* the start button is depressed *or* the contact M is closed.

2.3 TRUTH TABLES

To prove the validity of a Boolean expression, a truth table can be used. The table is erected as follows:

1. Write out the expression:

$$L = A + B \qquad (2.7)$$

2. Set up a table:

A	B	L = A + B

$$(2.8)$$

3. Let each input variable take its two binary values; this determines the output (e.g., $0 + 0 = 0$).

A	B	L = A + B
0	0	0
0	1	1
1	0	1
1	1	1

$$(2.9)$$

4. Only when $L = 1$ will the device be energized or active, that is, when contact A or B or both are closed, as shown by the table.

5. Repeat the procedure for $L = A \cdot B$, and the truth table will appear as follows:

A	B	L = A · B
0	0	0
0	1	0
1	0	0
1	1	1

$$(2.10)$$

Principally, truth tables are used as a means of verifying two or more Boolean expressions; that is, if similar entries in the table give identical results for each and every entry, then the expressions are the same.

For example, it is claimed that

$$AC + A\overline{B} + BC = A\overline{B} + BC \qquad (2.11)$$

To verify, set up a truth table complete with entries, and compare the table entry by entry. Columns 1, 2, and 3 constitute all possible combinations of A, B, and C.

Columns	1	2	3	4	5	6	7	8	9	
	A	B	C	$\bar{B}$	AC	$A\bar{B}$	BC	$AC + A\bar{B} + BC$	$A\bar{B} + BC$	(2.12)
	0	0	0	1	0	0	0	0	0	
	0	0	1	1	0	0	0	0	0	
	0	1	0	0	0	0	0	0	0	
	0	1	1	0	0	0	1	1	1	
	1	0	0	1	0	1	0	1	1	
	1	0	1	1	1	1	0	1	1	
	1	1	0	0	0	0	0	0	0	
	1	1	1	0	1	0	1	1	1	

Combine columns 1 and 3 to make up column 5.
Combine columns 1 and 4 to make up column 6.
Combine columns 2 and 3 to make up column 7.
Combine columns 5, 6, and 7 to make up column 8.
Combine columns 6 and 7 to make up column 9.

Output columns 8 and 9 are identical, which proves that $AC + A\bar{B} + BC = A\bar{B} + BC$. The reader should draw out the relay circuit and try by mental argument to verify that the parallel path of contacts AC is indeed redundant.

2.4 BOOLEAN POSTULATES

The symbolic and mathematical representations of the Boolean postulates are given in Figure 2.5.

2.5 BOOLEAN THEOREMS

This section will present a number of techniques that can be used in the reduction of the complexity and size of a Boolean expression. The section is divided into four parts:

1. De Morgan's theorem and dual
2. Included-term theorem
3. Optional-products theorem
4. Optional-sums theorem

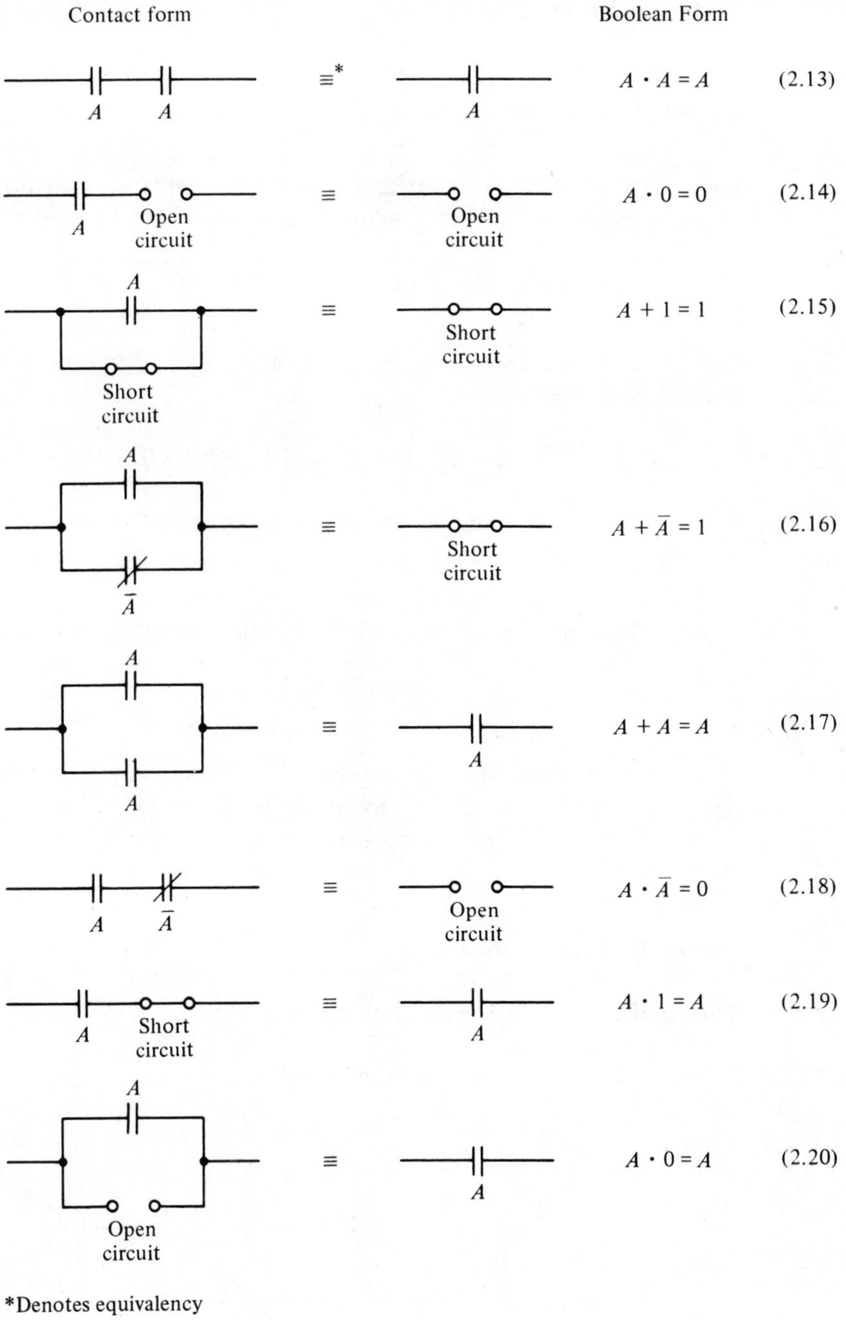

Contact form   Boolean Form

$$A \cdot A = A \qquad (2.13)$$

$$A \cdot 0 = 0 \qquad (2.14)$$

$$A + 1 = 1 \qquad (2.15)$$

$$A + \overline{A} = 1 \qquad (2.16)$$

$$A + A = A \qquad (2.17)$$

$$A \cdot \overline{A} = 0 \qquad (2.18)$$

$$A \cdot 1 = A \qquad (2.19)$$

$$A \cdot 0 = A \qquad (2.20)$$

*Denotes equivalency

Figure 2.5 Boolean Postulates

De Morgan's Theorem

The theorem developed by De Morgan is used in the complementation of OR and AND expressions and is illustrated next.

Step 1. Bracket all products and indicate all operations; also place the complementing sign over the whole expression. For example,

$$(A + \overline{B}C)D\overline{E} \quad \text{becomes} \quad \overline{[A + (\overline{B} \cdot C)] \cdot (D \cdot \overline{E})} \qquad (2.21)$$

Step 2. Complement the literals and change all operators, but do not change the brackets.

$$\overline{[A + (\overline{B} \cdot C)] \cdot (D \cdot \overline{E})} \quad \text{becomes} \quad [\overline{A} \cdot (\overline{\overline{B}} + \overline{C})] + (\overline{D} + \overline{E}) \quad (2.22)$$

Remember that the complement of the complement is the original (i.e., $\overline{\overline{B}} = B$).

Step 3. Remove unwanted brackets and the answer appears as follows:

$$\overline{A}(B + \overline{C}) + \overline{D} + E \qquad (2.23)$$

Dual. The dual of a Boolean expression is obtained by changing only the operators, that is, the $+$ and $\cdot$; for example, $AB + A\overline{B}$ becomes $(A + B) \cdot (A + \overline{B})$. This theorem can be used for reduction purposes, as will be shown later.

Included-Term Theorem

The Boolean expression for the circuit of Figure 2.6 is

$$L = A + AB = A \qquad (2.24)$$

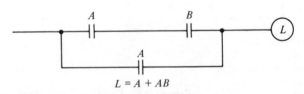

$$L = A + AB$$

Figure 2.6

The lamp will be on if contact A or contacts A and B are closed. Clearly, by inspection, it can be seen that the lamp will be on if contact A only closes. Thus contact B is redundant and serves no useful purpose. The theorem states that, if one term is totally contained in another term, then the larger term is redundant.

Example 1

$$L = AB + ABCD = AB \qquad\qquad (2.25)$$

However, if $L = ABC + ABD$, no term is totally contained in another term; therefore, the theorem is not satisfied and the Boolean statement cannot be reduced.

Optional-Products Theorem

A product is said to be optional if its presence or absence in a Boolean expression does not affect the outcome of the expression. See Figure 2.7, which is equivalent to Figure 2.8. Now from postulate (2.16)

$$(B + \overline{B}) = 1$$

and from postulate (2.19)

$$A \cdot 1 = A$$

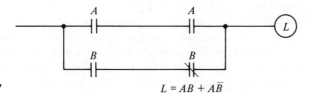

Figure 2.7 $\qquad\qquad L = AB + A\overline{B}$

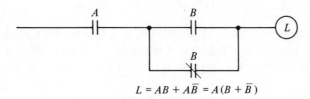

$$L = AB + A\overline{B} = A(B + \overline{B})$$

Figure 2.8

The circuit represented by $L = A(B + \overline{B})$ can be written as $L = A$, and this can be seen to be true. However, the first circuit did not show this quite so clearly, and had other contacts been included, this redundancy could have been missed.

To use this theorem, write out the equation as a sum of products, that is, as $AB + CD$, and not as a product of sums, which is of the form $(A + B) \cdot (C + D)$; then examine each product to see if the complement of any of the elements of that product is to be found in any of the other terms. From the example, $L = AB + A\overline{B}$, the term AB is compared to the next term $A\overline{B}$, and it is seen that B is in its inverted form. But A remains in the same form in both terms; therefore, the theorem may be applied.

A simple graphical technique exists to simplify the use of the theorem and at the same time acts as a reminder of what was actually done to the expression.

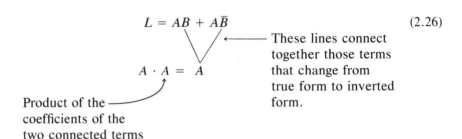

$$L = AB + A\overline{B} \tag{2.26}$$

These lines connect together those terms that change from true form to inverted form.

$A \cdot A = A$

Product of the coefficients of the two connected terms

The term created (A) by the use of the theorem is called an *optional product;* it is optional because it was created by the two parent members of the expression, and these two parents can together create the effect of the optional term. It is called a product because it was created by the product of the coefficients of the term that changed its form. Once the optional product has been created, it can be included in the original expression without affecting the outcome as an additional term, and usually it is written as follows:

$$L = (A) + AB + A\overline{B} \tag{2.27}$$

The parentheses are there to remind the user that the term is optional. By using the included-term theorem it is plainly seen that (A) is contained in AB and $A\overline{B}$ so that (A) can replace both of these terms, its parents.

$$\therefore \quad L = AB + A\overline{B} = A \tag{2.28}$$

Example 2

$$L = AB + A\bar{B}C \qquad (2.29)$$

$$AC$$

$$L = (AC) + AB + A\bar{B}C$$

Clearly, (AC) is included in term $A\bar{B}C$; thus (AC) replaces one of its parents, that is, $A\bar{B}C$.

$$\therefore \quad L = AB + AC \qquad (2.30)$$

Example 3

$$L = AB + \bar{B}C + AC \qquad (2.31)$$

$$AC$$

The term AC is included in the original expression, and AC is also created as an optional product by terms AB and $\bar{B}C$; therefore, AC may be left out of the original expression as its effect is created by the remaining terms $AB + \bar{B}C$.

$$\therefore \quad L = AB + \bar{B}C \qquad (2.32)$$

The term AC is optional *only* as long as the terms that can create it remain in the expression.

Example 4

$$L = AB + \bar{B}C + AC + B\bar{C} \qquad (2.33)$$

$$L = AB + \bar{B}C + AC + B\bar{C}$$

$$AC \qquad AB$$

This example produces two optional products, and if the optional product AC is removed from the original expression, then

$$L = AB + \bar{B}C + B\bar{C} \qquad (2.34)$$

Now it can be seen that the term AB is *no longer optional;* that is, it cannot be created by the terms that remain, $\overline{B}C$ and $B\overline{C}$. However, the term AC can still be created by AB and $\overline{B}C$ and so AC is optional. Therefore, when applying the optional products theorem, care must be taken not to get "carried away" with reduction; otherwise, the expression left may not be the equivalent of the original. The reader at this point should verify that the expression $AB + \overline{B}C + B\overline{C}$ must be equivalent to $\overline{B}C + AC + B\overline{C}$, where the optional product AB was deleted. The optional products presented thus far are called *first optionals,* and these may not always produce useful results immediately.

Example 5

$$L = AB + \overline{A}\overline{C} + C\overline{D} + \overline{B}\overline{D} \qquad (2.35)$$

No first optionals can be created that on their own will produce any reduction in the size of the complexity of the original expression. Therefore, higher optional products must be created. Going back to expression (2.35)

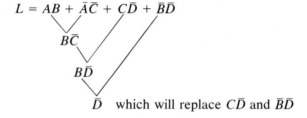

$$\therefore \quad L = AB + \overline{A}\overline{C} + \overline{D} \qquad (2.36)$$

Example 6

$$L = AB + \overline{A}\overline{B} \qquad (2.37)$$

No reduction can take place in this expression because there is a change of two variables, and the optional-products theorem can only be applied to expressions that exhibit a change of one variable, as illustrated in the examples given.

Optional-Sums Theorem

A sum is said to be optional if its presence or absence in a Boolean expression does not affect the outcome of the expression.

$$(A) \cdot (A + B) = A \qquad (2.38)$$

This theorem can be proved by multiplying the expression.

$$(A) \cdot (A + B) = AA + AB = A + AB = A \qquad (2.39)$$

This optional sum theorem can be applied to products of sums.

Example 7

$$L = (A + B)(A + \bar{B}) = A \qquad (2.40)$$
$$A$$

Example 8

$$L = (A + B)(A + \bar{B} + C) \qquad (2.41)$$
$$(A + C)$$

$(A + C)$ replaces the larger term $(A + \bar{B} + C)$.

$$\therefore \quad L = (A + B)(A + C) \qquad (2.42)$$

Example 9

$$L = (A + B)(A + \bar{B} + C) \qquad (2.43)$$

The dual of (2.43) is $AB + A\bar{B}C$, which reduces to $AB + AC$. To convert back, the dual must be applied again, and $AB + AC$ becomes $(A + B)(A + C)$.

$$\therefore \quad L = (A + B)(A + C) \quad \text{see Example 8} \qquad (2.44)$$

2.6   LOGIC GATES

Logic gates are electronic switches with one output and a number of controlling inputs. For example, a three-input gate has three controlling input lines. The various logic gates discussed in this section are shown in Figure 2.9 with their symbols, each of which will convey to the reader the type of logic operation being performed by the gate. Each of the gates presented represents a unique operation, and each will be dealt with in turn.

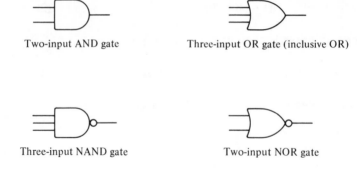

Two-input AND gate   Three-input OR gate (inclusive OR)

Three-input NAND gate   Two-input NOR gate

Two-input exclusive OR gate

Figure 2.9

Because logic gates are electronic switches, they must be supplied with good-quality direct current, typically +5 volts and ground.

AND Gate

The AND gate has a minimum of two inputs; the symbol, Boolean expression, and truth table representing a two-input AND gate are as follows:

A	B	X = AB
0	0	0
0	1	0
1	0	0
1	1	1

(2.45)

where $X = 1$ if and only if *all inputs* are 1 (similarly for three or more input gates).

Sometimes not all the input lines to a gate are utilized; for example, only two inputs of a three-input gate may be required, and the third line is left "hanging"; it is not controlled. This is not a good practice; unused leads should either be connected to another controlled lead or tied to the +5-V line supplying the gate. If a lead is left "hanging," the gate will operate as if the line were hi (1), and in some cases it acts as an antenna and produces noise.

OR Gate (Inclusive OR)

The OR gate has a minimum of two inputs; the symbol, Boolean expression, and truth table representing a two-input inclusive OR gate are as follows:

A	B	$X = A + B$
0	0	0
0	1	1
1	0	1
1	1	1

$$X = A + B \tag{2.46}$$

where $X = 1$ when *one or more* of its inputs are 1; inclusive means all the inputs can be included in the output, as shown by the truth table.

Exclusive OR Gate

This gate has only two inputs; the symbol, Boolean expression, and truth table representing an exclusive OR gate are as follows:

A	B	$X = A \oplus B$
0	0	0
0	1	1
1	0	1
1	1	0

$$X = A\bar{B} + \bar{A}B \tag{2.47}$$

where $X = 1$ if either A or B but *not* both are 1; thus the name exclusive OR.

This gate usually has both input lines controlled; however, it is sometimes used with one input tied hi (1), that is, to the $+5$-V supply, in which case the gate will act as an inverter. For example, if line B is tied hi ($B = 1$), then

$$X = \bar{A} \cdot 1 + A \cdot \bar{1} \quad (\bar{1} = 0) \tag{2.48}$$

$$\therefore \quad X = \bar{A}$$

This configuration is sometimes found in industrial circuits and can cause concern for the unwary.

NAND Gate

This gate has a minimum of two inputs; the symbol, Boolean expression, and truth table representing a two-input NAND gate are as follows:

A	B	$X = \overline{A \cdot B} = \bar{A} + \bar{B}$
0	0	1
0	1	1
1	0	1
1	1	0

(2.49)

$X = \overline{A \cdot B} = \bar{A} + \bar{B}$
(by De Morgan's theorem)

where $X = 1$ *when any input is zero* and will be zero only when *all* inputs equal 1. The term NAND is an abbreviation of NOT AND, and it is noticed that the inputs appear at the output in the inverted OR form.

Signal inversion is very often required in circuit operation and a two-input NAND gate can be used for this, with both inputs tied together.

NOR Gate

This gate has a minimum of two inputs; the symbol, Boolean expression, and truth table for a two-input NOR gate are as follows:

A	B	$X = \overline{A + B} = \bar{A} \cdot \bar{B}$
0	0	1
1	0	0
0	1	0
1	1	0

(2.50)

$X = \overline{A + B} = \bar{A} \cdot \bar{B}$

where $X = 0$ *when any input is 1* and $X = 1$ only when *all* inputs are zero. The term NOR is an abbreviation of NOT OR, and it is noticed that the inputs appear at the output in inverted AND form.

A gate with a "hanging" input will operate as if the line were equal to 1; this will cause the output to be zero. Therefore, to be able to operate this gate the unused input must be tied to the ground supply line. This gate can be used as an inverter similar to the NAND gate.

2.7   **LOGIC CIRCUITS**

Logic gates are connected together to form logic circuits that will generate the required Boolean expression at their output. This section will be presented in two parts: (1) NAND logic circuits, and (2) NOR logic circuits. The major emphasis will be toward industrial control systems and motor control.

NAND Circuits

Before designing complex circuits, the reader should memorize the basic OR and AND configurations, which are shown in Figure 2.10. A more complex example is given in Figure 2.11. This circuit requires a two-input OR (AB or $C\overline{D}$) preceded by the two-input AND ($A \cdot B$ and $C \cdot D$). Gates 1 and 2 constitute a double inversion and can be omitted. The redrawn circuit is shown in Figure 2.12.

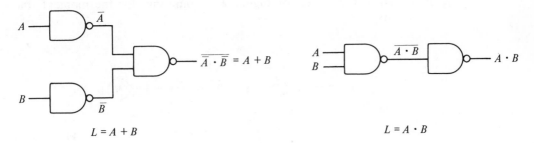

$$\overline{\overline{A} \cdot \overline{B}} = A + B$$

$$L = A + B$$

$$A \cdot B$$

$$L = A \cdot B$$

Figure 2.10

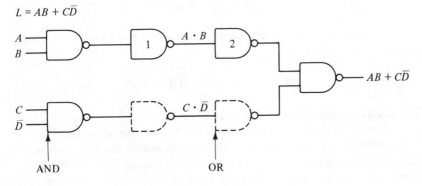

$$L = AB + C\overline{D}$$

$$A \cdot B$$

$$C \cdot \overline{D}$$

$$AB + C\overline{D}$$

AND

OR

Figure 2.11

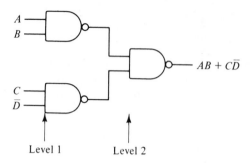

Figure 2.12

This method of laying out a circuit is now almost standard practice. It enables the designer to rapidly verify that it will generate the required expression, and it enables the technician to find circuit faults in a much shorter time.

Example 10. Design a logic circuit using NAND gates to derive the motor start/stop circuit of Figure 2.13, omitting, for the moment, the overloads.

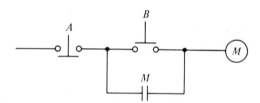

Figure 2.13

The Boolean expression is

$$M = \overline{A}(B + M)$$

Multiply out to get the expression into the sum of products form, which is required for NAND logic.

$$M = \overline{A}(B + M) = \overline{A}B + \overline{A}M$$

The circuit arrangement is shown in Figure 2.14.

To meet industrial requirements the stop button must be N.C. (normally closed); otherwise, a poor contact will inhibit the stop button from stopping the motor. Since a line that is "hanging" in NAND logic is equivalent to a 1, the push buttons must switch a 0 (ground) (see Figure 2.15), and therefore an inverter gate is necessary for proper operation of motor-control circuits.

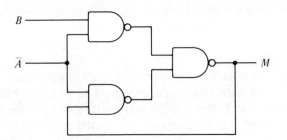

Figure 2.14

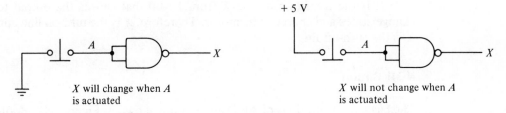

X will change when A
is actuated

X will not change when A
is actuated

Figure 2.15

In all industrial motor-control circuits *stop must override start,* which means that the stop button will be connected such that, when actuated, the output M will be 0, independent of the state of the start button (see Figure 2.16). The operation of the circuit is as follows: assume that

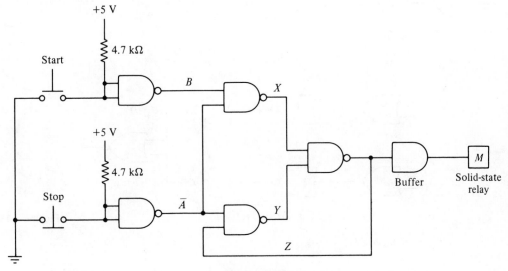

Figure 2.16

$M = 0$ (motor deenergized), line $Z = 0$, therefore line $Y = 1$. The push buttons are in their rest or quiescent position; thus line $B = 0$ and line $\overline{A} = 1$. Lines X and Y are therefore both 1, causing the output M to be 0.

Depress the start button; line $B = 1$ and, because line $\overline{A} = 1$, line X goes to 0, which causes output M to go to state 1; the motor starts. Line Z is now at state 1 and, because line $\overline{A} = 1$, line Y becomes 0, holding the output M at 1. Release the start button and $\overline{M}$ remains at 1.

If the stop button is now activated, line $\overline{A} = 0$, which in turn sets M to 0. If the start button is activated while the stop button is pressed, then the 0 on line $\overline{A}$ will hold lines X and Y at state 1, maintaining M at 0.

It is the transition of line X from 1 to 0 that causes the output to change states and energize the motor. Therefore, X is the turn-on line and Y is the turn-off line.

NOR Circuit

The basic configurations of AND and OR are shown in Figure 2.17. Boolean expressions in the form of products of sums are used with NOR gates, and relay circuits.

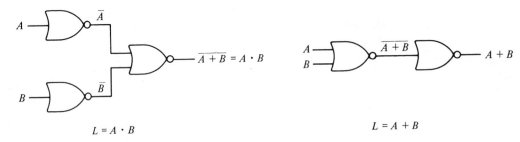

$$\overline{\overline{A} + \overline{B}} = A \cdot B$$
$$L = A \cdot B$$

$$\overline{A + B}$$
$$A + B$$
$$L = A + B$$

Figure 2.17

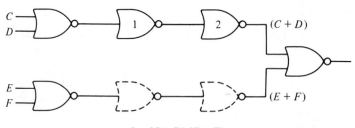

$$L = (C + D)(E + F)$$

Figure 2.18

Example 11

Gates 1 and 2 constitute a double change and can be omitted.

Example 12. Redraw Figure 2.16 using NOR gates only. The Boolean expression is $M = \overline{A}(B + M)$.

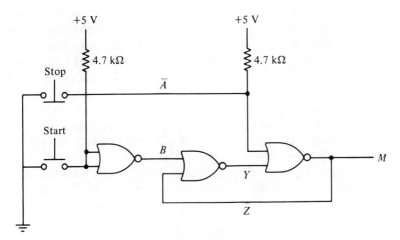

Figure 2.19

Circuit operation: Assume that $M = 0$, line $Z = 0$, and, since the start button is not depressed, line $B = 0$, and, because B and $Z = 0$, line $Y = 1$. A level 1 at any input of a NOR gate results in a 0 as an output; therefore, $M = 0$. The stop button is not depressed ($\overline{A} = 0$).

Depress the start button and line B goes to state 1; thus line $Y = 0$ and, since $\overline{A} = 0$, output M changes state from 0 to 1; the motor is energized and line $Z = 1$, which holds line Y at 0. Release the start button and line B goes to 0. If the stop button is now depressed, line $\overline{A} = 1$, which causes the output M to change state and deenergizes the motor. Line Z goes to 0, which holds line Y to 1, since line B is 0; thus the motor is deenergized and will remain so until the start button is depressed again.

It is the transition of line Y from 0 to 1 that causes the output to change state and energize the motor. Therefore, Y is the turn-on line and $\overline{A}$ is the turn-off line.

2.8 INDUSTRIAL PRACTICES

In industry, the logic panel is normally in an MCC room (motor-control center); the push buttons, on the other hand, are usually some distance away in the plant. Connecting the two together are cables or conduit and wires, which are vulnerable; that is, they can become shorted, grounded, or opened.

Industry, however, demands that the motor be deenergized if a stop wire becomes open circuited; in standard industrial magnetic circuits, it will. This also occurs in the logic circuits developed, because an open line to a NAND gate is a hanging line, which is 1, and in the way that the logic circuit has been designed, this is equivalent to depressing the stop button. The same answer occurs in the NOR circuit for similar reasons.

The logic circuits of Figures 2.16 and 2.19 have many other similarities to industrial magnetic circuits. The output M will not change from 0 to 1 if the start line breaks, and if in state 1 (motor energized), it will change to state 0 if the memory line (Z) is opened (i.e., the holding line is defective). Control is lost, however, if the stop or start buttons are shorted via the lines leading back to the MCC, but this loss of control can be prevented by high-security circuits.

The principal source of concern with logic circuits is the failure of a gate, causing a signal to change and energize a motor. While the concern is recognized and appreciated, gate failure rate is extremely low, and in the case of high-security circuits misoperation of equipment due to gate failure is virtually nil.

Examples of simple but effective security systems will now be offered.

Double Voting System

This is a higher-security start/stop motor-control system. Two sets of control leads leave the control station to connect the logic center. Each set of these two stop/start control wires goes by different routes and is never in the same enclosure except at the control end and the logic end.

The system operates as shown in Figure 2.20. To start the motor, both control lines A and B must go low; the motor will *not* start with only one line going low. The alarm will sound instead. To stop the motor, *one control line, A or B, must go high,* but the alarm will sound if only one line is activated.

Shielded or Dummy Cable Method

An alternative method of lower security than the first method is to run a dummy wire in the cable or conduit, or use the ungrounded shield if the cable is shielded, as shown in Figure 2.21. The motor starts and stops in the normal manner, but if the shield or dummy wire potential goes low (grounded), the motor will stop, if running, and will *not start* if the start button is activated. The alarm A will sound when the shield or dummy potential goes low.

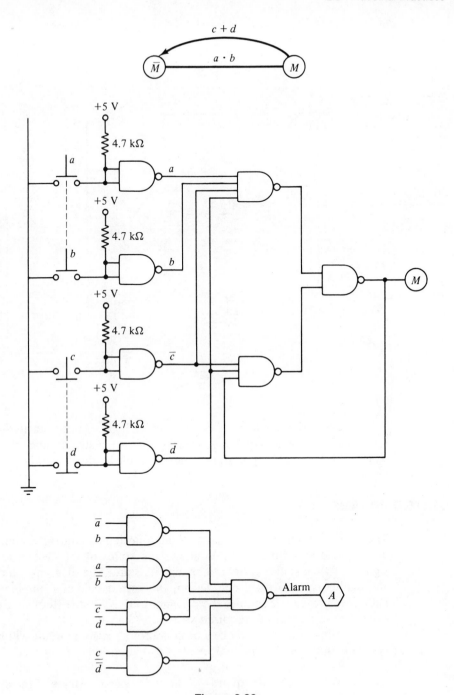

Figure 2.20

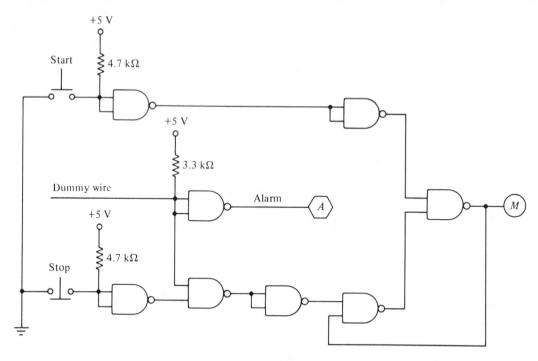

Figure 2.21

There are other methods of providing high-security start/stop systems, such as check back coded message format systems. Since these are usually high-technology systems, they will not be dealt with here.

2.9 STATE DIAGRAMS

The state diagram is a graphical method of designing control circuits. It is a cause-and-effect diagram that shows the states of the system and the signals that cause the system to change its state. The diagram is a design aid that helps to clarify circuit design, and with the aid of a simple mathematical expression it enables the designer to develop the Boolean expressions that represent the required system.

To illustrate the method, a basic motor start/stop circuit will be developed similar to Figure 2.13. A step-by-step procedure is used.

1. Draw circles or squares to show the various states of the system and inside write enough to convey the state of the system (see Figure 2.22).

Figure 2.22   Motor deenergized   Motor energized

2. The states (Figure 2.22) are now connected together by transition lines, which will cause the system to change its state. An arrow on the transition line shows the direction of the transition, that is, from $M = 0$ to $M = 1$, or vice versa (see Figure 2.23).

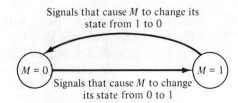

Figure 2.23

3. Write on the transition lines the identity of the Boolean variable that will cause the change of state (see Figure 2.24). The diagram now shows that if device B (push button) is actuated, the system will change its state; that is, the motor will be energized. If device A (push button) is actuated, the motor will be deenergized.

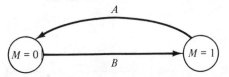

Figure 2.24

4. The following mathematical expression is used for industrial control circuits; it will always cause the stop to override the start.

$$M = \overline{\text{turn off}}\ (\text{turn on} + \text{memory})$$

This expression is divided into three parts.
a. The complement of that which causes M to change its state from 1 to 0.
b. The sum of that which causes M to change state from 0 to 1 and a signal from M of its *present* state (memory).
c. The product of items a and b.

Summary

Device B causes M to change state from 0 to 1, resulting in the turn-on set.

Device A causes M to change state from 1 to 0, resulting in the turn-off set.

$$M = \overline{A}(B + M) \qquad (2.51)$$

See Figure 2.13.

Example 13. Design a motor-control circuit with two start and two stop buttons; use s and r for stop, x and y for start (see Figure 2.25). Device M will change its state from S_0 to S_1 if device x or device y is actuated. Similarly, M will go from S_1 to S_0 if device s or device r is actuated. If the mathematical expression is now written, it will appear as follows:

$$M = (\overline{s + r})(x + y + M) = \overline{s} \cdot \overline{r}(x + y + M) \qquad (2.52)$$

which is realized into the relay circuit of Figure 2.26.

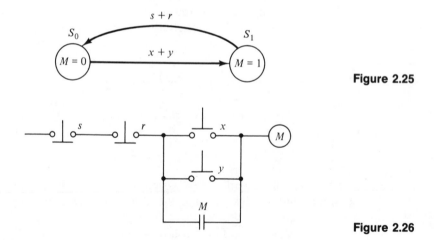

Figure 2.25

Figure 2.26

2.10 SEQUENTIAL CIRCUITS

Up to this point all the circuits that have been developed and discussed have been *combinational* circuits, that is, circuits in which the sequence of the input signals did not matter. For example, $C = A \cdot B$; as soon as A

and B are 1, C will be 1. The order in which the signals arrived did not affect the outcome. However, in sequential circuits, order is important. Sequential circuits have memory, and the events of the past affect the outcome of the future.

To demonstrate this principle a four-speed motor driving a cooling fan will be used (only three speeds are used). In Figure 2.27, signal x represents the low-speed start button, signal y, the medium speed, and signal z the high speed. Signal s stops the motor. These signals are called *primary* signals.

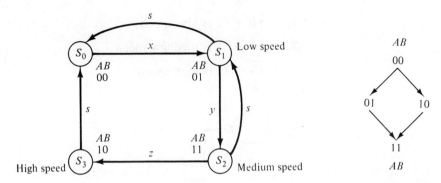

Figure 2.27   Figure 2.28

To remember the previous state, coding of the states S_0, S_1, and so on, is necessary. This is done by *secondary* signals A and B, which are coded in such a way that when moving from one state to another *no double change takes place*. If the system is in state $\bar{S}_0(\bar{A}\bar{B} = 00)$, then it is impossible to go to state $S_2(AB = 11)$ in one transition (see Figure 2.28); either A will change first or B will change first, since the outcome due to the simultaneous change of signals A and B is not considered.

According to the state diagram, it will be impossible to start the motor in high speed without first having gone through the low and the intermediate speeds. The next step is to write down the turn-off and turn-on sets of A and B respectively.

The turn-off set of A is from S_2 to S_1 when $B = 1$ and signal s causes the transition. The Boolean statement for this transition is Bs. A is not included since it changes state. Similarly, A is also *turned off* from S_3 to S_0, which is $\bar{B}s$; combining the two statements results in $\bar{B}s + Bs = s$. Thus if device s is actuated while A is set, it will cause A to reset. Similarly *turned-on* set of A is from S_1 to $S_2 = By$. *Turn-off* set of B is from S_1 to S_0 or from S_2 to $S_3 = \bar{A}s + Az$. *Turn-on* set of B is from S_0 to $S_1 = \bar{A}x$.

$$\therefore \quad A = (\bar{s})(By + A) = \bar{s}By + \bar{s}A \tag{2.53}$$

$$B = (\overline{\overline{As} + A\overline{z}})(\overline{A}x + B)$$
$$= (A + \bar{s})(\overline{A} + \overline{z})(\overline{A}x + B)$$
$$= (A\overline{z} + \overline{A}\overline{S} + \bar{S}\overline{Z})(\overline{A}x + B) = (A\overline{z} + \overline{A}\bar{s})(\overline{A}x + B)$$
$$= \overline{A}\bar{s}x + AB\overline{z} + \overline{A}B\bar{s}$$

The circuit for these secondary signals A and B is shown in Figure 2.29, which follows from Figures 2.11 and 2.12. Secondary signals A and B control the motor, and from the state diagram of Figure 2.27 it can be seen that

$$\text{State } S_1 = \overline{A}B = 01 \text{ is low speed}$$
$$S_2 = AB = 11 \text{ is medium speed}$$
$$S_3 = A\overline{B} = 10 \text{ is high speed}$$

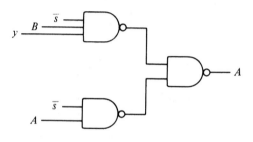

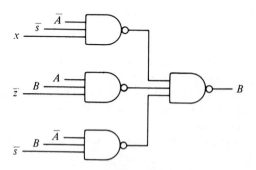

Figure 2.29

The result is shown in Figure 2.30.

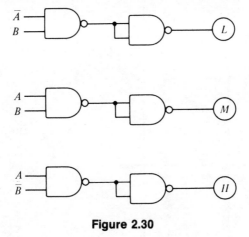

Figure 2.30

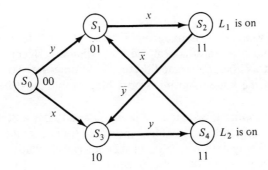

Figure 2.31

2.11 STATE REDUCTION

When developing a state diagram, it is important to use as few states as possible. This reduces the amount of secondary signals or auxiliary relays. This in turn simplifies the circuit and reduces the overall cost. It is, however, not always that easy to determine the minimum number of states necessary to do this. Consider the following circuit, which will give a visual indication of a sequence of events.

L_1 will be energized when two buttons y and x are pressed in that order, and L_2 will come on when the order in which the buttons are pressed is reversed. The state diagram of Figure 2.31 illustrates this. If button y is pressed, a change in state will occur from S_0 to S_1, and if button x is now activated the system moves to S_2 and L_1 comes on; loss of

	$\overline{x}\,\overline{y}$ 00	$\overline{x}\,y$ 01	xy 11	$x\overline{y}$ 10
	①	②	③	④
S_0	S_0	S_1	–	S_3
S_1	S_0	S_1	S_2	–
S_2	–	S_1	L_1 S_2	S_3
S_3	S_0	–	S_4	S_3
S_4	–	S_1	L_2 S_4	S_3

All the combinations of primary signals x and y

All states that appear on the state diagram

Transfer the data from the state diagram (Figure 2.31) to this table

Figure 2.32

signal y causes a change from S_2 to S_3. The reader is urged to check all the possible combinations on the state diagram. Keep in mind that the final circuit will be determined by this state diagram. To check whether reduction of the state diagram is possible, a table must be constructed (Figure 2.32).

Assume the system to be in S_0 and that x and y are both 0; then the system remains in S_0. This is indicated by entering S_0 in square 1. If, however, the system is in S_0 and y is activated, that is, $x = 0$, $y = 1$, then a change in state will occur; it will change to S_1. Enter S_1 in square 2 (row S_0, column $\overline{x}y$). Square 3 is left blank, because the system cannot change from S_0, where $x = 0$ and $y = 0$, to a state where $x = 1$ and $y = 1$, which constitutes a double change. Square 4 indicates that $x = 1$ and $y = 0$; this means a move to S_3.

When the table has been completely filled out, row by row will then be compared to check if any of them are identical, for these rows can be merged. Blank states can be merged with any states. In Figure 2.32, rows S_0, S_1, S_2 are identical and so are rows S_3 and S_4; therefore, these rows can be merged. A merged table is shown in Figure 2.33.

The five states have been reduced to two, and this circuit can be implemented with only one auxiliary relay A. Therefore S_0, S_1, S_2 can be called $\overline{A}$, and S_3, S_4 will then be called A (see Figure 2.33). Relay A is an auxiliary relay only; its function is to remember which button was pressed first. The main circuit must still be developed.

The first row $(\overline{A})$ of Figure 2.33 indicates that a change in $\overline{A}$ occurs only in the last entry of the row, that is, $x\overline{y}$, and the second row (A) indicates that a change will occur when the signals are $\overline{x}\,\overline{y}$ or $\overline{x}y$ $(\overline{A})$.

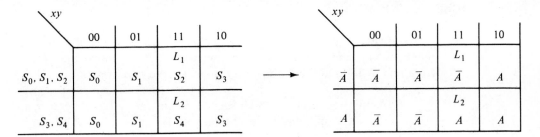

Figure 2.33

Figure 2.34

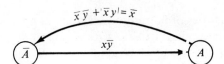

Figure 2.35

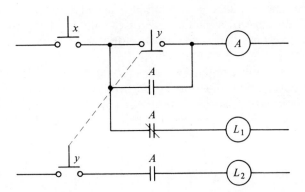

The table in Figure 2.33 can be represented by a simple state diagram as in Figure 2.34.

$$\therefore \quad A = (x)(x\bar{y} + A) = x\bar{y} + xA = x(\bar{y} + A) \qquad (2.54)$$

Figure 2.33 also indicates that $L_1 = \overline{A}xy$ and $L_2 = Axy$, which can intuitively be further reduced to $L_1 = \overline{A}x$, because A has already detected the presence of signal y; and $L_2 = Ay$, because A detected the presence of signal x.

The schematic of Figure 2.35 shows the final relay circuit.

2.12 CLOCKED SEQUENTIAL CIRCUITS

Clocked sequential circuits are pulse-actuated systems. A pulse, in conjunction with combinational signals or alone, acts to move the system from one state to another. Basically, pulses have two origins:

1. A pulse generator, which supplies identical pulses at regular intervals (oscillator, multivibrator, etc.).
2. Some electrical or electronic device that supplies a pulse when certain conditions are met. The pulses may be irregular in duration and frequency (packages passing an electric eye, limit switch operating, etc.).

Both types of pulses cause similar action: when specified conditions are met and a pulse arrives, a transition of state occurs.

A flip-flop is an electronic switch with two stable complementary outputs and a number of controlling input lines; the number varies with the type of flip-flop being considered.

Because a flip-flop has two stable outputs, it is called a bistable device, and a change in the output is usually coincident with the leading or trailing edge of an actuating pulse, commonly called a *clock pulse*. We shall assume that a change in output occurs at the trailing edge of the clock pulse.

A number of pulses connected together is called a *pulse train;* a typical pulse train is illustrated in Figure 2.36. The terms *mark* and *space* are a holdover from the keyed transmissions of the early telegraph systems and are used today to identify the high and low levels of a pulse.

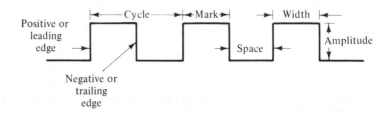

Figure 2.36

Two basic types of flip-flops will be considered here in turn with their state diagrams and a brief description of their operation. The D type of flip-flop is shown in Figure 2.37. State S_0 is called the reset state of the flip-flop. The transition from S_0 to S_1 occurs when $D = 1$ and the trailing edge of the clock pulse occurs; in this state $Q = 1$ and $\overline{Q} = 0$, that is, it is

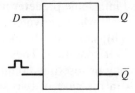

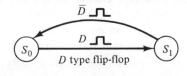

Figure 2.37

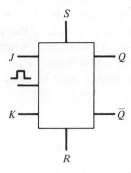

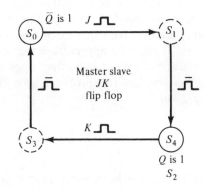

Figure 2.38

said to be in the set condition. The flip-flop will remain in this state, independent of the number of clock pulses that arrive, until $D = 0$, and then on the *next* negative-going edge the device will switch from S_1 to S_0 and will remain in this state until D changes.

The *JK* flip-flop (Figure 2.38) has some unique properties, which are used in code-generating circuits, counter circuits, and the like; these properties are as follows:

1. When $J = 1$ and $K = 1$, the flip-flop will switch back and forth on the trailing edge of every clock pulse.
2. When $J = 1$ and $K = 0$, the flip-flop will set on the trailing edge of the next clock pulse.
3. When $J = 0$ and $K = 1$, the flip-flop will reset on the trailing edge of the next clock pulse.
4. When $J = 0$ and $K = 0$, the flip-flop will lock in its present state independent of clock pulses.
5. When $J = \overline{K}$, the flip-flop behaves in the same manner as a D type.

There are two more control terminals on the *JK* flip-flop that have not been discussed yet; these are "set direct" and "reset or clear direct."

These terminals enable the flip-flop to be forced into some predetermined state independent of the clock. The set terminal, when activated, will force the flip-flop into the set condition. (Q is high); similarly, the clear or reset terminal will force the flip-flop into the reset condition. These two control terminals are used, usually, for initializing or resetting counters.

A typical application for a JK flip-flop is in the case of two sump pumps controlled by two float switches in such a way that motor 1 starts first when float switch 1 is closed and motor 2 starts later if the second float switch calls for additional pumping action. The second time motor 2 starts first, followed by motor 1 if necessary (see Figure 2.39). One problem encountered with this circuit would be contact bounce of the limit switches. It must be appreciated that every switch "bounces" when its contacts are closed; that is, they open and close a few times before settling. To overcome this problem, an antibounce switch must be used (see Figure 2.40).

2.13 COUNTERS

JK flip-flops have been used quite extensively to build synchronous or asynchronous counters. These can now be replaced by "chips," such as

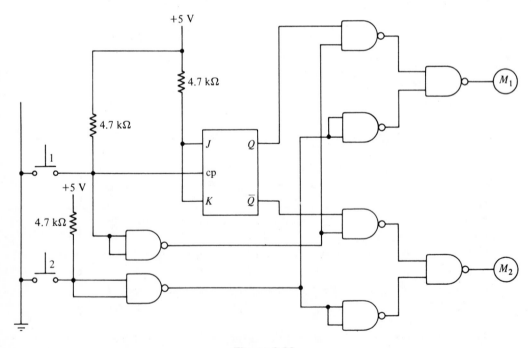

Figure 2.39

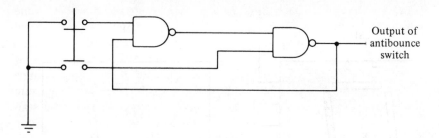

Figure 2.40

the 74190 Up/Down Decade and Binary Counter, which together with 7446 BCD driver and an IEE 1372 R, 7-segment display can be connected to form decimal counting and display units. Each display will show the decimal number between 0 and 9 corresponding to the binary-coded-decimal output of the counter. The counter output will show the number of input pulses. As the least significant counter changes from 9 to 0, its output point 13 increments the next counter by 1 (see Figure 2.41).

2.14 TUTORIAL EXAMPLES

Example 1.   Design a control circuit for a stop/start jog motor circuit according to Figure 2.42. Use NAND logic.

Solution

Turn on $F = x$	Turn on $J = y$
Turn off $F = s + y$	Turn off $J = s + \bar{y}$
$F = (\overline{s + y})(x + F)$	$J = (\overline{s + \bar{y}})y + J)$
$F = (\bar{s}\bar{y})(x + F)$	$J = (\bar{s}y)(y + J)$
$F = \bar{s}\bar{y}x + \bar{s}\bar{y}F$	$J = \bar{s}y$

Do not reduce the Boolean statement after the terms have been multiplied.

$$M = F + J$$

$$M = \bar{s}\bar{y}x + \bar{s}\bar{y}F + \bar{s}y$$

See Figure 2.43.

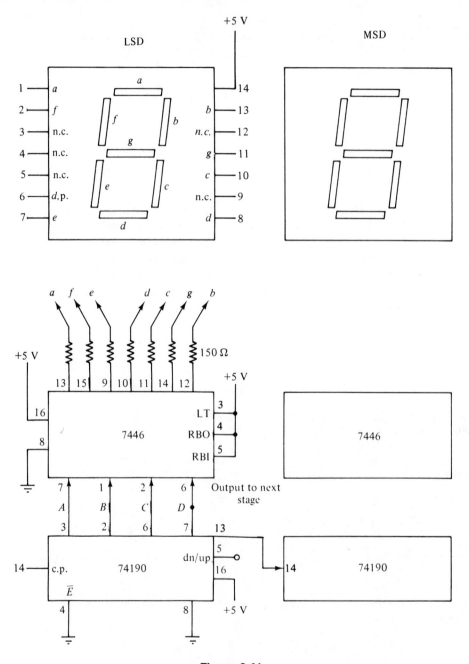

Figure 2.41

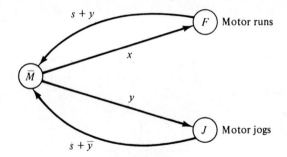

Figure 2.42

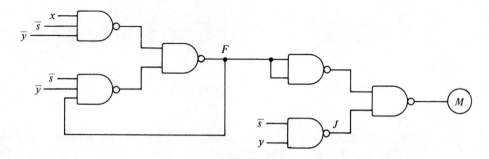

Figure 2.43

Example 2. Design a control circuit for a forward-reverse motor operation. Use the following push buttons and limit switches:

Stop button, s Forward limit, c
Forward start, f Reverse limit, d
Reverse start, r

Solution. Refer to Figure 2.44.

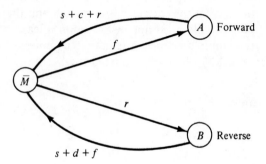

Figure 2.44

Turn on set of $A = f$ Turn on set of $B = r$

Turn off set of $A = s + c + r$ Turn off set of $B = s + d + f$

$A = \overline{(s + c + r)}(f + A)$ $B = \overline{(s + d + f)}(r + B)$

$A = (\bar{s} \cdot \bar{c} \cdot \bar{r})(f + A)$ $B = (\bar{s} \cdot \bar{d} \cdot \bar{f})(r + B)$

See Figure 2.45.

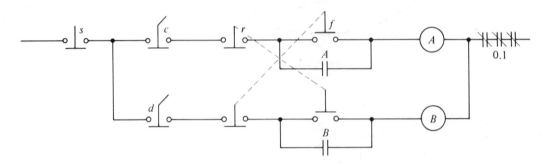

Figure 2.45

Example 3. Design a logic circuit to control the operation of a planer table. The forward and backward movements of this table are obtained by forward and reverse control of a suitable motor. The following push buttons and limit switches are used:

     Stop button, s Retentive start button, m

     Forward start button, f Limit switches h and k

     Reverse start button, r

Automatic control of the reciprocating action of the planer table is obtained by the use of momentary-contact limit switches located at each end of the planer-table bed. If, after stopping, the button m is pressed, the planer table resumes its previous direction of travel.

    The stop must override *all other* functions; for example, if the planer table is on the forward limit switch and the stop button is depressed, the planer table must not restart upon release of the stop button. Also, if the two limit switches are actuated simultaneously (one limit switch got "stuck"), the motor must stop.

Solution. Refer to Figure 2.46.

     Turn on set $A = B(f + h)$

     Turn off set $A = B(r + k)$

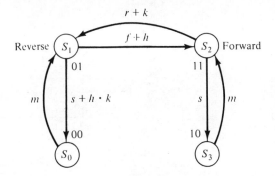

Figure 2.46

$$A = \overline{B(r + k)}[(B(f + h) + A]$$
$$A = (\overline{B} + \overline{rk})(Bf + Bh + A)$$
$$A = \overline{rk}Bf + \overline{rk}Bh + \overline{rk}A + A\overline{B}$$
Turn on set of $B = \overline{A}m + Am = m$
Turn off set of $B = \overline{A}(s + h \cdot k) + As = \overline{A}hk + s$
$$B = (\overline{A}hk + s)(m + B)$$
$$B = [(A + \overline{h} + \overline{k})\overline{s}][m + B] = A\overline{s}m + AB\overline{s} + \overline{hm}\overline{s} + B\overline{h}\overline{s}$$
$$\quad + \overline{ksm} + \overline{k}\overline{s}B$$

$$\text{Fwd} = AB, \qquad \text{Rev} = \overline{A}B$$

Example 4. Design a circuit to control a two-speed motor in accordance with the following specifications:

> The motor can only be started in speed 1.
> The motor can then be switched from speed 1 to speed 2.
> The motor cannot be switched from speed 2 to speed 1.

If excessive vibration occurs, the motor will stop and lock out. If this happens, the motor cannot be restarted until the control circuit is reset by means of a suitable reset button. If the stop button is pressed when the motor is running in either speed, the motor will stop, but will not lock out.

Solution. Refer to Figure 2.47.

A on $= B(y + \overline{v})$	B on $= \overline{A}x$
A off $= Bs + \overline{B}r$	B off $= \overline{A}s + Av$
$A = AB\overline{r} + B\overline{s}y + B\overline{s}\overline{v} + AB\overline{s}$	$B = AB\overline{v} + \overline{A}\,\overline{s}x + \overline{A}B\overline{s}$
Low speed $= \overline{A}B$	High speed $= AB$

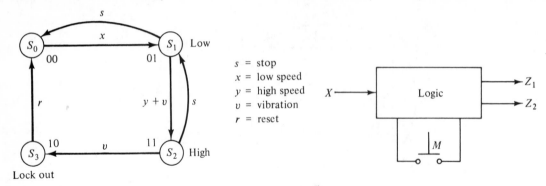

s = stop
x = low speed
y = high speed
v = vibration
r = reset

Figure 2.47

Figure 2.48

Problems

2-1. Reduce the following Boolean functions:
a. $(A + B)(\overline{A} \cdot \overline{B})$
b. $(A + B)(\overline{A} + \overline{B})$
c. $(\overline{A} + \overline{B})(A \cdot B \cdot \overline{C})$
d. $A + \overline{B} + \overline{A}B + \overline{C}$
e. $(A + \overline{B} + \overline{A}B)(C + \overline{D})(\overline{C}D)$

2-2. Reduce the following Boolean functions:
a. $A + \overline{B} + \overline{A}B + D(A + B)$
b. $A + \overline{A}B + AB$
c. $(AB + C)(AB + D)$
d. $AB + (\overline{A} + \overline{B} + \overline{C}\overline{D})E + \overline{E}F + FG$
e. $(A + B + CD)(\overline{A}\overline{B} + D)$

2-3. Reduce the following Boolean expressions:
a. $\overline{BC\overline{D}}$
b. $\overline{\overline{A}BC} + \overline{A}\overline{B}C$
c. $B + C(\overline{AB + AC})$
d. $AB + \overline{A(B + C)}$
e. $(A + \overline{BC})(\overline{AB} + \overline{\overline{A}BC})$

2-4. Redraw Figure 2.26 using NOR gates only.

2-5. Redraw Figure 2.26 using NAND gates only.

2-6. Design a pulse train switch according to the following specifications (use NAND gates). Given that signal X in the diagram is a pulse train, design a circuit that will cause the pulse train to appear at output Z_1 if button M is not activated, and at output Z_2 if button M is activated. In either case, the output is to consist of complete pulses only. It can be assumed that at least one input pulse will occur while button M is either activated or not. (See Figure 2.48 above.)

Chapter 3

The Microprocessor

The microprocessor was first developed in 1970 as a 4-bit unit, with a few discrete components, resistors, capacitors, and transistors compressed into an IC chip (integrated circuit). A few years later, MSI (medium-scale integration) was introduced and the number of components was increased to about 100 per IC. The power of the processor was increasing. This was followed by large-scale integration (LSI) with 5,000 to 10,000 transistors per IC chip, and today very large scale integration (VLSI) features 25,000 to 70,000 transistors per chip, which is within one order of magnitude of that used in the large computers.

A microprocessor is a clock-driven time sequential circuit, which corresponds to the central processing unit (CPU) of a digital computer. A microprocessor, some random-access memory (RAM) for storage of data, some programmable read-only memory (PROM) to hold instructions, a power supply, and interfacing circuitry for communication to the outside world (peripheral devices) comprise the elements of a microprocessor system or, as it is sometimes called, a minicomputer.

The key to using a microprocessor system is to treat it as a logic element rather than a "number crunching computer." This is the approach taken here, to use a microprocessor system in dedicated control applications in which the power is turned on and the system does its job.

The microprocessor chosen to illustrate and to demonstrate dedicated control is the Motorola MC 6800 and later the MC 6805. A 6800 evaluation kit is used to test the programs and the systems developed; this kit, as its name implies, is used only to evaluate the microprocessor and its family of chips, as well as to test out programs and routines. It is not made to be placed in a cabinet and used as an industrial tool, although it could be. Rather, the 6800 is wired into a circuit with its support chips and used in industrial systems.

3.1 OVERVIEW OF A MICROPROCESSOR SYSTEM

All digital computers from the very large machines down to the smallest microprocessor system contain three basic sections:

1. Central processing unit (CPU) or microprocessor unit (MPU)
2. Memory: PROM, ROM, RAM, and so on
3. An input/output section (I/O) for communication with its peripherals

Each of these will be considered in some detail, after a brief summary of how a microprocessor system operates.

A microprocessor system is basically a black box with a number of inputs from, and a number of outputs to, the outside world. It can make decisions, store data, do timing cycles, do simple arithmetic, convert codes, and so on. The basic difference between this black box and a hard-wired logic system using IC chips is that a specific coded message is stored in an area called *program memory,* which is a PROM or ROM IC chip. PROM or ROM is used to store the coded message or program because they are *nonvolatile;* that is, they retain the program, which was put into them before being plugged into the system, after the power is turned off. RAM memory is used to store collected data and also act as a temporary storage area for data (''scratch pad''); all information is lost when the power is turned off.

The system operates through the interaction of the processor and the program memory. When the power is turned on to the system, the processor reads the first code word (instruction) stored in memory and acts upon this instruction; when completed it goes back to the memory for the next instruction, and so on until the task is complete. This is a read/do cycle and is exactly the same process as is followed in the assembling of a do-it-yourself kit; read the instruction, perform the task, read the next instruction, and so on.

The speed with which each task is performed, read instruction and execute, is dependent upon the clock frequency of the system and the task to be performed (e.g., go to memory, fetch a piece of data, a number, and add it to another number); the code is 9B in the MC 6800 instruction set and the execution time is three clock cycles.

The processor communicates with the outside world via input gates and output latches. The gates and latches are packaged in an IC chip, which is one of the chips surrounding the MPU forming the processor system.

A typical processor system is drawn in skeleton form in Figure 3.1 which illustrates its basic form; this will be followed by a description of the MPU, memory, and I/O devices.

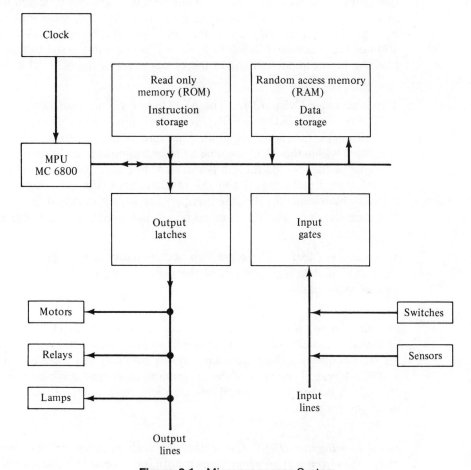

Figure 3.1 Microprocessor System

3.1.1 Microprocessor Unit (MPU)

The MPU (microprocessor) is usually packaged in a dual-in-line (DIP) plastic or ceramic container. Dual in line means the connections that lead from the integrated circuit inside to the world outside are arranged in two rows on either side of the long axis; they have the same appearance as an IC chip and are numbered in the same way. The MPU consists of a number of basic elements, an arithmetic logic unit (ALU), several accumulators (storage registers for work in process), general-purpose registers (temporary storage devices), program counter (a register that stores the address of the *next* instruction), instruction register (contains the instruction being executed), a condition codes register (contains specified flags, or information bits, which indicate that predetermined conditions have been met, e.g., overflow, carry, etc.), and a control unit (directs the operation of the processor). Each of these will now be discussed to give the reader a better understanding of the components of the MPU chip.

Arithmetic Logic Unit (ALU). The ALU uses logic manipulation, such as AND, OR, exclusive OR (EXOR), and complements and shifts to perform the arithmetic functions required. The condition code flags, which are generated within the ALU, assume a predetermined state according to the outcome of the *last* instruction performed. For example, if two numbers are compared (subtracted) and the result is zero, the zero flag will be raised, which indicates that the numbers are identical; raised flags manifest themselves by setting specified bits in the condition codes register.

Accumulators (Acc). There are two accumulators in the 6800; they are used to pass data into and out of the MPU, as well as to hold data for further manipulation.

Program Counter (PC). The program counter keeps track of the location of the *next* instruction to be executed from program memory. After each fetch from memory the program counter is automatically incremented by 1. The address of the start of the program to be executed is placed in the PC, which is then incremented one step at a time as the processor executes the program.

Instruction Register (IR). The instruction register receives each current instruction from memory and holds it until it has been executed; then the next instruction is loaded in, which simultaneously destroys that which is already in the register.

Condition Codes Register (CC). The condition codes register contains the flags that are used to indicate the outcome of the *previous* operation. The contents of this register are used as part of the decision-making process of

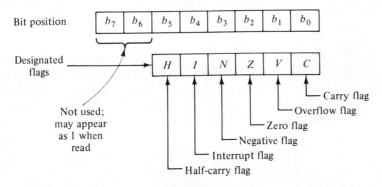

Bit position

Designated flags

Not used; may appear as 1 when read

Carry flag
Overflow flag
Zero flag
Negative flag
Interrupt flag
Half-carry flag

Figure 3.2

the processor (e.g., was the result of the last operation zero?). If the answer is yes, then the zero flag will be set and action based upon this result can now take place. The format of the register is shown in Figure 3.2 and a description of each bit follows.

Bit 0 is the carry flag and when set is used to indicate the following:

1. That a carry out of the b_7 position of the designated accumulator or memory location has occurred; for example,

$$
\begin{array}{r}
0\ 1\ 1\ 1\ 1\ 1\ 0\ 1 \\
1\ 1\ 1\ 0\ 1\ 1\ 1\ 1 \\
\hline
1\ \ 0\ 1\ 1\ 0\ 1\ 1\ 0\ 0
\end{array} +
\qquad (3.1)
$$

carry out of the b_7 position

2. That a borrow has occurred as a result of the preceding subtraction operation; for example, subtract 78 from 57:

	Decimal		*Binary*			
	57		0 0 1 1 1 0 0 1	minuend	(3.2)
	78	−	0 1 0 0 1 1 1 0	subtrahend	
borrow 1	79		borrow 1 1 1 0 1 0 1 1		
			1		

A borrow has occurred and the carry flag is set *because the absolute value of the subtrahend is larger than the absolute value of the minuend.*

3. That a particular bit in an accumulator or memory location is set; for example, assume that accumulator A is loaded with an un-

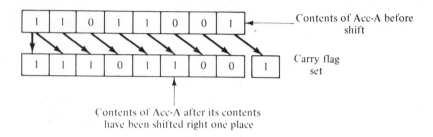

Figure 3.3

known binary number, and that the next operation of the processor is based upon whether bit 0 of the number is set or reset. A decision based upon the status of the carry flag will be utilized, the diagrams of Figure 3.3 show the contents of Acc-A.

Bit 1, the V bit, represents the status of the overflow flag; it is used with 2's complement arithmetic to illustrate that the result currently being held in an accumulator or memory location has overflowed the device's storage capacity. A subtraction will be carried out via an 8-bit processor and the 2's complement representation.

Subtract 40_{10} from -100_{10}

Decimal		*Binary*								
-100	$-$	1	0	0	1	1	1	0	0	$+$ 2's complement representation of -100 (3.3)
40		1	1	0	1	1	0	0	0	2's complement representation of 40
-140		0	1	1	1	0	1	0	0	

The 0 in the b_7 position indicates that the number is positive, and therefore the result would be read as $+116_{10}$ and not -140_{10}, as it should be. What has happened to cause this error? The capacity of the accumulator has been exceeded, and to inform the processor that something is in error, the *overflow* flag is set. Continued testing of the V bit is important after arithmetic operations involving 2's complement numbers.

Bit 2, the Z flag, is set *any* time the result of the last data manipulation, arithmetic or logic operation, caused the result to be zero. This flag finds its major use in the exact determination of the magnitude of a number. For example, assume that the processor is being used to count the number of objects passing a point. A counter is loaded with a prescribed number, and each time an object passes the contents of the counter will be decremented by 1. Ultimately, its contents will be 00 and at that time the zero flag will set, indicating that the desired count has been reached. Since the accumulators and memory locations are only 8 bits wide, the total count must be less than or equal to 255_{10}.

Bit 3 is the N bit; it is set if bit 7 of the specified accumulator or memory location is 1. This means that the resident number is negative, in signed numbers arithmetic.

Bit 4 represents the status of the interrupt mask bit; if I = 1, interrupts to the microprocessor via the $\overline{IRQ}$ line (interrupt request line) are inhibited. The bit may be set or cleared under program control via the instructions SEI or CLI, respectively. An interrupt is a request to the MPU by a peripheral for service, extra to the normal program.

Bit 5 is the half-carry bit H; it is set when a carry occurs from the lower 4 bits to the upper 4 bits in an accumulator or memory location; for example, $A = 00001001$, $B = 01001001$.

$$\therefore \quad A + B = \begin{array}{c} 0\ 0\ 0\ 0\ |\ 1\ 0\ 0\ 1 \\ 0\ 1\ 0\ 0\ |\ 1\ 0\ 0\ 1 \\ \hline 0\ 1\ 0\ 1\ |\ 0\ 0\ 1\ 0 \end{array} + \qquad\qquad (3.4)$$

half-carry from lower 4 bits (nibble)

The principal use of the half-carry bit is in the decimal adjust instruction DAA.

Control Unit (CU). The control unit directs the operation of the MPU and is actually a computer inside a computer; that is, it has all the functional elements of a classical computer, including read-only memory (ROM). The CU executes the user's program instructions by executing a series of its own "microinstructions," which control data transfers and all the functions from the computed results. The control signals form the "personality" of the microprocessor; if the microprogram could be altered in some way, the personality of the processor is changed. Therefore, the MPU is directly controlled by the microprogram in the control unit; in other words, it is a microprogram-controlled processor, hence the name "microprocessor."

3.1.2 Memory Units

Memory is used to store the user's program and/or information or data and is of two basic types: (1) volatile (stored information is lost when the power to the memory is switched off) and (2) nonvolatile (stored information is retained when power is removed, i.e., permanent memory). Each memory chip is partitioned into individual memory locations with each location having a unique address and the ability to store data. The information is stored in the form of 1's and 0's (bits), and the number of bits

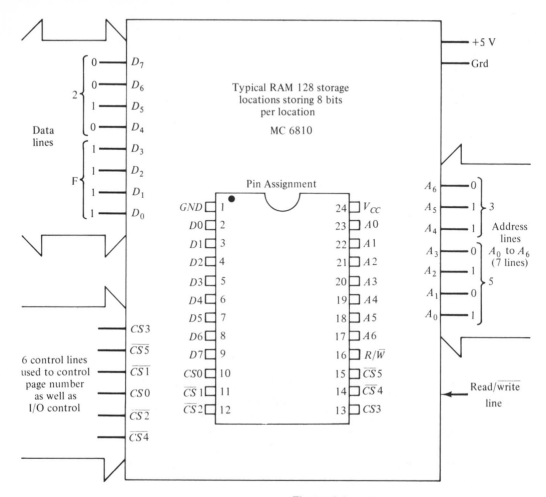

Figure 3.4

stored per memory location is dependent upon the memory or chip capability; that is, some chips store 4 bits per location, others store 8 bits, and still others store 1 bit per location. However, the technique of using (accessing) the memory is the same for all and will now be discussed. All numbers from this point on will be in hexadecimal notation unless otherwise stated.

A block diagram of a RAM chip is illustrated in Figure 3.4 and shows the data lines, address lines, and control lines. Precise details, timing diagrams, and so on, will be deferred until Figure 3.5; only the basic operation will be offered at this time. Each memory location, that is, each line of the memory element, has a unique address starting at 00 and going to

7F and may be accessed at any time, as follows. Assume that the processor has a piece of data it wants to save, say 2F at location 0035; the process required is a write cycle, and the sequence is

1. Set up the address lines to address 35.
2. Set up the control lines to access page 00 and, at the same time enable the control lines on the memory.
3. Set up the data lines to contain 2F.

Now the chip is set up to receive the data (see Figure 3.4), which will *not* be written into memory until the processor sets the R/$\overline{W}$ line low; the processor is now in the write or store mode. At that time, the data, which are on the data lines, are written into the memory location whose address is on the address lines. The R/$\overline{W}$ line is then tristated (set into a high-impedance state; it is neither at a high voltage level, binary 1, or at low voltage level, binary 0); the address lines, data lines, and so on, are then all changed to enable the processor to perform its next operation.

A read or load cycle is similar except that the processor wants to read the data held at a specified memory location. Assume that the processor is to read the data stored at memory location 0247; the process is as follows:

1. Set up the address lines to address 47; the page number 02, is controlled by the control lines.
2. When the $R/\overline{W}$ line is set high, the contents of the selected memory location are placed on the data lines; the $R/\overline{W}$ line is then tristated, and the system clock switches or clocks the data into the accumulator in the MPU and thence it is processed.

A read-only memory, as its name implies, cannot be written into. However, the read cycle is identical to that just outlined. When the processor writes to memory, an STA (Store Acc) instruction can be utilized; similarly, when a read cycle is required, a LDA (Load Acc) instruction can be used.

Most microprocessor systems use a combination of RAM and ROM. RAM allows the processor to store and retrieve data from that portion of the memory. However, the data in ROM are programmed in during manufacture by the semiconductor manufacturer according to specifications set up by the user; the program *cannot* be altered. Programmable read-only memory is semipermanent; it can be programmed in the field, but only with special equipment. EPROM (Erasable Programmable ROM) can have the complete contents erased by subjecting the EPROM to ultraviolet light; however, to write in the new contents requires special equipment.

Information read from RAM is only copied onto the data lines; it is not destroyed by the read process. Alternatively, information written into RAM destroys any other information present in the location at the time. Details for the timing of a data read from memory are contained in Section 3.2.

3.1.3 Input/Output Section

The input/output (I/O) section is used to connect the processor to the external data sources (switches, sensors, etc.) and data destinations (relays, lamps, etc.), all of which form part of the peripheral devices connected to the system. Data are transferred to and from peripherals in one of two ways: parallel and serial.

1. Parallel I/O ports (Section 3.5.1) usually have 8 lines to the connected peripheral, which allows 8 bits of data to be transferred in or out at a time under processor control. Data transfer rates can be as fast as the processor can operate.
2. Serial I/O ports (Section 3.5.2) have one input line from and one output line to the peripherals. Data transfers fom the MPU are still in 8-bit form; however, the port outputs or inputs data 1 bit at a time under the control of its own logic circuitry. The data transfer rate is slow compared to parallel I/O.

Input/output of data involves both hardware (IC chips and the like) and software (any means by which the microprocessor is instructed to perform a specific task) to transfer data between the MPU and the peripherals. The physical connection of the processor's data bus (the lines that are used to transfer data between memory and MPU and I/O ports) to the external hardware implies several other functions than an I/O port must satisfy:

1. It must isolate the data bus from external devices when not in use.
2. It must isolate external devices from one another when not being used by the processor.

Input/output systems can be categorized into two basic types: memory-mapped I/O and isolated I/O.

Memory-Mapped I/O Systems. A memory-mapped I/O system is simple to operate since there are no special I/O instructions and no special control lines. Each I/O device must contain registers that respond to the read/write commands of the processor.

The system designer assigns a section of memory address space as I/O addresses and does *not* allow any memory device (RAM, ROM, etc.) to share that space. Thus, a program-controlled data transfer to or from a peripheral is performed in the same manner as a read (LDA instruction) or a write (STA instruction) to memory.

Isolated I/O System. In the isolated I/O system the I/O ports are accessible only by using special input/output instructions; there are always at least two:

1. IN to transfer data from an input port to the MPU.
2. OUT to transfer data from the MPU to the output port.

The difference between the two systems is the method of addressing the port. (The IC chips can be identical in both cases; furthermore, a memory-mapped I/O system may be present in an isolated I/O based system.)

Advantages/Disadvantages

1. Isolated I/O systems have shorter instructions (less number of bytes) to actuate a data transfer than memory-mapped I/O and therefore operate faster.
2. Memory-mapped I/O has the advantage of using *all* the memory reference instructions, whereas isolated I/O has a limited number of instructions. There are several others, but since the 6800 and 6805 are memory-mapped I/O, they will not be considered here.

3.2 MPU TIMING

The timing of the 6800 is based upon a two-phase nonoverlapping clock, which outputs square waves called ϕ_1 and ϕ_2. The ϕ_1 clock is principally responsible for incrementing the PC and at the same time placing the address of the next instruction required onto the 16-bit address bus. The ϕ_2 clock is responsible for clocking the data out of the memory location accessed by placing it on the 8-bit data bus and then clocking it into the MPU, a load instruction; when $R/\overline{W} = 0$, data will be written by two master signals, both of which occur during ϕ_2. They are as follows:

1. VMA (valid memory address): This signal line is high when the address currently resident on the address bus is valid.

2. R/$\overline{\text{W}}$ (read/$\overline{\text{write}}$ line): The signal level on the line informs the memory location which way data are to be moved; that is, when R/$\overline{\text{W}}$ = 1, data will be read from the memory location *into* the MPU, a load instruction; where R/$\overline{\text{W}}$ = 0, data will be written from the MPU *into* the memory location, a store instruction.

The timing diagram of Figure 3.5 is for a small part of the 6800 program of Figure 3.6.

What happens inside the MPU is under the control of the program and the external control lines. However, if the processor chip fails in operation, there is nothing that can be done to repair it; therefore, we will concern ourselves with writing programs to control the external process and the interface between the input/output (I/O) port and the elements of the process (i.e., motors, relays, etc.).

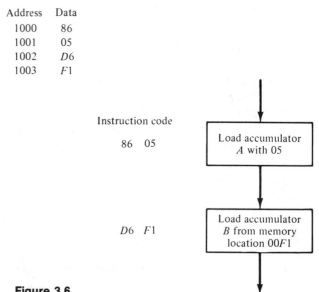

Address	Data
1000	86
1001	05
1002	D6
1003	F1

Instruction code

86 05 → Load accumulator A with 05

D6 F1 → Load accumulator B from memory location 00F1

Figure 3.6

3.3 THE 6800 ADDRESSING MODES

Immediate Addressing (IMM)

In this mode the data accompanies the instruction; the data are permanent in the program. For example, load accumulator A with number 37:

$$\text{LDA - A\#} \quad 37 \tag{3.5}$$

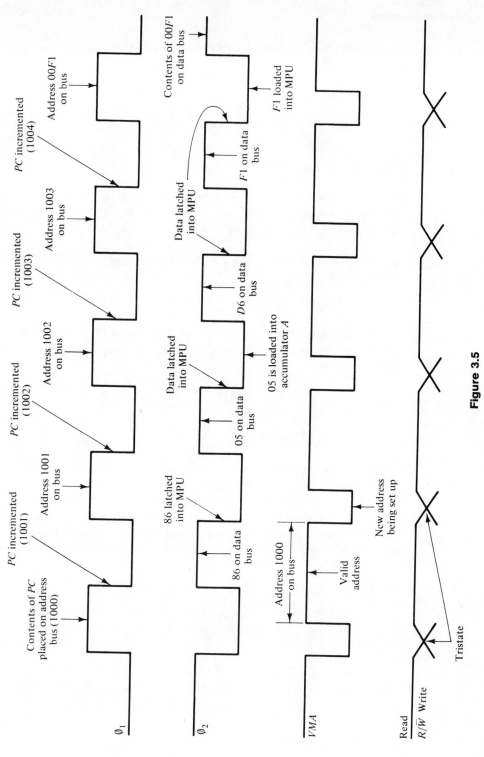

Figure 3.5

Inherent or Accumulator Addressing (INHERENT)

All the information is contained in the instruction; for example, add 1 to the contents of accumulator B:

$$INC - B \qquad\qquad (3.6)$$

Direct Addressing (DIR)

The instruction contains the address of the data to be operated on or transferred; the address *must* start with 00, that is, page 00. For example, load accumulator A from address 00F0:

$$LDA - A \quad F0 \quad \text{(00 is implied)} \qquad (3.7)$$

Extended Addressing (EXT)

The instruction contains the address of the data to be operated on; this address contains both the page and the line. For example, load accumulator B from address 0137:

$$LDA - B \quad 0137 \qquad\qquad (3.8)$$

Relative Addressing (REL)

Relative addressing is used to transfer the control position of the program to a point other than the next sequential location; it specifies a new memory location, *containing an instruction*, relative to the present position. For example, when a branch operation is executed, the program counter (PC) is two memory locations ahead of the current instruction. For example, *branch always* from 0036 to 003A; this constitutes a forward branch (see Figure 3.7).

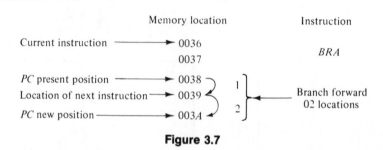

Figure 3.7

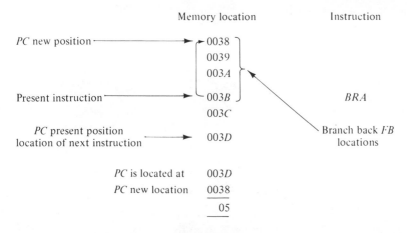

Figure 3.8

Consider now a backward branch, for example, branch always from 003B to 0038 (see Figure 3.8); since the branch is backward, the 2's complement must be taken of the result (05), which is FB.

The maximum move forward is 127_{10}, and 128_{10} the maximum backward relative to the PC's current position; this is because the number system used during relative addressing is the 2's complement system.

Indexed Addressing (IND)

This mode is a method of modifying addresses by fixed increments, relative to the contents of the index (X) register. The contents of the X register are added, in a temporary register, to a *positive* number called an *offset* to make up a new address. The *actual contents* of the X register are *not* changed. As an example, consider the following:

Store the contents of Acc-A at 02 locations from the address stored in the X register (see Figure 3.9). The result is that the contents of Acc-A are stored at memory location 0079.

3.4 PROGRAMMING THE 6800

A microprocessor will *not* operate without a set of sequential instructions. It has already been stated that the processor fetches an instruction from memory, performs the required operation, and then fetches the next instruction. It cannot stand still unless it is in the *wait* or *halt* state. Therefore, the microprocessor must have a continual supply of instructions, called a *program*. The first byte of each instruction, called the operation

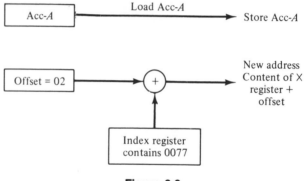

Figure 3.9

code (op-code), is *always* obtained from the instruction set (Appendix A). A typical instruction is shown in Figure 3.10 and will be used to illustrate how op-codes are used.

Example 1. Write the complete instruction codes for the operations shown in the flow diagram of Figure 3.11. An example of writing the program is given in Figure 3.12; the program itself is meaningless.

Load Accumulator   **LDA**

Operation:	$ACCX \leftarrow (M)$
Description:	Loads the contents of memory into the accumulator. The condition codes are set according to the data.
Condition Codes:	H: Not affected.
	I: Not affected.
	N: Set if most significant bit of the result is set; cleared otherwise.
	Z: Set if all bits of the result are cleared; cleared otherwise.
	V: Cleared.
	C: Not affected.

A	**B**	**C**	**D**	
Addressing Modes	Execution Time (No. of cycles)	Number of bytes of machine code	Coding of First (or only) byte of machine code	
			HEX.	
A IMM	2	2	86	
A DIR	3	2	96	
A EXT	4	3	B6	
A IND	5	2	A6	
B IMM	2	2	C6	
B DIR	3	2	D6	
B EXT	4	3	F6	
B IND	5	2	E6	

Figure 3.10

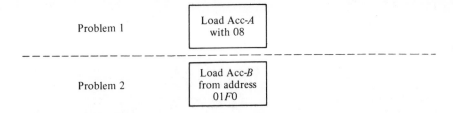

Problem	Address mode column *A*	Op-code column *D*	Data	Mnemonic	Remarks
1	*A IMM*	$\dfrac{86}{\text{1st byte}}$	$\dfrac{08}{\text{2nd byte}}$	LDA-A	Execution time 2 cycles (column *B*) Bytes of machine code 2 (column *C*)
2	*B EXT*	$\dfrac{F6}{\text{1st byte}}$	$\dfrac{01\ F0}{\begin{array}{c}\text{2nd and}\\\text{3rd bytes}\end{array}}$	LDA-B	Execution time 4 cycles Bytes of machine code 3

Figure 3.11

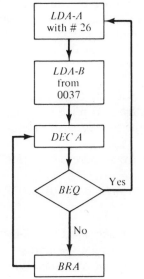

1. Load Acc-*A* with a number 26; therefore, addressing mode is *IMM*. OP-code 8̲6̲, Date 2̲6̲, 2-byte instruction.

2. Load Acc-*B* from address 0037, i.e., page 00 line 37, addressing mode *DIR*. OP-code *D̲*6, Data 3̲7̲, 2-byte instruction

3. Decrement *A* (subtract 1 from contents of Acc-*A*) addressing mode INHERENT. OP-code 4̲A̲, 1-byte instruction.

4. Check Acc-*A* (previous instruction dealt with Acc-*A*) to see if the contents are zero and branch back to start of program, addressing mode INHERENT. OP-code 2̲7̲, Data (to be determined), 2-byte instruction.

5. Branch always addressing mode INHERENT. OP-code 2̲0̲, Data (to be determined), 2-byte instruction.

Figure 3.12

Table 3.1

Address	Machine Code Op-Code	Data	Mnemonic	Remarks
2000	86	26	LDA-A #	
2002	D6	37	LDA-B	
2004	4A		DEC-A	
2005	27	F9	BEQ	7 steps back (-7 = F9)
2007	20	FB	BRA	5 steps back (-5 = FB)

The program will now be written in machine code (Table 3.1), starting at an arbitrary address (e.g., 2000). Each byte of the instruction code (op-code and data) is located at separate but adjacent addresses; that is, 86 is stored at address 2000 and 26 is stored at address 2001, and so on.

A mnemonic is a symbol that represents a valid machine instruction; mnemonics will be used for the remainder of the text to improve the readability of the programs. A system of abbreviations will also be used to identify two of the addressing modes, immediate and indexed: # is used to indicate immediate addressing mode, a 2-byte instruction; (N,X) is used to indicate the indexed addressing mode, where N is the positive offset, a 2-byte instruction.

If a comparison is made of the number of bytes of machine code in an instruction, it will be found that the number is either 1 (for inherent, accumulator, or relative addressing), 2 (for direct, indexed, or immediate addressing), or 3 (for extended addressing). Two of the 2-byte instructions are identified; the third 2 byte instruction, direct addressing, is not identified but can be easily distinguished from the extended addressing mode, which is also not identified, by the number of bytes in the instruction i.e. 3:

86	26	LDA-A#	Load Acc-A with 26 (IMM mode)
A6	03	LDA-A (3,X)	Load Acc-A from 03 relative to the contents of the index register (IND mode)
96	15	LDA-A	Load Acc-A from address 0015 (DIR mode)
B6	0173	LDA-A	Load Acc-A from address 0173 (EXT mode)

Thus, extended addressing will be apparent, as it is only used with 3-byte instructions. Similarly, with inherent accumulator or relative addressing, they are the only addressing modes used with 1-byte instructions. Which mode is in use at the time will be obvious by the instruction.

The instruction set of the 6800 contains 72 executable instructions as well as their description of operation, condition codes affected, execution time, and so on. The processor can *only* perform those tasks that have codes assigned to them by the manufacturer, which, collectively, in the 6800, form 197 valid machine codes.

Basically, the instruction set can be divided into four specific instruction groups; details of these groups and each instruction within the group are contained in Appendix A. A good understanding of the processor's language, addressing modes, system timing, interrupt handling, and so on, is essential for the design and operation of trouble-free systems.

Programming the microprocessor in machine language (hexadecimal bytes) requires that the programmer direct each step of the operation to be performed. If a step is omitted or forgotten, the program will *not* run correctly; *a processor cannot think for itself.* It does exactly what it is told via the program.

If something goes amiss with the program, such as 1 byte too many or too few in an instruction, the program will malfunction. So extreme care must be exercised when assembling programs to ensure that the correct number of bytes appears in each instruction. Furthermore, when transfers of control (relocation of the PC) are being used, care must be exercised in calculating the displacement (branch backward or forward), since control may be transferred to a memory location containing data instead of the correct op-code. The processor will interpret these data as op-code and again the operation will malfunction.

3.5 PROGRAMMABLE PERIPHERAL INTERFACES

A programmable peripheral interface (PPI) is an LSI device that contains registers, I/O ports, and an interface (a shared boundary between the parts of a processor system through which information is conveyed) to the microprocessor's data bus. The PPI that will be described here for parallel data transfers is the peripheral interface adapter (PIA), the MC 6820. The peripheral interface for serial data transmission is the asynchronous communications interface adapter (ACIA), the MC 6850. This device translates and buffers data between the parallel byte format of the microprocessor and the asynchronous serial format used by most data communications equipment. Finally, the programmable timer module (PTM), the MC 6840, will be described; this device can be used to achieve very accurate timing operations, without the need to employ software timing loops.

3.5.1 Peripheral Interface Adapter

The PIA is designed to operate with the MC 6800, the MC 6805, MC 6802D5, and others, and without any additional logic, other than address decoding, it can be used in any microprocessor system with correct translation of the control signals. A block diagram of the MC 6820 and the pinout diagram of the MC 6800 are shown in Figure 3.13; the pin out is included at this time to try to give the reader a feeling for the connections that will have to be made between these two devices in the future.

Each PIA chip contains two separately controlled I/O ports A and B; each I/O port comprises a data direction register (DDR), a peripheral data register (PDR), a control register (CR), and two interrupt/control lines, one of which can also act as an output control line. If the appropriate data are loaded into the DDR and the CR upon initialization, one or more of the following characteristics for that port can be defined:

1. I/O direction: each bit in the DDR defines the data direction of the corresponding I/O line, a 0 bit for input lines, a 1 bit for output lines.
2. I/O mode: each port can be programmed as direct I/O, while port A, only, can be programmed to *receive* data in the handshake mode or pulse mode, and port B, only, can be programmed to *output* data in the handshake or pulse mode.
3. Interrupt mode: each port can be programmed as interrupting or noninterrupting.
4. Interrupt control: whether a positive or negative transition of the interrupt/control lines will cause the processor to enter the interrupt cycle can be specified under program control.
5. Output control: control/interrupt lines C_{A2} and C_{B2} can act as an interrupt input line or as a program-controlled output line.

In addition, the control register can be interrogated under program control to reflect the status of the interrupt flags bits 7 and 6, even when the interrupts are masked (interrupt flag sets but processor does not recognize the interrupt).

The Motorola MC 6800 microprocessor is a memory-mapped I/O based system with the PIA located in the address space, at a position selected by the system designer.

The peripheral data register and data direction register share the same address; the register that is actually being serviced when that address is accessed is determined by bit 2 of the control register. In the MEK 6800 D2 evaluation kit, the memory locations selected for the PIA are as follows:

Port A   Address 8004   Peripheral data register or data direction register

Address 8005   Control register A (CRA)

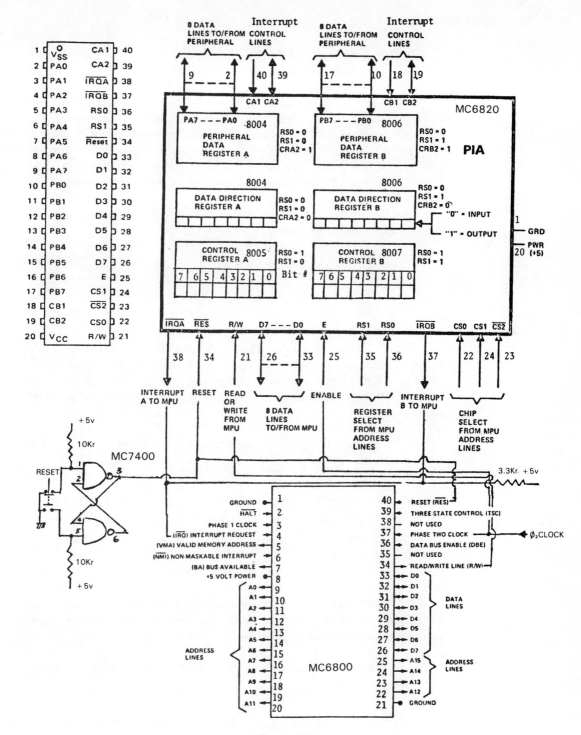

Figure 3.13

Port B   Address 8006   PDR or DDR
                Address 8007   Control register B (CRB)

It was stated previously that a PIA has five basic characteristics and that one or more may be selected at any time under program control. The I/O direction is determined by the 8 data bits written into the data direction register, whereas the I/O mode, interrupt mode, interrupt control, and input/output control are determined by the data written into the control register. Each bit controls some function of the ports' operation; the function of each bit will now be examined.

The bits in any register are always indentified from right to left as 0 to 7; an 8-bit register is shown as follows:

Bit No.:	7	6	5	4	3	2	1	0

Bits 1 and 0 of control register A and control register B are used to determine whether a positive or negative transition of the signal on the interrupt/control line C_{A1} or C_{B1}, respectively, will or will not cause the processor to interrupt its normal operation. Bit 0 controls the interrupt mask, when

$$b_0 = 0 \quad \text{the interrupt is masked (inhibited)} \qquad (3.9)$$

$$b_0 = 1 \quad \text{the interrupt is nonmasked}$$

Bit 1 controls which edge (a positive or negative transition of the signal on line C_{A1} or C_{B1}) will cause the interrupt flag to set:

$$b_1 = 0 \quad \text{interrupt on a negative edge} \qquad (3.10)$$

$$b_1 = 1 \quad \text{interrupt on a positive edge}$$

Bit 2 is the control bit that determines whether the processor is addressing the peripheral data register or the data direction register, when using address 8004 (port A) or 8006 (port B). When

$$b_2 = 1 \quad \text{processor can communicate with PDR} \qquad (3.11)$$

$$b_2 = 0 \quad \text{processor can communicate with DDR}$$

Bits 5, 4, and 3 of control registers A and B control the lines C_{A2} and C_{B2}, respectively:

Bits	5	4	3	Remarks	
	0	0	0	interrupt on a negative edge, interrupt masked	(3.12)
	0	0	1	interrupt on a negative edge, interrupt non-masked	
	0	1	0	interrupt on a positive edge, interrupt masked	
	0	1	1	interrupt on a positive edge, interrupt non-masked	

Bit 5 = 0 signifies that lines C_{A2} or C_{B2} are in the *input or interrupt mode;* mask or nonmask, positive or negative transition interrupts, are determined by bits 4 and 3 as shown in (3.12).

Bits	5	4	3	Remarks	
	1	0	0	handshake mode	(3.13)
	1	0	1	pulse mode	
	1	1	0	following mode	
	1	1	1		

Bit 5 = 1 signifies that the interrupt/control lines C_{A2} or C_{B2} are in the *output* mode; that is, control signals will pass out of these lines. Further various operating modes are initiated, and a detailed description of each will follow.

Handshake Mode, Port A.   When bits 5, 4, and 3 of control register A are coded 100, the handshake mode is initiated. The *output* data lines from a peripheral are tied to the P_{A0} to P_{A7} data lines of port A, and *two* control lines from C_{A1} and C_{A2} are tied to the peripherals' control lines. When the peripheral has data ready to transmit to the processor, it places the data on the data lines and at the same time sends an interrupt signal, either a positive or negative transition via line C_{A1}. The *PIA* responds automatically by raising line C_{A2}; this action informs the peripheral that the interrupt flag is raised and that the processor is aware that data are available, and it will read the data as soon as possible. When the processor loads the data in from the PIA port by means of an LDA statement, the port automatically sets line C_{A2} low, which now informs the peripheral that the data presented have been read. The peripheral will repeat the operation until all data have been transferred. A block diagram and timing diagram of the system are shown in Figure 3.14. In this operating mode, the peripheral presents data and informs the processor of their presence by creating an interrupt. The mode is used when data flows are irregular and could be used to transfer important data (e.g., furnace temperatures).

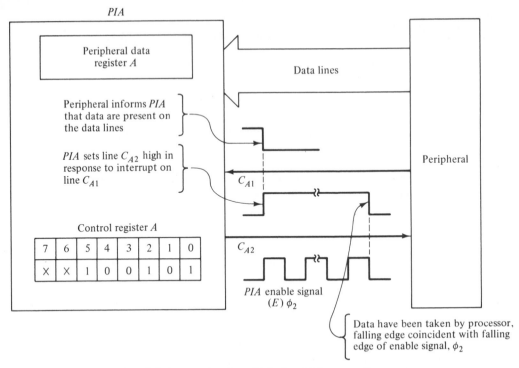

PIA

Peripheral data register *A*

Data lines

Peripheral informs *PIA* that data are present on the data lines

PIA sets line C_{A2} high in response to interrupt on line C_{A1}

Peripheral

C_{A1}

Control register *A*

7	6	5	4	3	2	1	0
X	X	1	0	0	1	0	1

C_{A2}

PIA enable signal $(E)\,\phi_2$

Data have been taken by processor, falling edge coincident with falling edge of enable signal, ϕ_2

Line C_{A1} is coded via control register *A* bits 1 and 0 for a negative transistion interrupt.

Figure 3.14 Handshake Mode Port A

Pulse Mode, Port A. When bits 5, 4, and 3 of control register A are coded 101, the pulse mode is initiated. The *output* data lines from a peripheral are tied to the data lines of port A, and *one* control line from C_{A2} is tied to the peripherals' control line; this is normally high. The peripheral places data on the data lines, but does *not* signal the processor that it has done so. The processor will read the data when ready; the PIA will then signal the peripheral that the processor is reading the data by automatically setting line C_{A2} low. When the data have been read, the line returns high and the peripheral now knows that the data presented have been read. A block diagram and timing diagram are shown in Figure 3.15. The signal on line C_{A2} goes low on the first negative edge of the enable signal ϕ_2 *after* the data are read by the processor (LDA instruction) and remains low for 1 cycle of the ϕ_2 clock.

In this operating mode, the processor does not know when data are presented, and reads those that are currently present. This mode could

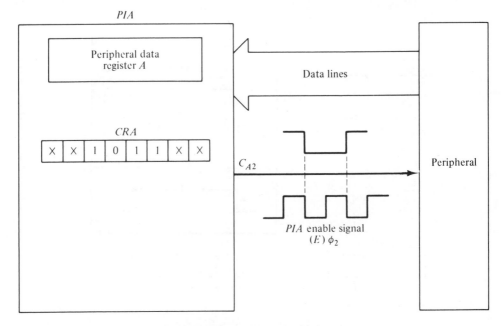

Figure 3.15 Pulse Mode Port A

be used to read a meter periodically. When the data are read, the meter up-dates the information.

Handshake Mode, Port B.   When bits 5, 4, and 3 of control register B are coded 100, the handshake mode is commissioned. The P_{B0} to P_{B7} data lines from port B are tied to the peripherals' *input* lines, and *two* control lines from C_{B1} and C_{B2} are tied to the peripherals control lines. When the peripheral wants data, it signals the PIA by sending an interrupt signal, either a positive or negative transition via line C_{B1}. The PIA responds automatically by raising line C_{B2}; this informs the peripheral that the processor is aware of its request and will place the data on the lines as soon as possible. When the processor places data on the output lines, the PIA will set line C_{B2} low, which informs the peripheral that data are available. The line stays low until the next request for data. A block diagram and timing diagram are given in Figure 3.16. The signal on line C_{B2} goes low on the first positive edge of enable signal ϕ_2 *after* the MPU has stored the data (STA instruction) in the peripheral data register B.

In this mode, the peripheral requests data by raising an interrupt. This mode is used when data requests are irregular, and could be used for a process where a specified starting production rate is given to the peripheral. When that rate is achieved, the peripheral informs the processor by

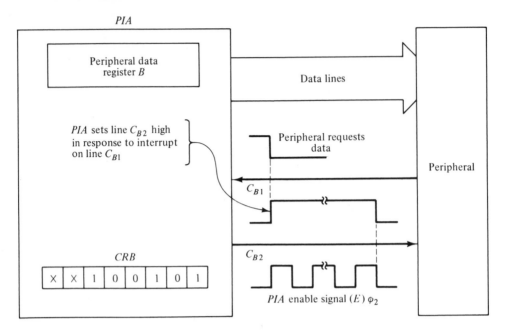

Figure 3.16 Handshake Mode Port B

requesting new data. For example, when a steam turbine is to be started from cold, the case of the turbine must be preheated by allowing quantities of steam into the chamber; as the case temperature rises more steam is admitted.

Pulse Mode, Port B. When bits 5, 4, and 3 of control register B are coded 101, the pulse mode is programmed in. The data lines from port B are tied to the peripherals *input* lines, and *one* control line from CB_2 is tied to the peripherals control line, which is normally *high*. The processor places data on the output data lines and at the same time the PIA sets the CB_2 line low. This indicates to the peripheral that new data have been presented, and it can read it when ready. The line CB_2 goes high after a predetermined time. Whether the data have been read by the peripheral is not known to the processor. A block diagram and timing diagram are shown in Figure 3.17. The signal on line CB_2 goes low on the first positive edge of enable signal ϕ_2 *after* the MPU has stored the data (STA instruction) in peripheral data register B; the line stays low for 1 cycle of the ϕ_2 clock.

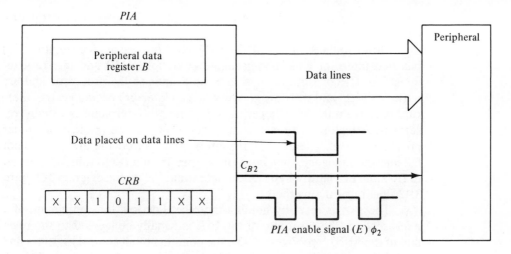

Figure 3.17 Pulse Mode Port B

In this mode, the processor places data on the data lines to the peripheral and informs the peripheral of this fact by transmitting a negative-going pulse. This mode could be used in a speed loop control system; the processor sends out a demand speed to the peripheral, which maintains that speed until a new demand is received.

Following Mode, Ports A and B. Line C_{A2} or C_{B2} is a program-controlled output line and follows the state of bit 3 of control register A or B, respectively, whenever bits 5 and 4 of the relevant registers are coded 11. Bit 3 is changed by an MPU write to control register command (STA instruction). Table 3.2 illustrates the relationship of bits 5, 4, and 3 and the output line. This mode could be used in a single-line control system such as control of a solenoid or relay or a starting pulse for a programmable timer.

Table 3.2 Bit 3 Following Mode

b_5	b_4	b_3	Output line C_{A2}/C_{B2}
1	1	0	0
1	1	1	1

Bits 7 and 6 of each control register are *read only* lines and reflect the status of received interrupts on lines C_{A1} and C_{A2}, respectively, for port A and C_{B1} and C_{B2}, respectively, for port B. Bits 7 and 6 will be set whenever the active transition for which that line is programmed occurs, even if the interrupt is masked; that is, bit 0 is zero. Bits 7 and 6 *can only* be cleared, under program control, by reading the respective peripheral data register.

How are these bits used? The MC 6800 processor chip (Figure 3.13) has two interrupt pins: nonmaskable interrupt ($\overline{\text{NMI}}$) and interrupt request line ($\overline{\text{IRQ}}$). Both of these are input lines to the processor and are normally high, and will only go low when an interrupt occurs via the interrupt lines from the MPU's peripherals. The $\overline{\text{NMI}}$ terminal is usually reserved for high-priority interrupts since it cannot be masked off under program control. Therefore, if the interrupt lines from the PIA port do not fall into this category, then lines $\overline{\text{IRQA}}$ (pin 38) and $\overline{\text{IRQB}}$ (pin 37) must go to $\overline{\text{IRQ}}$ line in a hard-wired OR configuration; the wiring diagram of Figure 3.13 illustrates the method.

The PIA is reset by holding the reset line low for a minimum of 1 microsecond (the reset line of the PIA is usually connected to the reset line of the MPU; see Figure 3.13); the contents of all control registers and data direction registers are cleared to zero. This means that *all* I/O lines are defined as inputs, and since all the bits in the control registers are zero, the MPU will service the *data direction registers* when the peripheral data register/data direction register address (8004 for port A, 8006 for port B) is accessed, because bit 2 of the control register is 0; see Equation (3.11).

Lines C_{A1}, C_{B1}, C_{A2}, and C_{B2} are all defined as masked interrupt lines negative edge sensitive, and should an interrupt occur during the period that the PIA is being initialized, bits 7 and/or 6 of the relevant control register can become set; the processor will *not* be interrupted since the interrupts are masked (control register bit 0 = 0). This can be a cause of concern, and special action must be taken during PIA initialization to avoid erroneous program operation; this will be illustrated in the tutorial examples.

The method of setting up a PIA will now be illustrated. Port A will be used in the examples to follow; however, to access and set up port B the same routine would be followed, except addresses 8006 and 8007 would be used for the PIA in lieu of 8004 and 8005.

Assume the PIA is in the *reset* state and that port A is to be initialized to perform the following:

1. Lines 0 to 3 (P_{A0} to P_{A3}) as data input lines.
2. Lines 4 to 7 (P_{A4} to P_{A7}) as data output lines.

Port A is located at addresses 8005 (control register A) and 8004 (data direction register/peripheral data register). The I/O lines of port A (pins 2 to 9) reset high, approximately +5 V; the port B I/O lines (pins 10 to 17) will reset to a high impedance state, approximately +2.5 V. The block

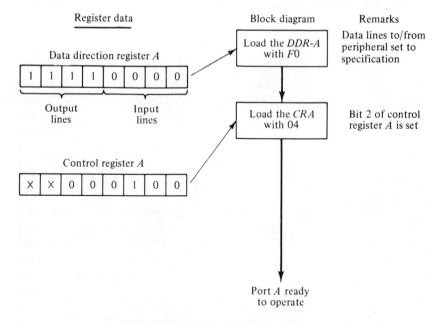

Figure 3.18   Initialization Procedure

diagram of Figure 3.18 will now be converted into the language of the microprocessor, that is, the machine language code of the 6800 as shown in Figure 3.19.

Port A is now ready to operate, and the next time that the MPU addresses memory location 8005, it will be in contact with the peripheral data register. The MPU can now communicate with the outside world and read data in via lines 0 to 3 and output data on lines 4 to 7. To alter the contents of the DDR (data direction register) when bit 2 of the CR (control register) is set will necessitate resetting this bit in the control register and then proceeding as if the PIA were being initialized. For example, the previous example set up port A with lines 0 to 3 as input lines; assume the device is currently in operation.

Set up the contents of the data direction register so that lines 0, 2, 4, and 6 are now input lines (see Figure 3.20).

3.5.2 Asynchronous Communications Interface Adapter (ACIA)

The ACIA is designed to operate with the MC 6800, the MC 6805, and others, without any additional logic other than address decoding. It can be used with any microprocessor system with the correct translation of the control signals. A block diagram of the MC 6850 is shown in Figure 3.21.

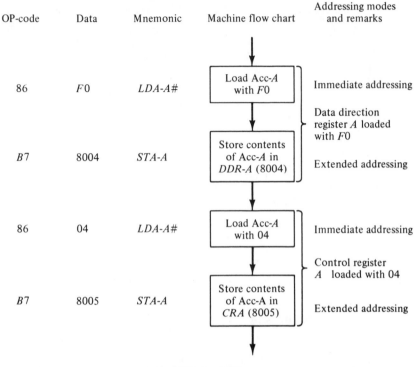

OP-code	Data	Mnemonic	Machine flow chart	Addressing modes and remarks
86	F0	LDA-A#	Load Acc-A with F0	Immediate addressing
B7	8004	STA-A	Store contents of Acc-A in DDR-A (8004)	Data direction register A loaded with F0 · Extended addressing
86	04	LDA-A#	Load Acc-A with 04	Immediate addressing
B7	8005	STA-A	Store contents of Acc-A in CRA (8005)	Control register A loaded with 04 · Extended addressing

Figure 3.19

The ACIA is used for the transmission and reception of serial data. Primarily, the chip is used as part of the interface between the microprocessor and a teletype or CRT. It is sometimes referred to as a UART (universal asynchronous receiver transmitter) since this device performs the same function as the ACIA. This section will describe the action of the device, its registers and its alarms or error flags.

It must be pointed out that the data are transmitted with the least significant bit first. The characters are usually transmitted in 7-bit ASCII code; a table of ASCII characters is contained in Appendix C.

The MPU addresses the ACIA via two different, but usually adjacent, addresses; in the MEK 6800D2 kit they are 8008 and 8009 (see Figure 3.21). These addresses will be used throughout when required in a program. Other addresses could be used as set up by a system designer.

Each ACIA chip contains six registers, two of which, status register (SR) and receive data register (RDR), are read only; two others, the control register (CR) and the transmit data register (TDR), are write only. The remaining two are the transmit (Tx) shift register and the receive (Rx) shift register. A processor interrupt line is also available, as well as sev-

Instruction	Mnemonic	Block diagram	Remarks

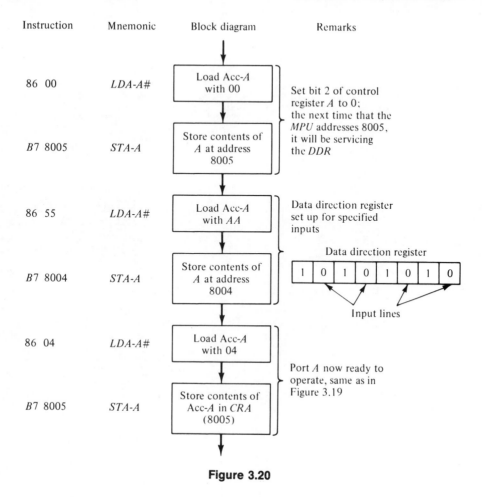

Figure 3.20

eral control lines and separate receive transmit clock lines. For peripheral or modem (modulator/demodulator) operation, three control lines are available, clear to send ($\overline{\text{CTS}}$), request to sent ($\overline{\text{RTS}}$), and data carrier detected ($\overline{\text{DCD}}$).

If the appropriate data are loaded into the CR on initialization, the program can define one or more of the following characteristics.

1. **Word length:** the transmitted character (data bits only) can be 7 or 8 bits long.
2. **Clock division:** the receive and transmit clocks can be divided by fixed ratios, ÷ 1, ÷ 16, and ÷ 64.
3. **Interrupt:** interrupts can be generated by the receive and transmit

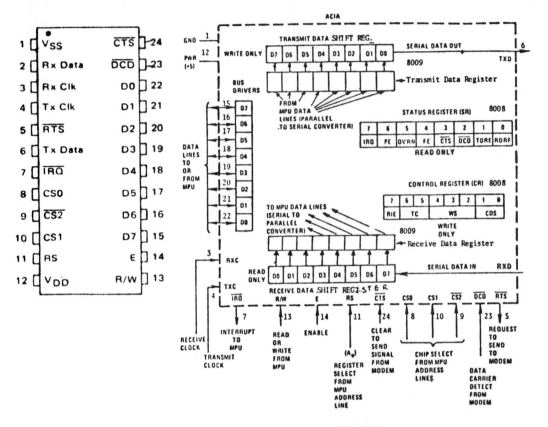

STATUS REGISTER

SR7	SR6	SR5	SR4	SR3	SR2	SR1	SR0
Interrupt Request (IRQ)	Parity Error (PE)	Receiver Overrun (OVRN)	Framing Error	Clear-to-Send	Data Carrier Detect	Transmit Data Register Empty (TDRE)	Receive Data Register Full (RDRF)
				Modem Status			

CONTROL REGISTER

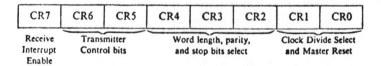

CR7	CR6	CR5	CR4	CR3	CR2	CR1	CR0
Receive Interrupt Enable	Transmitter Control bits		Word length, parity, and stop bits select			Clock Divide Select and Master Reset	

Figure 3.21

sections, each of which is separately controlled, but share the same interrupt line to the MPU.

4. Error checking: checking for errors in the received data is performed as the data are received when the following errors occur: framing error, parity error, or receiver overrun. The status of the error flags is available to the MPU under program control.

5. Parity selection: a horizontal parity bit is available, if required, under program control.

In addition to this, the status register can be interrogated under program control to determine if:

1. The receive data register is full.
2. The transmit data register is empty.
3. The data carrier signal is available.
4. It is clear to send new data.
5. There is a framing error in the currently received data.
6. There is an overrun in the currently received data (i.e., a data character has been overwritten).
7. There is a parity error.
8. There is an interrupt request, as the interrupt to the processor may be masked.

The status register and control register share the same address location 8008; the register actually being addressed by the MPU is determined by the state of the read/$\overline{\text{write}}$ (R/$\overline{\text{W}}$) line. If the R/$\overline{\text{W}}$ line is high, the MPU is in the read or LDA-instruction mode and the status register is being addressed. Similarly, if the R/$\overline{\text{W}}$ is low, data are being written into the transmit data register.

Furthermore, the receive and transmit data registers are double buffered, which means that a data character may be read from the receive data register at the same time that the *next* character is being received. Similarly, a new character may be written into the transmit register at the same time that the previous character is being transmitted; that is, the second character will be automatically transferred into the output shift register when the transmission of the first character is completed.

Clock Frequency. The clock input frequency is determined by the output baud rate (a measure of the data flow rate, the number of signal elements per second, based upon the duration of the shortest element) and the divide by ratio, and is calculated as follows:

$$\text{Input frequency} = \text{baud rate} \times \text{divide by ratio}$$

For example, a 300-baud transmission rate and a ÷ 16 ratio will require a supplied frequency of 4800 Hz (300 × 16). Separate lines are supplied for the receive and transmit clocks because the ACIA may be receiving and transmitting at different baud rates; also the received data must be synchronized *externally* when the ACIA is operating with the ÷ 1 ratio. Internal synchronization is used when operating with ÷ 16 or ÷ 64 ratios, and as such only one input clock is required when the input and output baud rates are the same. Therefore, the Rx and Tx terminals can be tied together and supplied locally with the correct frequency.

Bit synchronization is initiated by the ACIA when it identifies the start bit of the incoming character. The incoming data line is normally high and makes a negative transition at the onset of the start bit. Figure 3.22 illustrates the message format for 7-bit standard teletype ASCII code (American Standard Code for Information Interchange).

Two stop bits are used at the 110 baud rate; above this rate only 1 stop bit is used. This is a generally recognized rule. At the 110-baud rate, 110 character bits will be transmitted per second; this includes start bits, stop bits, parity bits, and the data bits. If a character is 11 bits long (see Figure 3.22), then 10 characters will be transmitted per second. Similarly, at the 300-baud rate with 10 character bits (only 1 stop bit), 30 characters will be transmitted per second.

It was stated previously that various word lengths, clock divisions, interrupts, and parity can be selected at any time under program control by the data written into the CR. Each bit controls some function of the ACIA operation. The function of each bit will now be examined.

Control Register. Bits 1 and 0 control the clock divide by ratios and the master reset, as Table 3.3 shows.

Bits 4, 3, and 2 select the word length, number of stop bits, and parity if required. Table 3.4 illustrates the eight different combinations and the number of bits in a transmitted/received character (including the start bit). When bits 4, 3, and 2 are changed, the change becomes effective immediately as they are not buffered.

Bits 6 and 5 are the *transmitter* interrupt control bits; they also control the request to send line (RTS), which is *only* used when the ACIA is in communication with a peripheral via a telephone link (see Figure 3.23). To achieve this communication link, an additional interface is required, that is, a modem (modulator/demodulator), typically an MC 6860 chip. This device provides fully automatic answering and disconnecting capability, as well as processes the data via modulation and demodulation of the carrier wave. The modem is *not* dealt with here, but a comprehensive treatment can be found in Motorola's Applications Manual Section 3.4.3, page 3-28.

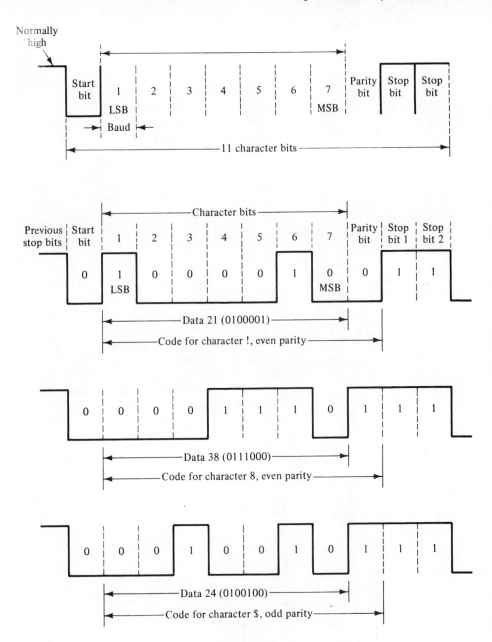

Figure 3.22

Table 3.3

Control Register Bits		
1	*0*	*Remarks*
0	0	$\div 1$, the received data are *externally* synchronized to its clock
0	1	$\div 16$
1	0	$\div 64$	Synchronization is performed internally
1	1	Master reset

Table 3.4

Bit				*Bits in a*
4	*3*	*2*	*Function*	*Character*
0	0	0	7 bits + even parity + 2 stop bits	11
0	0	1	7 bits + odd parity + 2 stop bits	11
0	1	0	7 bits + even parity + 1 stop bit	10
0	1	1	7 bits + odd parity + 1 stop bit	10
1	0	0	8 bits + 2 stop bits	11
1	0	1	8 bits + 1 stop bit	10
1	1	0	8 bits + even parity + 1 stop bit	11
1	1	1	8 bits + odd parity + 1 stop bit	11

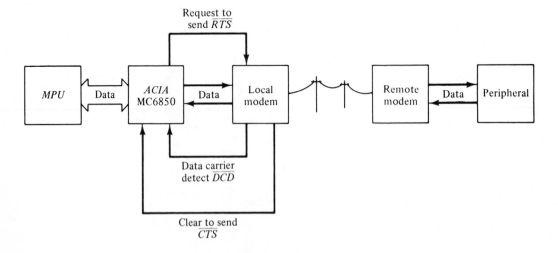

Figure 3.23

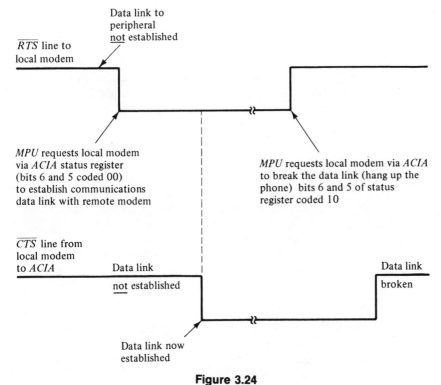

Figure 3.24

Table 3.5 will be used to illustrate the four combinations of bits 6 and 5. The timing diagram of Figure 3.24 illustrates how the $\overline{RTS}$ terminal is used.

Bit 7 is the receive interrupt enable bit. This bit will enable or disable interrupts from the receive section; that is, bit 7 = 1 interrupts enabled, and bit 7 = 0 interrupts disabled.

The method of setting up and operating the control register under program control will now be examined. The ACIA chip is selected when the select control lines C_{S0}, C_{S1}, and $\overline{C_{S2}}$ are coded 110; this is done via the address lines, address decoding logic, and the VMA signal. A fourth control line, a register select line, is usually connected to address line A_0; thus when $A_0 = 1$, the processor is addressing memory location 8009, which is the receive/transmit registers, and when $A_0 = 0$, the processor is addressing 8008, which is the control/status register.

There is no external reset line on the ACIA chip; therefore, the first thing that must be done when the chip is initially accessed in a program is to actuate the master reset. This is accomplished as shown in Figure 3.25. When power is applied to the ACIA (pins 1 and 12 of Figure 3.21), a circuit

Table 3.5

Bit

6	5	*Function*
0	0	$\overline{RTS}$ pin is held *low;* the transmit interrupts are disabled; this code is used as a request to the local modem that the data link be established; it is not clear to send data
0	1	$\overline{RTS}$ pin is held *low;* transmit interrupts are enabled, the data link has been established; it is now clear to send
1	0	$\overline{RTS}$ pin is set *high;* transmit interrupts disabled; this code is used when requesting the local modem to break the data link with the peripheral
1	1	$\overline{RTS}$ pin is held *low;* a break signal will be transmitted to the peripheral; the data link is maintained but data will not be transmitted; transmit interrupts are disabled

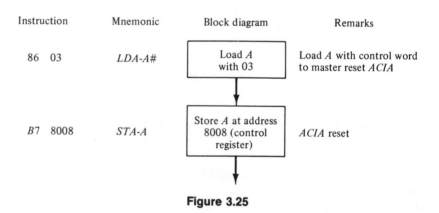

Instruction	Mnemonic	Block diagram	Remarks
86 03	*LDA-A#*	Load *A* with 03	Load *A* with control word to master reset *ACIA*
*B*7 8008	*STA-A*	Store *A* at address 8008 (control register)	*ACIA* reset

Figure 3.25

is initiated (power-on reset circuit) within the chip, which sets specified conditions upon several functions as follows:

1. Resets all the bits in the status register except bits 3 and 2 (these are controlled externally).
2. Sets the $\overline{RTS}$ output line high.
3. Sets the $\overline{IRQ}$ line to the microprocessor high.
4. Resets the receive interrupt.
5. Resets the transmit interrupt.

The power-on reset circuit is used to detect power turn-on transitions and to hold the ACIA in its reset state until initialization is complete; this is necessary to prevent erroneous outputs from occurring. Furthermore, the power-on reset logic will inhibit *any* change in control register bits 6 and 5 until *after* master reset is completed. Therefore, the control word that generates the master reset will not change the $\overline{RTS}$ output or transmit interrupt enable.

The master reset when actuated will:

1. Clear *all* registers.
2. Clear the status register except bits 3 (clear to send) and 2 (data carrier detect), which are controlled by external conditions.
3. Release control of bits 6 and 5 of the CR.

The master reset can also be used in conjunction with the establishment of a communications link without generating an interrupt from the receive or transmit sections. This is accomplished as shown in Figure 3.26. This will cause the $\overline{RTS}$ line to go low and at the same time enable the transmit interrupts. The $\overline{RTS}$ line going low will enable the local modem, which will clear the input line clear to send ($\overline{CTS}$) to the ACIA low when it has established a communication link with the remote modem via the telephone line. The communications link is now established and can be verified by the processor reading the ACIA's status register (which will be covered later in this section).

One final aspect of the power-on reset logic is that it is sensitive to the way in which the power supply to the chip rises when initially turned on. The power turn-on *must* have a positive slope throughout its transition; otherwise, the reset function will fail to operate.

After the master reset has been applied and the communications link has been established, the control register must be set up with the code word for the ACIA's transmit/receive operating mode (see Figure 3.27), for example, 7 bits + odd parity + 1 stop bit, and so on. The ACIA is now ready for operation.

There are five other registers in the ACIA: the status register, transmit data shift register, transmit data register (buffer), receive data shift register, and the receive data register (buffer).

Status Register

The status register (SR) is a read-only register and contains 8 bits or flags. The indications carried by each status bit will now be examined, as well as the ramifications of the bit being set (high) or reset (low).

Instruction	Mnemonic	Block diagram	Remarks

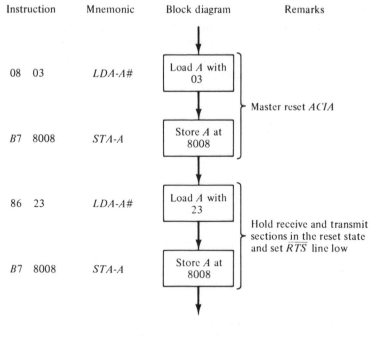

08 03	*LDA-A#*	Load *A* with 03	
B7 8008	*STA-A*	Store *A* at 8008	Master reset *ACIA*
86 23	*LDA-A#*	Load *A* with 23	Hold receive and transmit sections in the reset state and set $\overline{RTS}$ line low
B7 8008	*STA-A*	Store *A* at 8008	

Control register

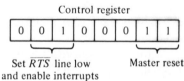

0	0	1	0	0	0	1	1

Set $\overline{RTS}$ line low and enable interrupts

Master reset

Figure 3.26

Instruction	Mnemonic	Block diagram	Remarks

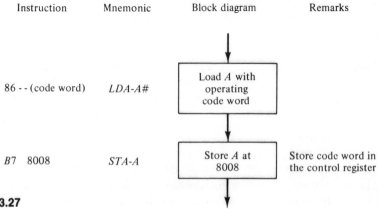

| 86 - - (code word) | *LDA-A#* | Load *A* with operating code word | |
| B7 8008 | *STA-A* | Store *A* at 8008 | Store code word in the control register |

Figure 3.27

Bit 0, receive data register full (RDRF), indicates the status of the receive data register (RDR), and if the bit is set, the register is loaded with *new* data. Reading the data will reset the bit. If the $\overline{DCD}$ *input* is *low,* the RDRF status bit reflects the current status of the receive data register. However, a high on the $\overline{DCD}$ input will force the RDRF status bit to the reset condition (the data could be incorrect) independent of the current status of the receive data register. The bit will remain in this state until the $\overline{DCD}$ input returns to the *low* state. Bit 0 is *not* affected by the $\overline{DCD}$ status bit.

Bit 1, transmit data register empty (TDRE), indicates the status of the transmit data register (TDR); if the bit is set, the contents of the TDR have been transferred to the transmit shift register, and *new* data may be loaded. New data *should not* be loaded into the TDR when the bit is reset as it will destroy the data currently residing there. An internally generated transfer signal will initiate the data transfer from the transmit data register to the transmit shift register. Bit 1 reflects the present status of the TDRE bit as long as it is clear to send, that is, as long as the $\overline{CTS}$ *input* to the ACIA is in the low state.

Bit 2, data carrier detect ($\overline{DCD}$) bit, generally reflects the status of the $\overline{DCD}$ input line from the local modem; it is normally low. When the $\overline{DCD}$ input goes high, it indicates that the data carrier signal has been lost, and the following occur:

1. The $\overline{DCD}$ status bit is set.
2. The RDRF status bit is inhibited and will be reset.
3. The receiver is initialized (the receive data register and the receive shift register are reset).

If the receive interrupt is enabled (i.e., bit 7 of the control register is set), then the loss of the carrier will also cause the following:

1. An interrupt on the processor.
2. Bit 7 of the status register to go high.

The ACIA has variations in its response to the $\overline{DCD}$ input changing its state dependent upon the conditions prevailing at the time. The six timing diagrams that follow will illustrate this.

1. A master reset operation will reset the interrupt status bit (bit 7 of the SR). The $\overline{DCD}$ status bit will follow the $\overline{DCD}$ input (Figure 3.28).
2. The $\overline{DCD}$ status bit will follow the $\overline{DCD}$ input if the $\overline{DCD}$ input goes "high" *during* a master reset (Figure 3.29).

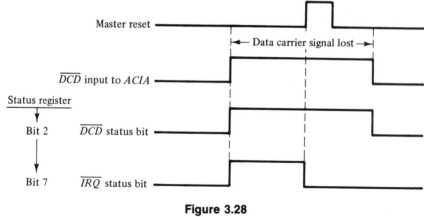

Figure 3.28

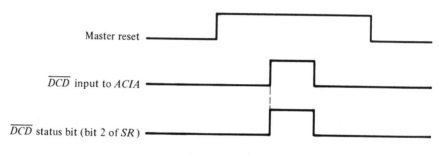

Figure 3.29

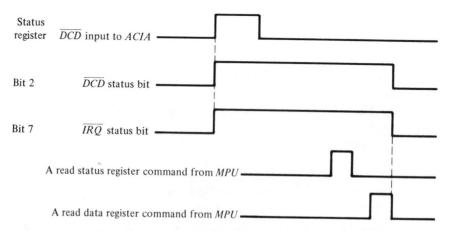

Figure 3.30

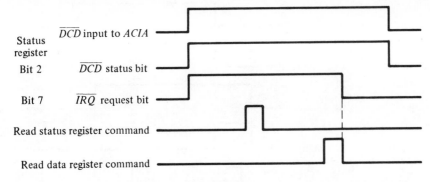

Figure 3.31

3. The interrupt and $\overline{DCD}$ status bits are reset by first reading the status register, followed by a read data command, providing the $\overline{DCD}$ input has returned to the low level.

4. The $\overline{DCD}$ status bit will follow the $\overline{DCD}$ input if a read status register, followed by a read data command, occurs and the $\overline{DCD}$ remains high. The $\overline{IRQ}$ request bit will be reset.

5. The received data are transferred into the RDR (receive data register), the RDRF (receive data register full) status bit goes high, and the interrupt request status bit goes high ($\overline{IRQ}$). The $\overline{DCD}$ input and status bit are both low. Before a read data register command can be executed, the $\overline{DCD}$ input goes high, the $\overline{DCD}$ status bit follows, and the RDRF status bit is immediately inhibited low. The read data command is executed, the data is loaded into the processor, and the interrupt status bit is *still* set. To reset this bit, a read status register followed by a read data command must be executed. The $\overline{DCD}$ status will follow the $\overline{DCD}$ input.

6. If the $\overline{DCD}$ input goes high during a read status register command or read data register command, it will *not* be recognized until the trailing edge of the read status register command, at which time the $\overline{DCD}$ status bit and the $\overline{IRQ}$ status will go high.

NOTE: If the $\overline{DCD}$ input is *not* used, it must be tied *low*.

Bit 3 ($\overline{CTS}$) continuously reflects the state of the $\overline{CTS}$ input from the peripheral. The $\overline{CTS}$ bit will be raised when the local modem signals that it is *not* clear to send data; it does this by raising the $\overline{CTS}$ line high. A high on the $\overline{CTS}$ input will inhibit the operation of the TDRE status bit and the associated interrupt status bit. The $\overline{CTS}$ input has no effect upon a character being transmitted or a character in the TDR; that is, the transmitter section is *not* initialized by the $\overline{CTS}$ input going high. If the $\overline{CTS}$ input is *not* used, it *must* be tied *low*.

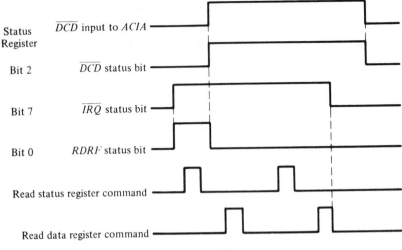

Figure 3.32

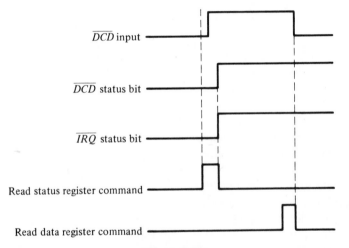

Figure 3.33

Bit 4 is the framing error status bit (FE), and when reset it indicates that the received character is correctly framed with its start and stop bits. A framing error indicates the absence of the *first* stop bit due to the following:

1. Loss of synchronication
2. "Break" in transmission (all spaces)
3. Faulty transmission

The framing error bit is updated when the data are transferred from the receive shift register into the receive data register. The timing diagram of Figure 3.35 will illustrate this. A high on the $\overline{\text{DCD}}$ input will disable and reset the FE status bit.

Bit 5 is the overrun error bit (OVRN) and, when low, indicates that the data in the receive data register (RDR) is valid. An overrun condition exists when a character has been received and transferred into the RDR and *not* read, and the *next* character has been received and is being held in the receive shift register. The overrun condition begins at the midpoint of the last bit of the *second* character received in succession without a read cycle having occurred.

The first read data register command forces the receive data register full status bit to remain high and the OVRN status bit to set, if an overrun condition exists. The *next* read command returns the RDRF and the OVRN status bits low. Figure 3.34 illustrates this. A high in $\overline{\text{DCD}}$ input will disable and reset the OVRN flag.

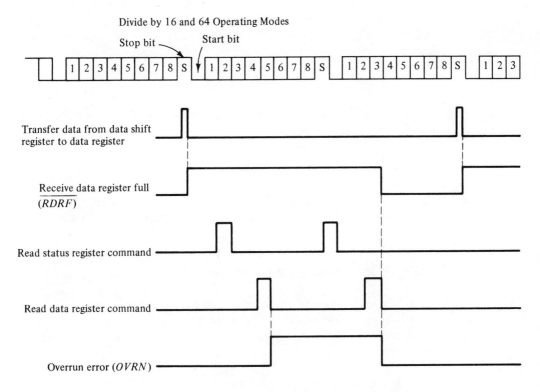

Figure 3.34 Divide by 16 and 64 Operating Modes.

Bit 6, parity error bit (PE), is used to indicate *incorrect* parity and will be set when a character with incorrect parity is transferred into the RDR. The status of the bit is updated during the data transfer cycle, as Figure 3.35 illustrates. A high on $\overline{\text{DCD}}$ input will disable and reset the PE bit.

Bit 7 is the interrupt request bit ($\overline{\text{IRQ}}$) and will be set when any of the following occurs:

1. When the transmitter data register is empty *and* the transmitter interrupts are enabled (bits 6 and 5 of the control register).
2. When the receive data register is full *and* the receive interrupt enable bit is set (bit 7 of the control register).
3. When the $\overline{\text{DCD}}$ input line goes high *and* the receive interrupt enable bit is set (bit 7 of the control register).

The state of bit 7 reflects the state of the $\overline{\text{IRQ}}$ line to the processor; that is, when the bit is set, an interrupt request is on the processor for service.

The bit is reset by a read RDR or a read status register followed by a write into TDR. It is possible that the status word will change before, during, or after the status register is read. This does not present a problem because the status register is *not* reset after a read cycle is executed, and the status change will manifest itself during the next read cycle.

The procedure to receive data is as follows:

1. Read the status register to determine if bit 0 = 1 (is the receive data register full).
2. Check for framing errors, parity errors, and overrun errors.
3. Read the data if they have no errors.

A logic flow chart can now be drawn, as shown in Figure 3.36, and its machine-dependent flow chart is shown in Figure 3.37. The program is located at starting address 00E0 for illustration only.

3.5.3 Programmable Timer Module PTM

The PTM is designed to operate with the 6800 family of components with no need for additional logic other than address decoding. The MC 6840 timer module is used in association with a microprocessor to achieve the following:

1. Accurate timing loops
2. Counting external events

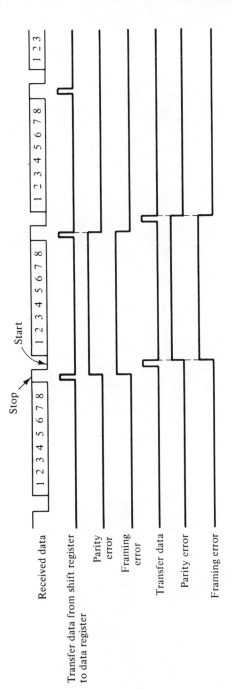

Figure 3.35

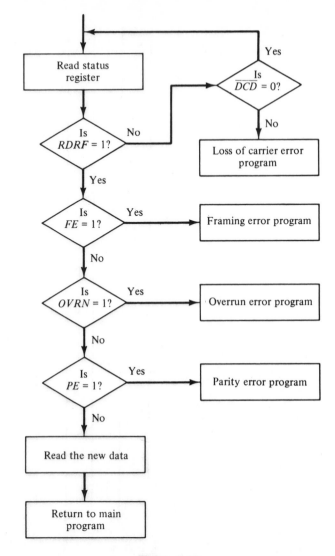

Figure 3.36

3. Frequency measurements
4. Square-wave generation
5. Single pulses of controlled duration
6. Pulse-width modulation
7. Numerous other related timing tasks

Address	Instruction	Mnemonic	Block diagram

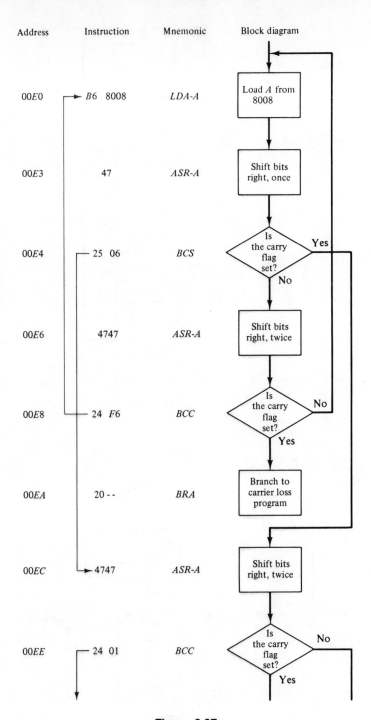

Figure 3.37

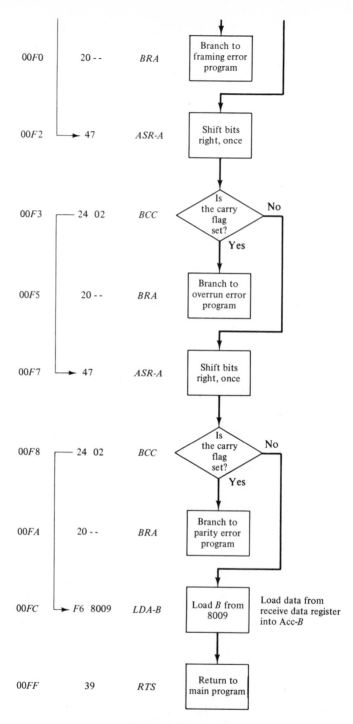

00F0	20 - -	BRA	Branch to framing error program
00F2	47	ASR-A	Shift bits right, once
00F3	24 02	BCC	Is the carry flag set?
00F5	20 - -	BRA	Branch to overrun error program
00F7	47	ASR-A	Shift bits right, once
00F8	24 02	BCC	Is the carry flag set?
00FA	20 - -	BRA	Branch to parity error program
00FC	F6 8009	LDA-B	Load B from 8009 — Load data from receive data register into Acc-B
00FF	39	RTS	Return to main program

Figure 3.37 *(cont.)*

102

The module contains three timers that can operate independently of one another and the microprocessor itself. Thus the processor is released to execute its primary tasks while the timer is executing its assigned functions. The timer has two independent outputs; either or both may be used.

1. An output terminal from each timer to external loads
2. An interrupt terminal that can interrupt the processor in its assigned tasks

The chip comprises three decrementing-type counter/timers, three 16-bit binary latches, three write-only control registers, one read-only status register, two chip select lines, C_{S1} and $\overline{C_{S0}}$, and three register select lines, R_{S2}, R_{S1}, and R_{S0}. A block diagram of the MC 6840 is shown in Figure 3.38.

Each timer has its own gating line $(\overline{G_1}, \overline{G_2}, \overline{G_3})$ to start the counter (timer) and its own output pulse line (O_1, O_2, O_3), which emits a pulse when the correct internal conditions exist; its mark/space ratio is determined by the mode in which the timer is operating. Furthermore, there is a choice of clocks that can be used to decrement the counter (i.e., the enable or internal clock), which is normally taken from the processor's ϕ_2 terminal or an external clock that can be fed into the selected counter via the specified terminal, $\overline{C_1}$ or $\overline{C_2}$ or $\overline{C_3}$. Each counter has the capability to interrupt the processor via the $\overline{IRQ}$ or $\overline{NMI}$ line when the preset count has been decremented to zero or time out (TO) has been reached. The counters are all reset simultaneously either externally via the reset line or internally via control register 1 (CR10).

The basic operation of a timer is as follows: data (the count to be decremented) are loaded into a latch from the MPU via the 8-bit bidirectional data bus, and on command, the initialization signal, the data are transferred from the latch into the counter, which is then decremented by 1 for each clock cycle, be it the external or the internal clock. When the preset count has been decremented to 00 + 1 count, a positive pulse may be transmitted via the output terminal and the appropriate action occurs in the connected system.

A description of how to initialize the programmable timer module from an initial condition and also how to change the mode of operation, and the like, will now follow:

Initially, after the reset line has been held low for a period exceeding 10.25 μs, five major events have occurred within the chip:

1. *All* the counter latches are preset to their maximum count; that is, $65,536_{10} = FFFF_{16}$ or the count currently being held in the latches.

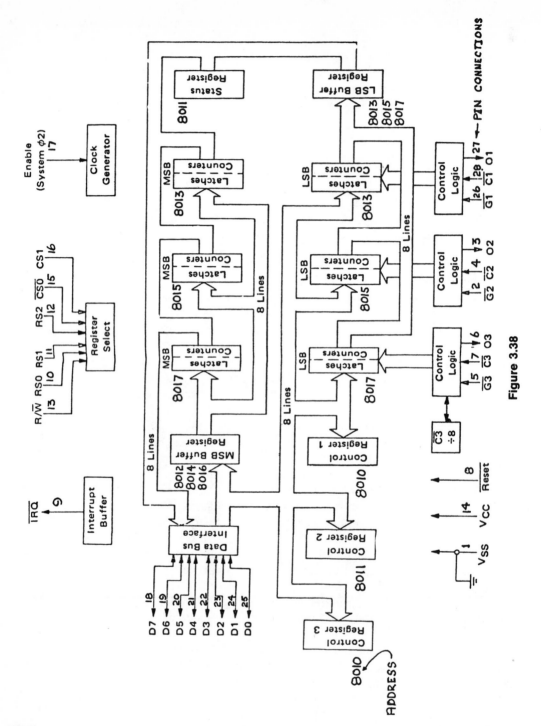

Figure 3.38

2. *All* control register bits in control registers 1, 2, and 3 are set to zero *except* bit 0 of CR1; this bit is set at 1.

3. *All* counter outputs are reset to zero.

4. *All* counter clocks are disabled.

5. *All* status register bits (they reflect the status of the interrupt flags) are cleared.

Under normal operating conditions, control register 2 (CR2) is the master register, and to alter the contents of CR1 and/or CR3, bit 0 of CR2 must be utilized since the addresses of CR1 and CR3 are the same; that is, the register select lines R_{S2}, R_{S1}, R_{S0} are coded 000, as Table 3.6 illustrates. These lines are usually tied to address lines A_2, A_1, and A_0, respectively. Therefore, control register 3 can be written into immediately following the positive transition of the reset line, since bit 2 of CR2 is cleared by the reset operation.

Table 3.6

$(A_2)\,R_{S2}$	$(A_1)\,R_{S1}$	$(A_0)\,R_{S0}$	*Bit 0 of CR2*	*Remarks*
0	0	0	0	Processor can write to CR3
0	0	0	1	Processor can write to CR1

Each control register is an 8-bit device; the function of each bit will be described, commencing with bit 0 of each register.

Control register 1: bit 0 is the timer module internal reset bit. All timers are allowed to operate when bit 0 = 0. When bit 0 = 1, all the timers are held in their preset state (the binary number contained in their latches is automatically loaded into the counters); that is, all timers are reset to their original counts.

Control register 2: bit 0 is a CR1 and CR3 address bit and has been previously dealt with.

Control register 3: bit 0 is timer 3 clock selection control bit. When the bit is cleared, the clock actuating timer 3 is divided by 1. When the bit is set, the clock is divided by 8.

The remaining bits of the control registers can be grouped together since identical bits perform similar functions with its related timer. The following comments relate to CR1, CR2, and CR3.

Bit 1 is the timer clock source selection bit.

Binary Value

0   The associated timer is actuated by an external clock applied to the specified terminal (i.e., $\overline{C_1}$, $\overline{C_2}$ or $\overline{C_3}$). The clock line is active low.

1   The timer uses the internal clock or the clock fed into the enable terminal (17), usually ϕ_2.

Bit 2 is the counting mode control bit.

0   The associated timer is set up to operate with a 16-bit binary number in its counter.

1   The timer is set up to operate with two 8-bit binary numbers in its counter, an upper and lower number.

Bits 5, 4, and 3 are the counter operating mode and interrupt control bits. These bits, in conjunction with bit 2, make up the 12 different operating modes of each timer; each will be discussed in detail later in this section.

Bit 6 is the timer interrupt enable bit.

0   The associated interrupt via the $\overline{\text{IRQ}}$ or $\overline{\text{NMI}}$ line is masked; the internal interrupt flag is still operative.

1   Interrupt line enabled.

Bit 7 is the timer counter output enable bit.

0   The output pulse is masked from appearing at its associated output terminal, O_1, O_2, O_3.

1   The output terminal is enabled and a positive pulse will be transmitted under the appropriate conditions.

Figure 3.39 is a condensed table of the preceding control register bits.
An example will now be given to illustrate the preceding outline.

Example 1. Set up an 8-digit number so that CR1 will control its timer to the given specification (see Table 3.7).
The format of control register 1 is

b_7	b_6	b_5	b_4	b_3	b_2	b_1	b_0
1	0	×	×	×	0	1	0

  (3.14)

The contents of the status register will now be examined; it is a read-only register and contains *only* the status of the interrupt flags.

Bit 0 reflects the status of timer 1 interrupt flag.
Bit 1 reflects the status of timer 2 interrupt flag.

CR10 Internal Reset Bit		CR20 Control Register Address Bit		CR30 Timer #3 Clock Control	
0	All timers allowed to operate	0	CR#3 may be written	0	T3 $\overline{\text{Clock}}$ is not prescaled
1	All timers held in preset state	1	CR#1 may be written	1	T3 $\overline{\text{Clock}}$ is prescaled by ÷8

CRX1*	Timer #X $\overline{\text{Clock}}$ Source
0	TX uses external clock source on $\overline{\text{CX}}$ input
1	TX uses Enable clock

CRX2	Timer #X Counting Mode Control
0	TX configured for normal (16-bit) counting mode
1	TX configured for dual 8-bit counting mode

CRX3 CRX4 CRX5	Timer #X Counter Mode and Interrupt Control (See Table 3)

CRX6	Timer #X Interrupt Enable
0	Interrupt Flag masked on $\overline{\text{IRQ}}$
1	Interrupt Flag enabled to $\overline{\text{IRQ}}$

CRX7	Timer #X Counter Output Enable
0	TX Output masked on output OX
1	TX Output enabled on output OX

*Control Register for Timer 1, 2, or 3, Bit 1.

Figure 3.39

Table 3.7

Specification	Bit No.	Binary Value
All timers must be able to operate	0	0
Timer must run on the enable clock	1	1
Counting mode to be 16-bit	2	0
Bits 5, 4, 3 not yet utilized	3	× Not
	4	× known
	5	×
Interrupt flag bit masked	6	0
Output terminal enabled	7	1

Bit 2 reflects the status of timer 3 interrupt flag.

Bits 6, 5, 4, and 3 are not used and appear as zero when read.

Bit 7 is a composite interrupt flag that will be set if any one of the following occurs:

1. An interrupt occurs on timer 1 provided bit 6 of its control register is set.
2. An interrupt occurs on timer 2 provided bit 6 of its control register is set.
3. An interrupt occurs on timer 3 provided bit 6 of its control register is set.

When bit 7 of the status register goes high, it causes the interrupt line from the timer module to the processor to go low, which interrupts the MPU. When the processor enters its interrupt routine, the status register is read to determine which timer caused the interrupt. The interrupt flag (bit 7 of the status register) can then be reset by one or more of the following methods:

1. External reset line set to zero for *at least* four enable (ϕ_2) clock pulses.
2. Internal reset via bit 0 of CR1.
3. The MPU reads the status register, followed by a read of the *timer counter that caused the interrupt;* the interrupt flag will not be cleared unless the correct timer counter is read.

Each timer in the chip has a counter and each counter has a latch, and the number to be decremented in the counter is first transferred from the MPU to the latch. However, the latch is 16 bits wide and the data bus from the MPU is only 8 bits wide; therefore, the data must be entered in two parts: the most significant byte (MSB) is entered first and is stored temporarily in a register or buffer; it is then transferred into the latch when the LSB is written in. This is accomplished as follows. To enter the MSB the address/register select lines $(A_2)R_{S2}$, $(A_1)R_{S1}$, $(A_0)R_{S0}$ must be coded 010 or 100 or 110; actually they all do the same thing since there is only one temporary storage buffer. The MSB and the LSB are entered into the relevant latch when the MPU presents the LSB to the programmable timer module; the latch and timer being addressed is determined by the coding of the address lines, as follows:

A_2	A_1	A_0	Remarks
0	1	1	MPU writes to timer 1 latch
1	0	1	MPU writes to timer 2 latch
1	1	1	MPU writes to timer 3 latch

For example, set timer 2 latch to F367 (see Figure 3.40). The memory map for this module is shown in Table 3.8.

The preset data are transferred into the counter, when one of the following events occurs:

1. Negative transition of the specified timer gate $(\overline{G_1}, \overline{G_2},$ or $\overline{G_3})$.
2. A write to timer latches command subject to specific timer modes (i.e., modes 1, 3, 5, 7).
3. Internal or external reset.

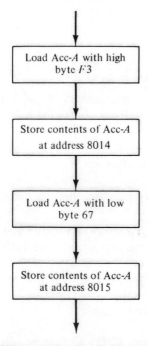

Figure 3.40 Block Diagram

Table 3.8

Address	Read ($R/\overline{W}$ line = 1)	Write ($R/\overline{W}$ line = 0)
8010	No operation	Processor writes to control registers 1 and 3
8011	Read status register	Processor writes to control register 2
8012	Read timer 1 counter	Processor stores MSB in temporary buffer
8013	Read LSB buffer register	Processor stores LSB and transfers MSB into timer 1 latches
8014	Read timer 2 counter	Processor stores MSB in temporary buffer
8015	Read LSB buffer register	Processor stores LSB and transfers MSB into timer 2 latches
8016	Read timer 3 counter	Processor stores MSB in temporary buffer
8017	Read LSB buffer register	Processor stores LSB and transfers MSB in timer 3 latches

The 12 different operating modes of a timer will now be examined, and they will be categorized into four basic groups:

1. Continuous operating modes (1 to 4).
2. Single shot modes (5 to 8).
3. Frequency comparison modes (9 and 10).
4. Pulse-width comparison modes (11 and 12).

The timer employs two different counting modes, which can give vastly differing results, as shown:

1. Sixteen-bit binary number: this is a standard 16-bit number (e.g., $0202_{16} = 514_{10}$).
2. Dual 8-bit binary number, an upper and lower 8-bit number: in this mode the most significant byte (MSB) and the least significant byte (LSB) are separate, and each time the LSB experiences a count down to 00 + 1 the MSB will be decremented by 1. Thus a number such as 0202_{16} is equal to 9_{10} counts, as compared to 515_{10} counts in 16-bit mode.

The dual 8-bit number 0202 will now be offered as an example (see Figure 3.41). The positive edge of the first clock pulse *after* the output lines goes low (initialization) causes the MSB to be loaded into the most significant counter, called the M counter, and the LSB to be loaded into the L counter. On the *next* positive transition of the clock, the start of pulse 2, the L counter is decremented by 1; on the next clock pulse (3) it is decremented again. Now the L counter contains 00. On the *next* clock pulse (4) the M counter is decremented by 1 and the L counter is automatically reloaded with 02. Ultimately, the contents of the L counter will again be 00 (positive edge of pulse 6), the M counter contains 01; on the next positive edge (pulse 7) the L counter is reloaded with 02 and the M counter is decremented to 00. On pulse 8 the output line rises to 1, where it remains for 2 counts, the contents of the L counter. The output line then falls to zero, the counters automatically reload from the latches, and the cycle continues as long as the associated gate line is held low. The individual interrupt flag is automatically raised at time out (TO) for a total of 9 pulses.

Operating modes 1 to 8 inclusive share two common methods of counter initialization:

1. A *negative* transition of the signal that is supplied to the gate line.
2. A timer reset, either external or internal.

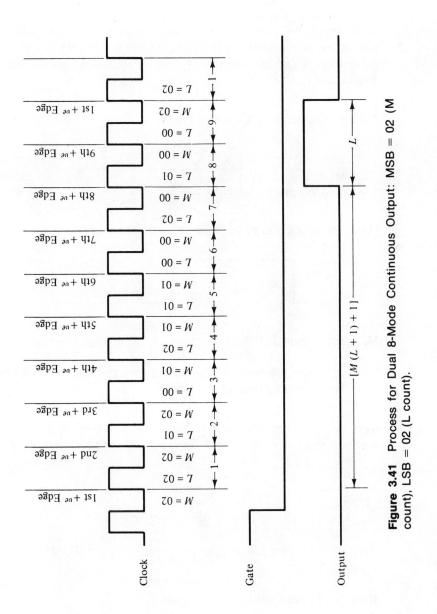

Figure 3.41 Process for Dual 8-Mode Continuous Output: MSB = 02 (M count), LSB = 02 (L count).

Furthermore, modes 1, 3, 5, and 7 share a third initialization method, an MPU write to timer command.

The first positive edge always loads the most significant byte and least significant byte from their latches into their respective counters. The output line goes low during initialization and will remain low until the M is *all* zeros plus 1 additional pulse; see the seventh and eighth + VE edges in Figure 3.41.

Mode 1. Sixteen-Bit Counting Mode

Control register format of the selected register

	b_7	b_6	b_5	b_4	b_3	b_2	b_1	b_0	Bit number
	1	×	0	0	0	0	×	×	× means not known

Pulse out of the output terminal ($b_7 = 1$):

Mark/space ratio 1:1:

N = decimal number contained in the 16-bit binary number
T = periodic time of the clock

Example 2. Determine the width and frequency of the pulse (in microseconds) when $n = 06$ and the frequency of the clock is 1 MHz.

$$T = \frac{1}{f} = \frac{1}{10^6} = 1 \ \mu s$$

$$\therefore \ \text{Width} = (6 + 1)1 = 7 \ \mu s$$

$$\therefore \ \text{Frequency} = \frac{1}{2 \times 7 \times 10^{-6}} = \frac{10^6}{14} = 71,428 \ \text{Hz}$$

Mode 2. Sixteen-Bit Counting Mode

Control register format of the selected register	b_7	b_6	b_5	b_4	b_3	b_2	b_1	b_0
	1	×	0	1	0	0	×	×

The output wave form is identical to mode 1. The only difference between modes 1 and 2 is that in mode 2 the count is *not* reset by a write to timer latches; that is, if the count in the latch is changed partway through a cycle, for example, timer 2 is operating in mode 1 (16-bit mode) with 0002(N) in the latch, the output line O_2 is enabled, and the timer is operating on external clock ÷ 1. Normally three pulses on the clock line will cause the output line to change its state; however, after two pulses of a new cycle have been entered, the latch is reset to 0004. The next output pulse is delayed, as Figure 3.42.

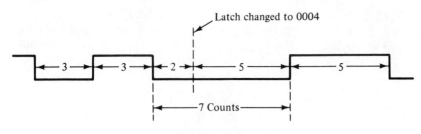

Figure 3.42

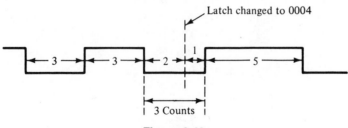

Figure 3.43

Had this same event occurred in mode 2, the output pulse would have appeared as in Figure 3.43. No counts have been lost.

Mode 3. Dual 8-Bit Counting Mode

b_7	b_6	b_5	b_4	b_3	b_2	b_1	b_0
1	×	0	0	0	1	×	×

Pulse out of the output terminal if L ≠ 0 is shown in Figure 3.44, where M = decimal number in the most significant byte and L = decimal number in the least significant byte.

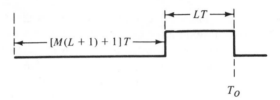

Figure 3.44

Example 3. Determine the width of the mark and the space when the number loaded into the latch equals $0F0A_{16}$, and the frequency of the clock is 2 MHz. Convert the hex number to two decimal numbers: $0F_{16} = 15_{10} = M$ and $0A_{16} = 10_{10} = L$.

$$T = 0.5 \ \mu s$$

$$\therefore \quad \text{Width of space} = [15(10 + 1) + 1]0.5$$
$$= 83 \ \mu s$$

$$\therefore \quad \text{Width of mark} = 10 \times 0.5 = 5 \ \mu s$$

If $L = 0$, the counter will revert to the 16-bit counting mode and the output wave form will appear as in mode 1. If $M = L = 0$, the output will toggle at half the clock frequency.

Mode 4. Dual 8-Bit Counting Mode

b_7	b_6	b_5	b_4	b_3	b_2	b_1	b_0
1	×	0	1	0	1	×	×

The output wave form is identical to mode 3. When a new number is written into the latch in mode 3, counts can be lost similar to mode 1.

Mode 5. Sixteen-Bit Counting Mode

b_7	b_6	b_5	b_4	b_3	b_2	b_1	b_0
1	×	1	0	0	0	×	×

The pulse out of the output terminal is illustrated in Figure 3.45. The width of the mark is NT; the output remains low until reinitialization. The counter continues to cycle at $(N + 1)T$ intervals.

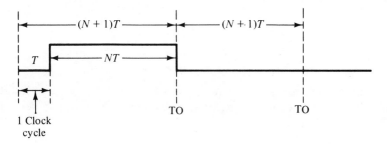

Figure 3.45

The state of the output does *not* depend upon the gate signal remaining low during the countdown operation; a gate pulse is good enough, as long as it has sufficient width, at least 4 enable clock cycles. If the high byte and low byte equal zero (i.e., $N = 0$), then the output line will go low and will stay low until a number is put into the latches.

Since bit 4 of the control register is 0, this mode *can* be initialized by a write to timer latches, and counts can be lost in the same manner as in modes 1 and 3.

Mode 6.   Sixteen-Bit Counting Mode

b_7	b_6	b_5	b_4	b_3	b_2	b_1	b_0
1	×	1	1	0	0	×	×

The output wave form is identical to mode 5 (Figure 3.45). Bit 4 being high means that a write to latches command will *not* initialize the timer.

Mode 7.   Dual 8-Bit Counting Mode

b_7	b_6	b_5	b_4	b_3	b_2	b_1	b_0
1	×	1	0	0	1	×	×

The pulse out of the output terminal is shown in Figure 3.46.

Mode 8.   Dual 8-Bit Counting Mode

b_7	b_6	b_5	b_4	b_3	b_2	b_1	b_0
1	×	1	1	0	1	×	×

The output wave form is identical to mode 7. Bit 4 is 1; thus mode 8 *cannot* be reinitialized by a write to timer latches (change of number in the latches).

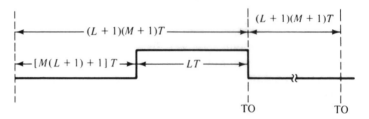

Figure 3.46

Modes 9 and 10 Frequency Comparison. These modes are used to generate an interrupt if the periodic time of the frequency of the signal *on the gate* terminal is less than the countdown time (mode 9) or greater than the countdown time (mode 10); whether the counters operate in 16-bit mode or dual 8-bit mode depends upon the state of bit 2 in the control registers (i.e., $b_2 = 0$, 16-bit mode; $b_2 = 1$ dual 8-bit mode). These time interval modes are used where more flexibility of interrupt generation and counter initialization are required. It is *not* the output that is utilized, but the interrupt; the individual interrupt flags are set as a result of both counter time out and the transitions of the gate input.

Mode 9

b_7	b_6	b_5	b_4	b_3	b_2	b_1	b_0
×	×	1	0	1	×	×	×

An interrupt will be generated *if* the gate input line goes negative *after* time out (TO). See Figure 3.47. However, if the gate makes its negative transition *before* the time out occurs, then an internal flip-flop is set (i.e., the counter enable flip-flop), and the counter *reinitializes* on the negative edge of the gate pulse, as illustrated in Figure 3.48. The counter will continue to reinitialize on the negative edge of the gate signal until TO occurs

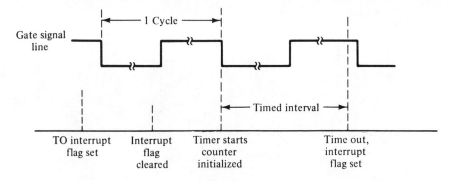

Figure 3.47

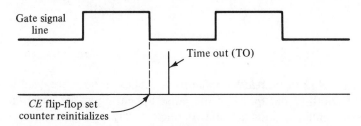

Figure 3.48

before the negative transition. At that time, the interrupt flag is raised and the internal flip-flop is reset.

Mode 10

b_7	b_6	b_5	b_4	b_3	b_2	b_1	b_0
×	×	0	0	1	×	×	×

An interrupt will be generated *if* the gate input line goes negative before time out (TO) occurs (see Figure 3.49). If TO occurs before the gate signal makes its negative transition, the internal CE flip-flop will set; this prevents an interrupt from occurring until after a new counter initialization cycle has been completed. If TO occurs *after* the negative transition of the gate line, then the interrupt flag is raised and the counter enable flip-flop is reset.

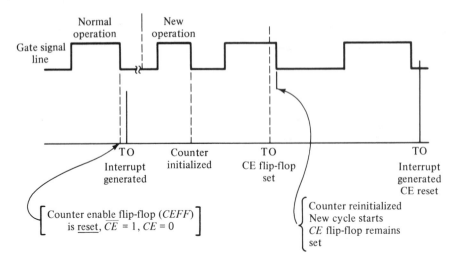

Figure 3.49

Modes 11 and 12 Pulse Width Comparison. These modes are used to measure the width of the *space* of the pulse appearing at the *gate line*. The modes are similar to modes 9 and 10, respectively, except that the counter is activated by the positive edge rather than the negative edge, as in the two previous cases.

Mode 11

b_7	b_6	b_5	b_4	b_3	b_2	b_1	b_0
×	×	1	1	1	×	×	×

An interrupt will be generated if the pulse width is greater than time out, as shown in Figure 3.50. If the gate signal makes its positive transition before TO, the counter will reinitialize on the next negative transition of the clock line, and comparison will continue until either an interrupt occurs or the time is changed.

Mode 12

b_7	b_6	b_5	b_4	b_3	b_2	b_1	b_0
×	×	0	1	1	×	×	×

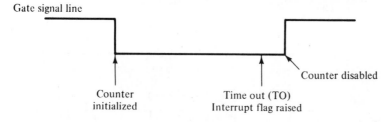

Figure 3.50

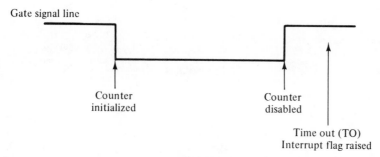

Figure 3.51

The interrupt flag will be raised if time out occurs after the positive transition of the gate line, as Figure 3.51 shows. If the gate line makes its positive transition after TO, the counter will reinitialize on the next negative edge and the comparison will continue.

A timer output is controlled by bit 7 of the control register, and if $b_7 = 0$, the output line will go low and stay low regardless of the operating mode. However, the timer will continue to operate even though no output results.

Gate Inputs: $\overline{G_1}$, $\overline{G_2}$, $\overline{G_3}$.   The gating signal that controls the initialization of the timer, in some modes, is *not* effective until *after* the first negative transition of the enable clock, after the set-up time has elapsed (four enable clock pulses). See Figure 3.52. If the gate goes positive during a timing interval, the change in gate status will *not* become effective until four enable gate pulses have elapsed. See Figure 3.53. The timing process will cease after the positive gate transition is recognized. The $\overline{G_3}$ gate signal set-up time is not affected by using the $\div 8$ presealer.

Input Clock Lines: $\overline{C_1}$, $\overline{C_2}$, $\overline{C_3}$.   These inputs are asynchronous and can be used to decrement the timers. The external clocks are clocked in by the

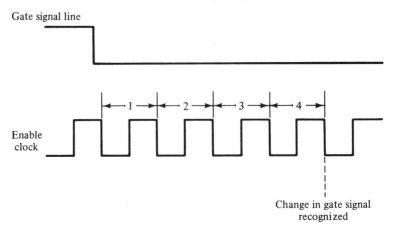

Figure 3.52

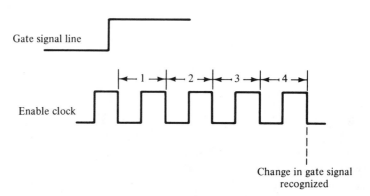

Figure 3.53

enable clock (ϕ_2), and it will take three enable pulses to synchronize and process the external clock. The fourth pulse will decrement the internal counters. Once the external clock is synchronized and its transitions recognized, the timer will be operated satisfactorily by this clock, subject to it falling into the specification laid down by the chip manufacturer. The output wave form will not be in phase with the input wave form because of the time delay due to synchronization and processing of the external clock.

An example of how to initialize the PTM from a reset condition and when operating will now be offered (see Figure 3.54).

The timer has just been externally reset; initialize timer 3 for mode 4 operation with a count of 0204. Disable the interrupts; enable output line

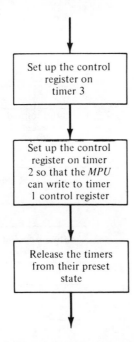

Figure 3.54 Block Diagram

0_3. The activating clock will be external $\div 1$. A machine-dependent flow chart is given in Figure 3.55.

Timer 3 control register will be set up as shown in Figure 3.56. Control register 2 must be accessed to set bit 0 so that the MPU can write to control register 1 (see Figure 3.57). Timer 2 is not required in this example; therefore, all the bits in its control register will be 0 except bit 0.

Control register 1 (Figure 3.58) must also be accessed to clear bit 0; this will release all the timers from their preset state.

The reader should verify the instruction codes of Figures 3.56, 3.57, and 3.58 from Figure 3.55.

Timer 3 is now operating; now initialize timer 2 to operate from the internal clock, interrupts disabled, output terminal 0_2 enabled, operating mode 2, count 3F16 (see Figure 3.59).

Timer 2 now operational. The programs illustrate the simplicity of setting up a timer; once the timer is set up it will operate independently of the processor, which will then release the processor to perform other tasks. The reader should verify the hexadecimal number written into CR2, instruction 0161.

Address	Instruction	Mnemonic	Block diagram	

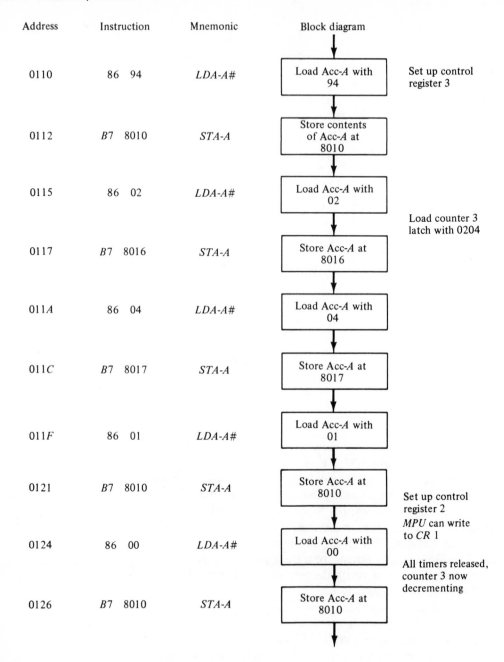

Address	Instruction	Mnemonic	Block diagram	
0110	86 94	*LDA-A#*	Load Acc-*A* with 94	Set up control register 3
0112	*B*7 8010	*STA-A*	Store contents of Acc-*A* at 8010	
0115	86 02	*LDA-A#*	Load Acc-*A* with 02	Load counter 3 latch with 0204
0117	*B*7 8016	*STA-A*	Store Acc-*A* at 8016	
011*A*	86 04	*LDA-A#*	Load Acc-*A* with 04	
011*C*	*B*7 8017	*STA-A*	Store Acc-*A* at 8017	
011*F*	86 01	*LDA-A#*	Load Acc-*A* with 01	
0121	*B*7 8010	*STA-A*	Store Acc-*A* at 8010	Set up control register 2 *MPU* can write to *CR* 1
0124	86 00	*LDA-A#*	Load Acc-*A* with 00	All timers released, counter 3 now decrementing
0126	*B*7 8010	*STA-A*	Store Acc-*A* at 8010	

Figure 3.55

7	6	5	4	3	2	1	0	Bit number
1	0	0	1	0	1	0	0	Binary state
9				4				Hexadecimal number

Figure 3.56

7	6	5	4	3	2	1	0	Bit number
0	0	0	0	0	0	0	1	Binary state
0				1				Hexadecimal number

Figure 3.57 Control Register 2

7	6	5	4	3	2	1	0	Bit number
0	0	0	0	0	0	0	0	Binary state
0				0				Hexadecimal number

Figure 3.58 Control Register 1

3.6 TUTORIAL EXAMPLES

Example 1. Subtract 121_{10} from 54_{10}; list the state of the following flags: N, Z, V, and C.

Decimal		*Binary*	
54 −	Minuend	0 0 1 1 0 1 1 0	+
121	Subtrahend	1 0 0 0 0 1 1 1	(-121)
− 67	Result	1 0 1 1 1 1 0 1	

N flag is set because bit 7, of the result, is set.

Z flag is clear because the result is *not* zero.

V flag clear because there is no 2's complement overflow.

C flag is set because the absolute value of the subtrahend is larger than the absolute value of the minuend.

Address	Instruction	Mnemonic	Block diagram

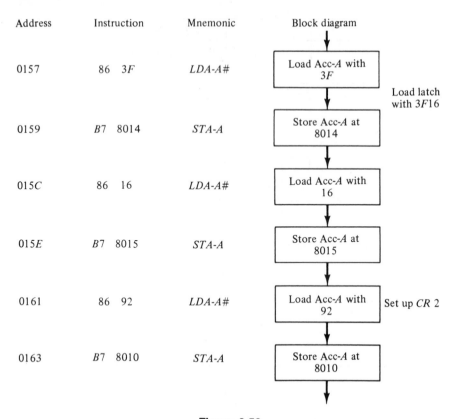

Figure 3.59

Example 2. Subtract 61_{10} from -67_{10}; list the state of the following flags: N, Z, V, and C.

Decimal		*Binary*
-67	Minuend	1 0 1 1 1 1 0 1
61	Subtrahend	1 1 0 0 0 0 1 1 (-61)
-128	Result	1 0 0 0 0 0 0 0

N flag set since bit 7 is set.

Z flag clear.

V flag clear because there is no 2's complement overflow; the register is full.

C flag clear since the absolute value of the subtrahend is less than the absolute value of the minuend.

Example 3. Subtract 62_{10} from -67_{10}; list the state of the following flags: N, Z, V, and C.

Decimal		*Binary*
-67	Minuend	1 0 1 1 1 1 0 1
62	Subtrahend	1 1 0 0 0 0 1 0
-129	Result	0 1 1 1 1 1 1 1

(with $-$ and $+$ operation marks)

N flag is clear since bit 7 is clear.
Z flag is clear.
V flag is set since a 2's complement overflow exists; -128 is the maximum that an 8-bit register can contain.
C flag is clear.
The contents of the register appear as $+127$, an obvious error.

Example 4. Add 58_{10} to 70_{10}; list the state of the following flags: N, Z, V, and C.

Decimal		*Binary*
58	Addend	0 0 1 1 1 0 1 0
70	Augend	0 1 0 0 0 1 1 0
128	Result	1 0 0 0 0 0 0 0

(with $+$ operation mark)

N flag is set since bit 7 is set.
Z flag is clear.
V flag is set since a 2's complement overflow exists; the result is $+128$. The indicated result is -128; an error has occurred.
C flag is clear since there is no carry from bit 7.

Example 5. Add 57_{10} to 70_{10}; list the state of the following flags: N, Z, V, and C.

Decimal		*Binary*
57	Addend	0 0 1 1 1 0 0 1
70	Augend	0 1 0 0 0 1 1 0
127	Result	0 1 1 1 1 1 1 1

(with $+$ operation mark)

All the flags are clear.

Summary of Examples 1 to 5

The arithmetic logic unit (ALU) always considers the numbers that it is currently manipulating as a member of the 2's complement number system. Thus the V flag, when set, indicates a 2's complement overflow, and if an overflow exists the result may be in error. The conditions under which the V flag, C flag, and so on, are raised is contained in Appendix A and require a good understanding of Boolean formulas, signed and unsigned number systems, and 2's complement number systems.

An 8-bit register when used with 2's complement numbers is only capable of handling numbers in the range of -128 to $+127$. Numbers larger than $+127$ (i.e., $+128$ and up) are considered as overflows. Therefore, when numbers are being manipulated extreme care must be taken to ensure that the numbers do not fall outside of the specified range, that is,

$$-128 \leq \text{Range} \leq +127$$

Failure to observe this rule could cause the following problem.

Decimal		*Binary Equivalent*			
183	X_7	1011	0111	(B7)	Minuend
$-\ 54$	M_7	0011	0110	(36)	Subtrahend
$+129$					

The ALU forms the 2's complement representation of (36), the subtrahend, that is, 1100 1010, and then adds this to (B7) the minuend as follows:

$$
\begin{array}{lll}
& 1011 \quad 0111 & \text{(B7)} \\
& \underline{1100 \quad 1010}\ + & \text{(2's complement of 36)} \\
\text{Result} \quad 1 & 1000 \quad 0001 & \\
& \nearrow & \\
& R_7 &
\end{array}
$$

Q: Which flags are set?

A: From Appendix A

$$N \text{ flag} = R_7$$

$$Z \text{ flag} = \overline{R_7} \cdot \overline{R_6} \cdot \overline{R_5} \cdot \overline{R_4} \cdot \overline{R_3} \cdot \overline{R_2} \cdot \overline{R_1} \cdot \overline{R_0}$$

$$V \text{ flag} = X_7 \cdot \overline{M_7} \cdot \overline{R_7} + \overline{X_7} \cdot M_7 \cdot R_7$$

$$C \text{ flag} = \overline{X_7} \cdot M_7 + M_7 \cdot R_7 + R_7 \cdot \overline{X_7}$$

Therefore,

$$N \text{ flag} = 1 \quad (\text{set})$$
$$Z \text{ flag} = \overline{1} \cdot \overline{0} \cdot \overline{0} \cdot \overline{0} \cdot \overline{0} \cdot \overline{0} \cdot \overline{0} \cdot \overline{1} = 0 \quad (\text{clear})$$
$$V \text{ flag} = 1 \cdot \overline{0} \cdot \overline{1} + \overline{1} \cdot 0 \cdot 1 = 0 \quad (\text{clear})$$
$$C \text{ flag} = \overline{1} \cdot 0 + 0 \cdot 1 + 1 \cdot \overline{1} = 0 \quad (\text{clear})$$

which indicates that the V flag is clear, and at first sight this could lead the reader to believe that no overflow exists. This is quite erroneous. An overflow does exist; also the result is incorrect.

Why is this error in overflow detection in effect? Because the minuend is outside of the bounds of the 2's complement number system. Therefore, if one of the binary numbers being manipulated is outside of the 2's complement range, the result may be in error.

Example 6. Draw a block diagram to illustrate how the addition of two unknown signed numbers should be executed so that overflow errors and negative numbers may be observed.

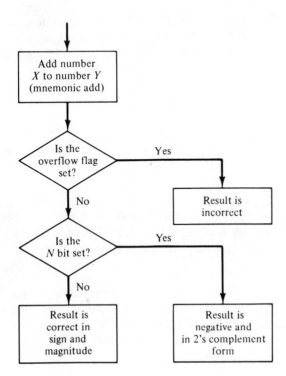

Example 7. Determine the code word to be written into control register B of the PIA to achieve the following: Port B, handshake mode, positive transition interrupt on line C_{B1}.

7	6	5	4	3	2	1	0	Bit number
X	X	1	0	0	1	1	1	Binary state
2				7				Hexadecimal number

Example 8. Determine the code word to be written into the control register of an ACIA to achieve the following: Divide by 16, 8 bits + 1 stop bit, $\overline{RTS}$ pin to be held low and transmit interrupts enabled, receive interrupts disabled.

7	6	5	4	3	2	1	0	Bit number
0	0	1	1	0	1	0	1	Binary state
3				5				Hexadecimal number

Example 9. Determine the code word to be written into control register 3 of the PTM to achieve the following: Output terminal enabled, interrupt disabled, operating mode 6, 16-bit counting mode, internal clock, ÷ 8.

7	6	5	4	3	2	1	0	Bit number
1	0	1	1	0	0	1	1	Binary state
B				3				Hexadecimal number

Example 10. Draw the flow chart of the operations to be performed to change the operating mode of timer 1. All three timers are operating in the PTM module, and bit 0 of control register 2 is cleared. Timers 2 and 3 must remain operating while 1 is being changed.

Example 11. Port A of a PIA is currently operating with FF in its data direction register and 04 in its control register. Change the I/O lines so that lines 7 and 6 are output, the remainder input. Draw the flow chart for this operation, and write out the instructions using Acc-B. The port is located at address 8004/8005.

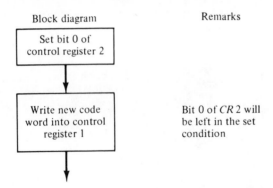

Block diagram   Remarks

Set bit 0 of
control register 2

Write new code   Bit 0 of *CR* 2 will
word into control   be left in the set
register 1   condition

Instruction   Block diagram

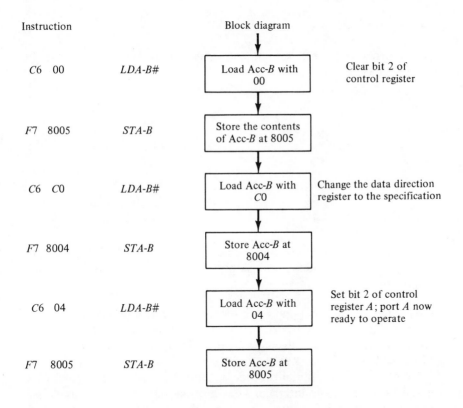

| C6 | 00 | LDA-B# | Load Acc-B with 00 | Clear bit 2 of control register |

F7 8005   STA-B   Store the contents of Acc-B at 8005

C6 C0   LDA-B#   Load Acc-B with C0   Change the data direction register to the specification

F7 8004   STA-B   Store Acc-B at 8004

C6 04   LDA-B#   Load Acc-B with 04   Set bit 2 of control register *A*; port *A* now ready to operate

F7 8005   STA-B   Store Acc-B at 8005

Example 12. A PTM timer is operating with timer 2 operational, draw a flow chart to internally reset the timer. Bit 0 of CR2 is set.

Example 13. A PIA has just been initialized; line C_{A1} is to be used as an interrupt line. The interrupt mask bit in the condition codes register is set.

Block diagram　　　　　　Remarks

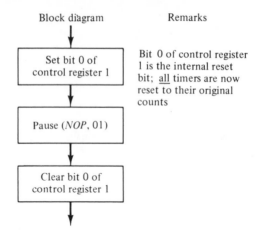

Bit 0 of control register
1 is the internal reset
bit; <u>all</u> timers are now
reset to their original
counts

Write a program to reset the interrupt flag in port A before clearing the interrupt mask bit.

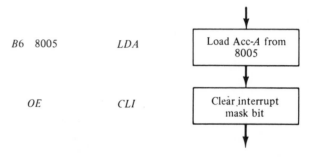

Assume that port A is located at address 8004/8005. The interrupt flag is reset, and the interrupt mask bit is cleared; the processor can now be interrupted.

Example 14. Write a program to initialize the PIA so that port A is used for input signals only and port B is used for output signals. No interrupt signals are to be acknowledged during initialization. The program starts at address 0020 (see Table 3.9).

When the PIA is initialized, the contents of the control registers are cleared to 00; thus the interrupt lines are masked and programmed for negative transition interrupts. Therefore, if a negative transition occurs on the lines C_{A1}, the relevant flag (bits 7 or 6) in the control register will set, which records the advent of an interrupt.

This example sets up port A for a positive transition interrupt, instruction 86 07 at address 0026. However, a negative transition interrupt may have previously occurred, in which case immediately after the CLI instruction is executed the program goes to the interrupt program even though a positive transition interrupt has not occurred. To eliminate this

Table 3.9

0020	0F	SEI (ensure no interrupts)
0021	86 00	LDA-A#
0023	B7 8004	STA-A
0026	86 07	LDA-A#
0028	B7 8005	STA-A
002B	86 FF	LDA-A#
002D	B7 8006	STA-A
0030	86 04	LDA-A#
0032	B7 8007	STA-A
0035	0E	CLI (port A set up to receive interrupts)
0036	Remainder of program	

LDA-A#, STA-A (0021, 0023) } All lines set up as input lines

LDA-A#, STA-A (0026, 0028) } Port A set up for positive transition interrupt

→ LDA-A#, STA-A (002B, 002D) } All lines set up as output lines

possibility, the interrupt flags must be cleared immediately prior to the point where interrupts will be recognized, for example, following the initialization of port A; → indicates the suggested point where the flags are cleared. The interrupt flags are cleared by a read peripheral data register, (i.e., load Acc-A from address 8004).

Problems

3-1. Write the code word to operate timer 2 as follows:
   a. External clock
   b. Dual 8-bit counting mode
   c. Operating mode 7
   d. Output terminal enabled
   e. Interrupt disabled
   f. CR1, CR3 address bit to CR1

3-2. Write the code word for control register A of a PIA to achieve the following; the PIA has just been reset.
   a. Negative transition interrupt on line C_{A1}
   b. Pulse mode operation
   c. MPU to be in communication with peripheral data register

3-3. What causes the following bits in the condition codes register to set?
   a. V bit
   b. N bit
   c. C bit

3-4. The following signal 5V ⌐ is applied to line C_{A2}. Determine the
                        0V⌐
   code word to be written into control register A to enable the C_{A2} interrupt and to place the MPU in communication with the peripheral data register.

3-5. How can the following errors be detected in received serial data, and what causes the error?

      a. Framing error
      b. Parity error
      c. Overrun error

3-6. Assume an ACIA is transmitting data; the transmit interrupts are disabled. How would you know when to load another character in the transmit data register?

3-7. How do you know when an ACIA, which is receiving data, has a character ready to load into the MPU?

3-8. A program instruction calls for a branch to 0036; the branch instruction is located at address 0075. What is the displacement?

3-9. Calculate the displacement for a forward branch from branch instruction location 0076 to 00A1.

3-10. Determine the addressing mode to load Acc-A from address 0137.

3-11. Determine the addressing mode to store the contents of Acc-B at address 0017.

3-12. Which counter is used to keep track of where the program is executing in memory?

3-13. Which instruction does nothing?

3-14. What is the storage capacity of an RAM chip that has 7 address lines?

3-15. What is the function of the R/$\overline{\text{W}}$ line?

3-16. State one advantage and one disadvantage of a memory-mapped I/O system as compared to an isolated I/O system.

3-17. What is the function of the ϕ_2 clock?

3-18. Draw the block diagram to master reset an ACIA.

3-19. Describe the reset logic circuit of Figure 3.13. What type of circuit is this?

3-20. Explain each of the following types of data transfers and list the instructions that fit into each category.
      a. Register to register
      b. Register to memory
      c. Immediate

3-21. Identify the register that contains the address of the next instruction to be executed.

3-22. What is the difference between parallel data output and serial data output?

3-23. Write a program to produce the following logic expression:

$$S = \bar{A}\bar{B}C + ABC + \bar{A}B\bar{C} + A\bar{B}\bar{C}$$

Hint: Use the EXOR instruction.

Chapter 4

Structured Programming

This chapter describes and implements a few of the *basic structures* used in the discipline of structured programming. It is important that the reader obtain an understanding of the function of each and how they are applied to a process control or motor control situation. A short list of texts and papers on structured programming and structure analysis is included at the end of this chapter; the list is by no means exhausted.

Structured programming is a set of programming constructs (such as sequence, repeat-until, do-while) for developing program code for computers, microprocessors, and the like. The constructs are the building blocks from which large programs emerge.

The technique of applying structured programming is to make a steady progression from a clear overview of the complete problem down to the detail of a program step. This technique is known as *top-down* design or stepwise refinement. The strategy is to progressively break the problem down into a series of smaller and smaller problems, which are easier to understand and handle.

To begin, the method of program development via structured programming will be used, and it will take precedence over high program efficiency and minimum program size. Later, methods of making the programs smaller and more efficient will be examined. The object of this chapter is therefore to develop a general procedure that can be used to design software programs to solve motor-control problems. The procedure will be machine independent. It does not depend upon the processor used and can be applied to any of the problems that are likely to be encountered.

The method or technique of solving control problems is basically in two parts: (1) *What* is the problem to be solved? and (2) *How* is the problem to be solved? Do *not* consider the *how* before having determined the *what*.

After the problem has been defined and understood, partition it into blocks; then describe within the block the action to be performed. It can be as complex as a complete motor-control situation or as simple as interrogating a push button. There is *no* best method. Consider, as an example, the following problem: an electric motor is to be started and stopped via a start/stop push-button station. Interrogate the push buttons to determine that they are in their correct quiescent condition as a pretest before start. The problem is shown partitioned in Figure 4.1.

This program could of course be modified in many ways; that is, instead of waiting for the stop or start buttons to take up their quiescent positions (as we could wait forever), an alarm light could flash and operator intervention could then correct the problem. This, however, was not asked for in the job definition, as it was poorly defined, but since the program of Figure 4.1 was a creative step a change can be made here quite easily by drawing in another block or rewording the description within the block. A program rewrite is *not* required as writing the program has not yet begun.

When the block diagram is complete, examine each block and the overall scheme to see that it meets all the specified conditions. Be sure that *all* inputs, outputs, data transfers, error conditions, and so on, are accounted for, and keep a list of these requirements. *Above all, understand the problem that is being programmed.*

When satisfied that everything is as it should be, proceed with the final step and draw the logic flow chart or machine flow chart. It is now that the *how* is achieved.

4.1 BASIC STRUCTURES OF A STRUCTURED PROGRAM

A basic structure is a small number of constructs that can be put together to implement a specified operation, such as adding two numbers currently in memory and placing the result at some specified location.

All basic structures have two things in common and that is that they have *one entry point and one exit point* and that they are composed of at least two connecting blocks.

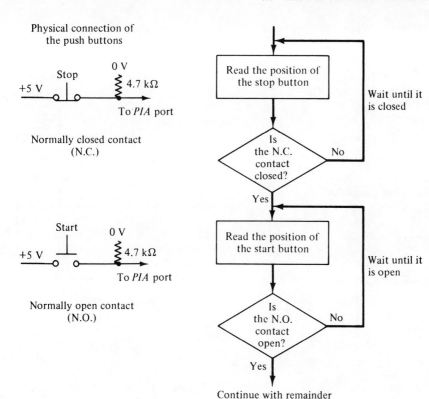

Figure 4.1

4.1.1 Sequence Structure

The sequence structure (Figure 4.2) is the most basic of all the structures. Control is transferred in, a process is performed, and control is transferred out to the next structure.

A sequence structure may be one or many microprocessor operations, such as output a positive pulse (Figure 4.3) to the clock terminal of a specified *J/K* flip-flop; it is being used in some process control scheme. The output line is normally held low; it will go high while the pulse is on the line. The output port is already initialized.

4.1.2 Decision Structure

The decision structure (Figure 4.4) is a true/false program-modifying instruction. Control passes into the decision structure and then the condition is tested (e.g., check to see if the zero flag is set). If the result of the

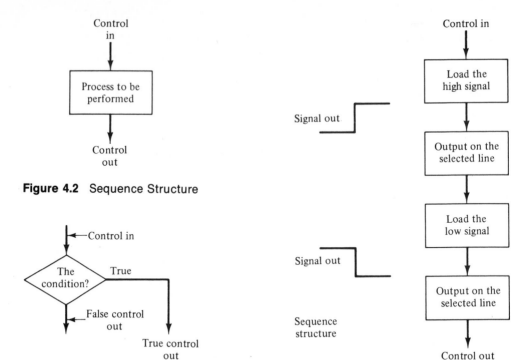

Figure 4.2 Sequence Structure

Figure 4.4

Figure 4.3 Sequence Structure

test is true, exit to the right. If it is false, pass straight through. This structure violates the rule of a basic structure, that is, one entry and one exit. Since this structure is never used alone but in combination with other structures, it is still regarded as a single-entry, single-exit structure.

True/false instructions use the processor's arithmetic/logic instruction set; these program decisions are of 2 types:

1. *Active test:* an arithmetic or logic operation that is performed sequentially for the purpose of testing some condition; for example, a data byte has been loaded into an accumulator from an external device and some of the flags are affected by this action. To actually determine the contents of the accumulator, a specific test must be performed, and from the results of this test the next action that the program is to perform will be determined. The instructions *COMPARE* (CMP, CBA) and *SUBTRACT* (SBA, SUB) are all active tests.

2. *Passive test:* a test of the result of an operation that is going to be performed as part of the program. For example, the contents of a

memory location have been decremented by 1. Is the location empty? If the contents of the memory location became 00 as a result of the decrement operation, then the zero flag will automatically be set. Thus, the status of the zero flag is tested each time the decrement operation is performed and a decision can be made without any extra instructions. *All* arithmetic/logic instructions *except* CMP, CBA, SBA, and SUB are passive instructions. Should it be necessary to know the exact contents of a memory location, an active test must be performed via the compare or subtract instructions. The terms passive and active are *not,* as yet, universally accepted standards.

4.1.3 If-Then-Else Structure

The structure (Figure 4.5) consists of a decision element and two sequence elements: if the condition is true, do this; *else* do that. As an example of its use, assume that the processor is controlling the temperature of an oven (Figure 4.6) by simply turning the heat on and off. The voltage across a positive thermally sensitive resistance device is measured periodically. It is then converted into a digital number that is compared to the demand via an active test.

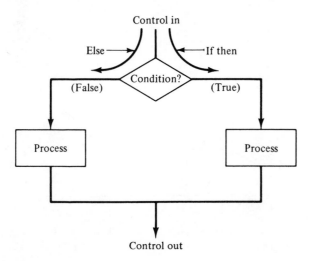

Control out   **Figure 4.5**

4.1.4 If-Then Structure

This structure (Figure 4.7) consists of a decision element and one sequence element. Basically, there are two types, if-then true and if-then false. If the condition is met, do this, and do nothing if it is not met. To

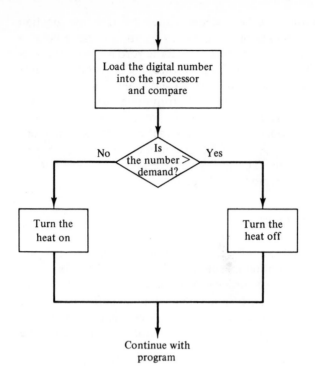

Continue with
program

Figure 4.6

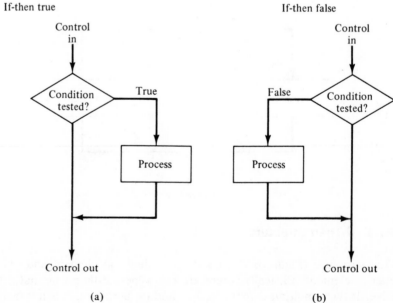

Figure 4.7   (a)   (b)

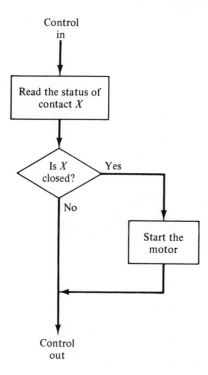

Control
in

Read the status of
contact X

Is X
closed?   Yes

No

Start the
motor

Control
out

Figure 4.8

illustrate the use of the if-then true structure, Figure 4.8 is now presented. The processor is reading the status of a contact X. If the contact is closed, start the motor, and if *not* closed, do nothing.

In all the preceding structures, control has passed into the structure; a decision was then made as to what process to perform next, and then control was passed out. These structures are called *open structures*. There are other structures in which the control is passed back to the entry point; these are called *closed structures* and they are considered in the next section.

4.1.5 Do-While Structure

This structure (Figure 4.9) consists of a decision element and one sequence element; basically, there are two types, do-while true and do-while false. If the condition is met, repeat the process; otherwise do nothing. It can be seen that the control is passed back to the entry point; thus this type of software structure is sometimes called a *loop*. A typical application of these structures is software timing loops (e.g., energize the circuit for X milliseconds).

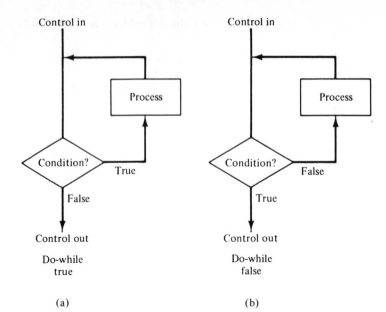

Control in

Process

Condition? — True

False

Control out

Do-while
true

(a)

Control in

Process

Condition? — False

True

Control out

Do-while
false

(b)

Figure 4.9

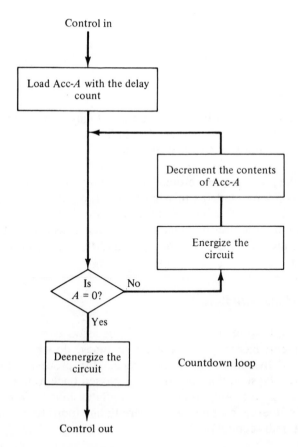

Control in

Load Acc-A with the delay
count

Decrement the contents
of Acc-A

Energize the
circuit

Is
$A = 0$? — No

Yes

Deenergize the
circuit

Countdown loop

Control out

Figure 4.10

When a do-while structure is used and the condition is met on the first pass through the structure, the loop will *not* be *executed,* not *even once.* Countdown loops and count-up loops are an application of a do-while structure and are used to meet different program requirements.

4.1.6 Repeat-Until Structure

This structure (Figure 4.11) contains the same elements as the do-while structure except that they are arranged differently. In these loops the *process is performed* even if the *loop is not executed.* As an example of its use, the push-button interrogation scheme of Figure 4.1 will be used again (Figure 4.12). This test is a signal-level test, not a signal-transition test. The key difference between the do-while and repeat-until structures is the entrance into the loop; that is, the do-while structure tests for the exit condition before performing any operations, whereas the repeat-until structure executes the process before it tests for the exit condition.

The major problem with the use of loops is correct termination; be sure that the loop does not execute one more or one less operation than is required. Correct operation can only be guaranteed when the loop variable (the number originally placed in the counter) is correctly selected.

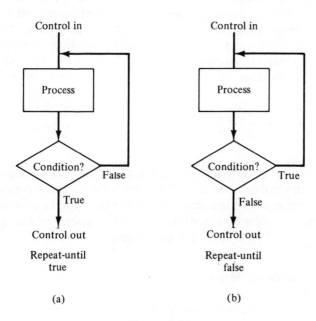

Figure 4.11

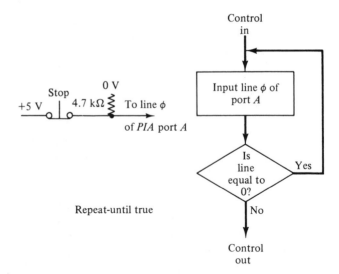

Figure 4.12

Example 1. Determine the loop variable for three operations of a specified process; this is a count-up loop, an increment before test (see Figure 4.11). Write the program, starting at address 0117, and leave 30_{10} memory locations for the process (see Figure 4.13).

Example 2. Determine the loop variable for three operations of a specified process; this is a count-up loop, an increment before test (see Figure 4.9). Write the program starting at address 0117; the program for the process starts at 0156 and requires 45_{10} memory locations. Accumulator B will be used as the operations counter (see Figure 4.14). The loop variable is 4. What is the loop variable if the process is *not* to be executed at all? (1)

4.2 MASKING

Masking is a technique that is used for testing and manipulating individual bits of a data word or byte; a *mask* is a *bit pattern*. The processor's logic instruction set is used when masking operations are to be performed. The three basic types of masking, *exclusive, inclusive,* and *exclusive OR,* will now be presented with several applications.

4.2.1 Exclusive Masking

Exclusive masking is used to set or clear *all* the bits that are *not* included in the mask.

Address	Instructions	Mnemonic	Block diagram

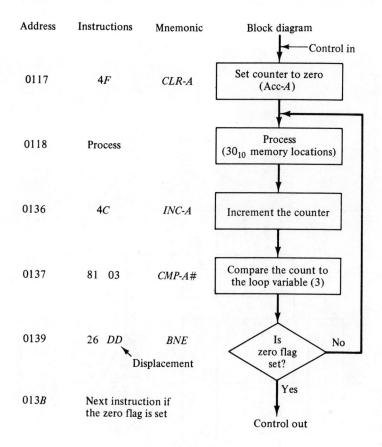

Control in

0117   4F   CLR-A   Set counter to zero (Acc-A)

0118   Process   Process (30_{10} memory locations)

0136   4C   INC-A   Increment the counter

0137   81 03   CMP-A#   Compare the count to the loop variable (3)

0139   26 DD   BNE   Displacement   Is zero flag set?   No

013B   Next instruction if the zero flag is set   Yes   Control out

The loop variable is 3

Remarks

Acc-A will be used as the operations counter; the process uses 30_{10} memory locations

The program enters the loop, performs the process, increments the counter, then performs an active test; if the result is <u>not</u> the one required, repeat the complete operation. When the desired result is achieved exit from the loop.

Figure 4.13

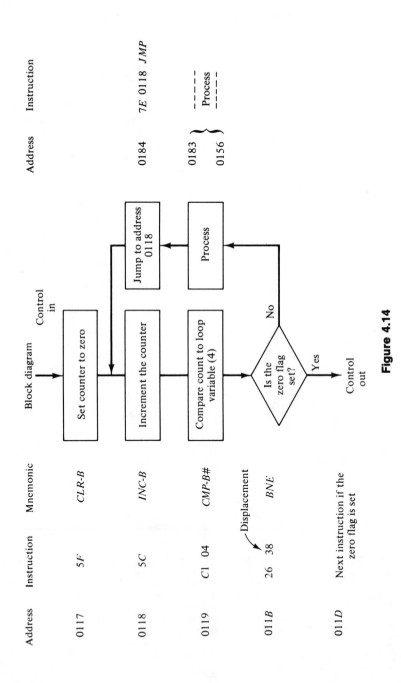

Figure 4.14

Example 3. A byte of unknown data is loaded into Acc-A; is bit 5 a 1 or a 0?

Assume that the data loaded into Acc-A are 01001101. To determine whether bit 5 is 1 or 0, the remaining data bits will be reset to 0. This is accomplished via the logic operation *AND* (6800 instruction AND, address mode IMM), since $1 \cdot 1 = 1$ and $0 \cdot 1 = 0$. Therefore, it is necessary to AND some number with the contents of Acc-A, which will leave bit 5 in its original condition and at the same time reset the remainder to 0. The number to be ANDed is called the *mask,* in this case (4.1) it will be 00100000 (20).

The AND Operation

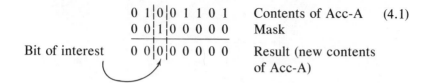

0 1 | 0 | 0 1 1 0 1	Contents of Acc-A (4.1)
0 0 | 1 | 0 0 0 0 0	Mask
Bit of interest 0 0 | 0 | 0 0 0 0 0	Result (new contents of Acc-A)

The result of the AND operation is zero, which causes the Z bit (bit 2 of the condition codes register) to set automatically; therefore, if a test is made of the state of the Z bit it can be determined whether bit 5 of Acc-A is set or is reset.

The block diagram Figure 4.15 illustrates such an operation. The data in the Acc-A were destroyed in the AND operation; however, had the *bit* instruction been selected, the same type of operation would have been performed. A similar flow chart would have been used but the data being tested would *not* have been destroyed. Similar operations can be performed on data stored in memory by utilizing the other addressing modes of the selected instruction; for example, is bit 3, in memory location 0076, set? Use the bit test instruction. See Figure 4.16.

4.2.2 Inclusive Masking

Inclusive masking is used to set or clear *individual* bits of a selected data byte; the remainder are *left unaffected.* For this reason, inclusive masking is usually used to add or remove selected bits from the data rather than as a bit-testing method. Inclusive masking is performed via the processor's logic AND operation or the logic OR operation, depending upon whether the data bits in the mask are 0 or 1.

Example 4. Assume that the data contained in Acc-A are known

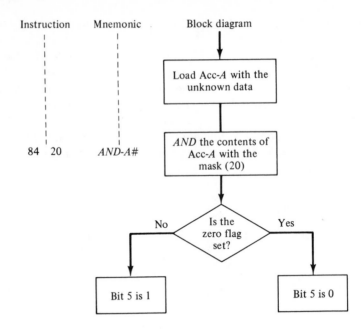

Figure 4.15

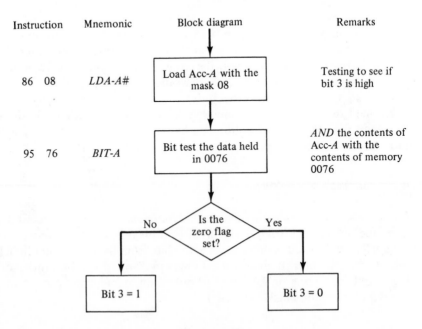

Figure 4.16

(01101010), bit 2 = 0; change it to 1 while leaving the remainder of the data bits unaffected.

This can be accomplished by using the logic operation OR (6800 instruction OR, addressing mode IMM), since 1 + 0 = 1 and 0 + 0 = 0. Therefore, it is necessary to OR the mask and the original data to achieve the desired result.

The OR Operation

$$
\begin{array}{lll}
0\ 1\ 1\ 0\ 1\ |0|\ 1\ 0 & \text{Original contents Acc-A} & (4.2)\\
0\ 0\ 0\ 0\ 0\ |1|\ 0\ 0 & \text{Mask (04)} & \\
\hline
0\ 1\ 1\ 0\ 1\ |1|\ 1\ 0 & \text{Result (new contents of Acc-A)} &
\end{array}
$$

— Bit 2 is set

Data contained in a memory location can be dealt with in a similar manner.

Example 5. The data contained in memory location 0137 are 11010110; reset bit 2, leaving the remainder unaffected.

This can be accomplished by using the logic operation AND as follows:

AND Operation

$$
\begin{array}{lll}
1\ 1\ 0\ 1\ 0\ |1|\ 1\ 0 & \text{Original contents of memory} & (4.3)\\
1\ 1\ 1\ 1\ 1\ |0|\ 1\ 1 & \text{Mask (FB)} & \\
\hline
1\ 1\ 0\ 1\ 0\ |0|\ 1\ 0 & \text{Result (new contents of memory)} &
\end{array}
$$

Bit 2 is changed from 1 to 0

Examples 4 and 5 have demonstrated how a data bit can be set or cleared under program control via inclusive masking. The status of the original bit may be known or unknown. However, a situation can arise where the status of a bit is *not known* but it must be inverted (if 1 change to 0, and vice versa), while the remainder must be unaffected. Logic operation *Exclusive OR* has some unique properties, as shown by the following table, which relates two variables, A and B, via the Exclusive OR (EXOR). See also Equation (2.53).

A	B	$A \oplus B$	
0	0	0	(4.4)
1	0	1	
0	1	1	
1	1	0	

which can be summarized as follows: the output is 1 when one or the other of the inputs is 1, but *not* both.

This logic operation will now be applied as a masking operation.

Example 6. The contents of memory location 0236 are not known; invert the status of bit 3.

Assume that the contents are 10110110 (bit 3 = 0); apply a mask via the EXOR (Exclusive OR) operation.

Exclusive OR Operation on Memory Location 0236

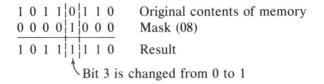

```
1 0 1 1 | 0 | 1 1 0      Original contents of memory
0 0 0 0 | 1 | 0 0 0      Mask (08)
1 0 1 1 | 1 | 1 1 0      Result
```
↖ Bit 3 is changed from 0 to 1

The block diagram of Figure 4.17 illustrates the program. It must be noted that contents of memory location 0236 are *not* changed by this operation; *the result is* being held *in accumulator A.* How the result will be utilized is now the concern of the programmer. *The bits that are Exclusive ORed with 1 are inverted; the remainder are unaffected.*

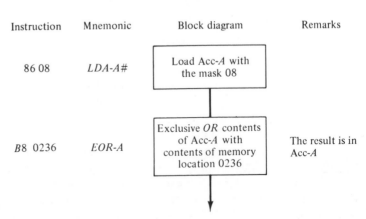

Instruction	Mnemonic	Block diagram	Remarks
86 08	*LDA-A#*	Load Acc-*A* with the mask 08	
B8 0236	*EOR-A*	Exclusive *OR* contents of Acc-*A* with contents of memory location 0236	The result is in Acc-*A*

Figure 4.17

4.3 COMBINED EFFECT OF DO-WHILE AND REPEAT-UNTIL STRUCTURES

When a program is implemented, it often happens that there are program loops within program loops (see Figure 4.18), for example, a counter that counts from 0 to 90. There are many variations to the loop, as Figure 4.19

Program Steps

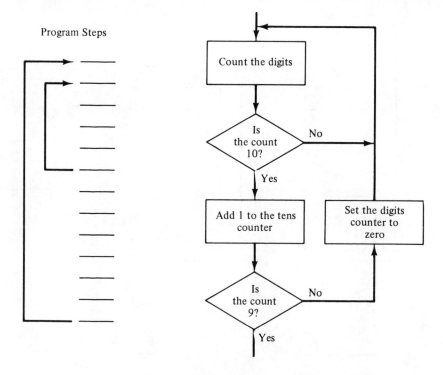

Figure 4.18

illustrates. Loops 1, 2, and 3 are *wholly* contained within other loops. These are loops that terminate in the reverse order from which they were initiated. This process is called *subroutine nesting*.

4.4   INDEX REGISTER

The index or X register has been mentioned previously under the heading of addressing modes; however, because the register has unique properties, it will now be discussed in more detail, with some illustrative examples.

The register (Figure 4.20) is a 16-bit device that forms into an upper and lower byte as illustrated. The 6800 instruction set contains a number of instructions that enable the contents of the X register to be manipulated and utilized, such as compare with a 16-bit number, increment, or decrement. Its contents may be stored in memory or the register may be loaded from memory, or the contents of other registers can be manipulated using the contents of the index register as an address reference point, base

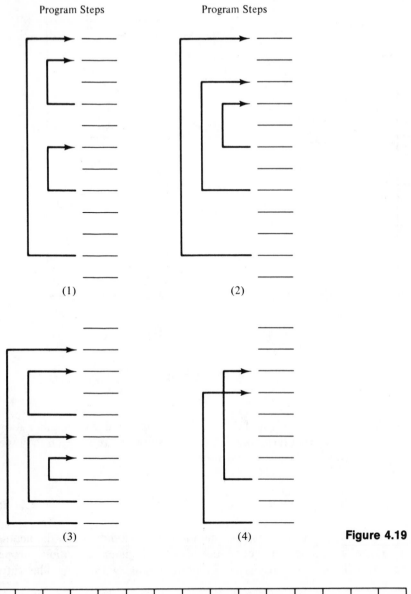

Program Steps   Program Steps

(1)   (2)

(3)   (4)

Figure 4.19

Bit number	15	14	13	12	11	10	9	8	7	6	5	4	3	2	1	0
X register contents	0	1	0	1	0	1	1	1	1	0	0	0	1	1	1	1

	5	7	8	F
	High byte		Low byte	

Figure 4.20

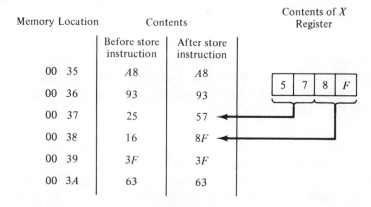

Figure 4.21

point, or pointer. Because the device is 16 bit and memory locations are only 8 bit, two memory locations must be used to store the data contained in the register or act as a source of data to load the register. As an example, assume that the contents of the X register (578F) are to be stored in memory starting at address 0037 [STX-(DIR), instruction op-code DF 37, store contents of X register at address 0037]. Figure 4.21 illustrates where the data are stored. Notice the high byte is stored at the specified address 0037, and the low byte is stored at the *next higher address* location; thus, the data are stored in two adjacent memory locations with a single instruction (STX). Similarly, with a load instruction (LDX), the data stored in two adjacent memory locations can be copied into the X register, as shown in Figure 4.22. Load the X register from memory location 0157 [LDX-(EXT), instruction op-code FE 0157]. Notice, the high byte is trans-

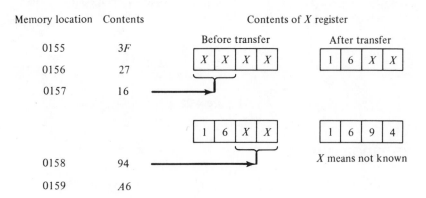

Figure 4.22

ferred from the specified memory location 0157, whereas the low byte is transferred from the *next higher address*.

A comparison will now be made using the X register and one of the accumulators (A) to initialize port A of a PIA, which is located at address 8004 / 8005. Assume that the PIA has just been reset; that is, all bits in the control registers and data direction registers are zero. The control register is located at address 8005.

Example 7. Set all peripheral lines as output lines, and inhibit the interrupts. A comparison-type solution will be offered. Compare the solution via the Acc-A (Figure 4.23) with the solution via the index register (Figure 4.24). It is readily seen that using the index register to load the registers of the PIA is more direct and takes less program steps and memory, *provided that* the memory locations accessed are adjacent. From this point on the X register will be used whenever possible.

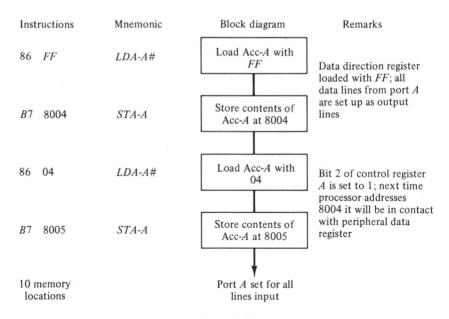

Instructions	Mnemonic	Block diagram	Remarks
86 *FF*	*LDA-A#*	Load Acc-*A* with *FF*	Data direction register loaded with *FF*; all data lines from port *A* are set up as output lines
B7 8004	*STA-A*	Store contents of Acc-*A* at 8004	
86 04	*LDA-A#*	Load Acc-*A* with 04	Bit 2 of control register *A* is set to 1; next time processor addresses 8004 it will be in contact with peripheral data register
B7 8005	*STA-A*	Store contents of Acc-*A* at 8005	
10 memory locations		Port *A* set for all lines input	

Figure 4.23

4.5 SUBROUTINE PROGRAMMING

Up to this point all the programs that have been developed have rigidly adhered to the principles of structured programming, that is, one entry point and one exit point. This means that one section of the program

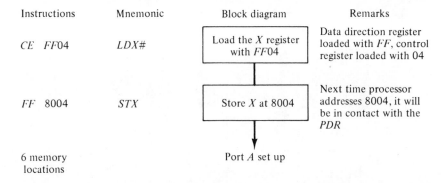

Instructions	Mnemonic	Block diagram	Remarks
CE FF04	LDX#	Load the X register with FF04	Data direction register loaded with FF, control register loaded with 04
FF 8004	STX	Store X at 8004	Next time processor addresses 8004, it will be in contact with the PDR
6 memory locations		Port A set up	

Figure 4.24

cannot make use of a function that is located in a different area of the program, and that each time the function is required all the programming steps necessary to generate that function must be repeated. For example, timing loops are used repeatedly in process applications and delays are usually generated as shown in Figure 4.25. The program will stay in the timing loop until the contents of the X register reach 0000; that is, decrement X, then check to see if the zero flag is set. If the answer is no, repeat the operation; if the answer is yes, continue with the program.

This function may be required many times in a program, and each time it is required it must be generated again. This type of programming is

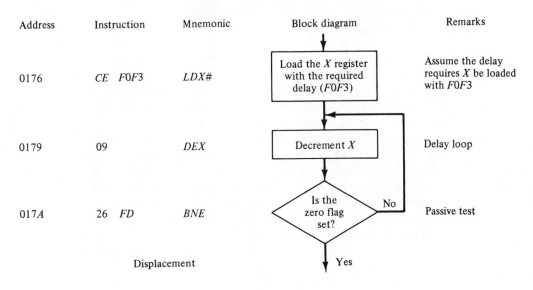

Address	Instruction	Mnemonic	Block diagram	Remarks
0176	CE F0F3	LDX#	Load the X register with the required delay (F0F3)	Assume the delay requires X be loaded with F0F3
0179	09	DEX	Decrement X	Delay loop
017A	26 FD	BNE	Is the zero flag set?	Passive test
	Displacement		Yes	

Figure 4.25

Block diagram   Remarks

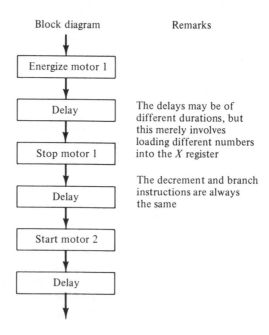

Energize motor 1

Delay   The delays may be of
different durations, but
this merely involves
loading different numbers
Stop motor 1   into the X register

The decrement and branch
instructions are always
Delay   the same

Start motor 2

Delay

Figure 4.26

called *linear* or *straight-line* programming; a typical program using such a process is given in Figure 4.26. Being forced to repeat the delay loop programming steps each time that the delay function is required would considerably increase the size of the program. To circumvent this, a technique known as subroutine programming is used. A subroutine is a program that can be used by other programs to perform some specified operation. Control is transferred from the main program to the subroutine; the subroutine executes its program and then transfers back to the main program to continue where it left off. For example, Figure 4.26 will be repeated using a subroutine for the delay loop (Figure 4.27). The transfer of control from the main program to the subroutine is called a *subroutine call*, and the transfer back is called a *return from subroutine*. The 6800 instruction set has two instructions for a subroutine call JSR (jump to subroutine) and BSR (branch to subroutine) and one instruction for the RTS (return from subroutine). The difference between the subroutine calls is that a BSR can only transfer control a maximum of -128_{10} or $+127_{10}$ steps relative to the current position of the program counter. Its area of influence is limited, whereas the JSR uses the indexed and extended modes of addressing; thus its area of influence is the whole memory space.

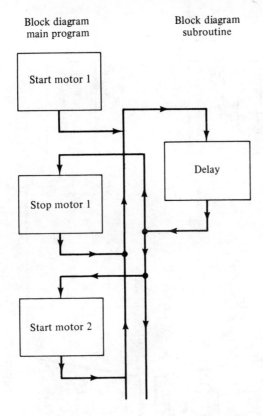

Block diagram
main program

Block diagram
subroutine

Start motor 1

Delay

Stop motor 1

Start motor 2

Figure 4.27

When a subroutine call is encountered, the processor will enter into a subroutine call sequence whereby it will automatically save the current contents of the program counter on the stack. It then loads the transfer address, which accompanies the subroutine call op-code, into the program counter and proceeds to execute the subroutine program. At the end of the subroutine program, there *must be* an RTS instruction; when it is encountered, the processor will reload the original contents of the program counter (these were saved when the subroutine call was encountered), which will destroy the program counter's current contents. The processor is now ready to resume executing the main program at what would have been the *next* instruction. An example will now be offered to illustrate the preceding.

Example 8.   A solenoid valve has just been energized; it must remain so for Z seconds (delay F376 in the register). Use a subroutine call for the delay (see Figure 4.28). This is one type of subroutine. It is called a *fixed subroutine* since the same operation will always be performed regardless

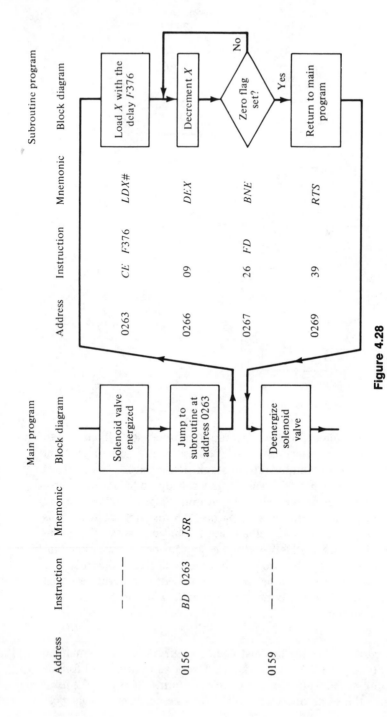

Figure 4.28

of the program that called it; that is, later in the main program the same delay will be used again, many times.

A second type, known as a *parameterized subroutine,* can perform different operations based upon the conditions of the calling program. This is made possible by using *pass parameters*. The data are sent from the calling program to the subroutine; that is, they are passed in. For example, assume that the delay in Example 8 was required to be fixed for some areas of the main program and yet be variable for other areas. The delay data in this case would be loaded into the X register before the processor transfers control to the subroutine program, as Example 9 will illustrate.

Example 9. A particular section of the process, of which Example 8 is a part, requires a delay of Y seconds (F067 in the X register). Use the subroutine already in existence (see Figure 4.29).

Parameters may also be passed back from the subroutine to the calling program via the flags of the condition codes register and/or the accumulators A and B.

The use of subroutines is not always a saving in memory and programming steps since the instruction JSR requires 3 bytes, which are contained in three memory locations, and the RTS instruction requires 1 byte. Thus a total of four memory locations is required per subroutine call, using the extended mode of addressing. Before setting up subroutines, therefore, an investigation must be made into the size of the subroutine and how many times it will be required in the program. For example, a subroutine requires three memory locations to store its program and it will be used twice. Would it be profitable to use a subroutine or should the three-step program be written twice in the main program? See Table 4.1.

Table 4.1

Subroutine Method		Main Program Method	
Instruction	*Memory Locations*	*Instruction*	*Memory Locations*
JSR	3	Program	3
JSR	3	Program	3
RTS	1		6
Program	3		
	10		

If a subroutine is used it will require four more memory locations than if the program had been written twice, a loss of four memory locations.

Figure 4.30 will assist the reader in determining the profit/loss of memory for subroutine calls. For example, assume the subroutine has

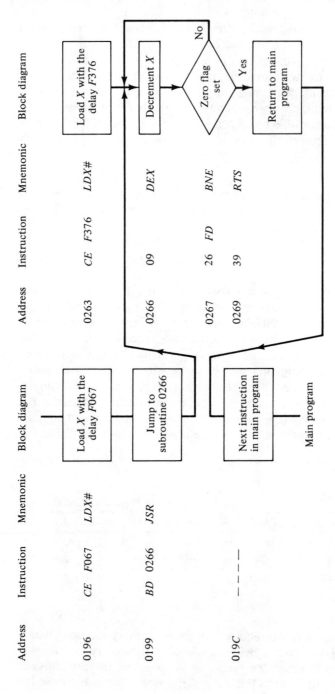

Figure 4.29

	No. of times subroutine occurs								
	1	2	3	4	5	6	7	X	
1	−4	−6	−8	−10	−12	−14	−16	−2X − 2	
2	−4	−5	−6	−7	−8	−9	−10	−X − 3	Loss of memory
3	−4	−4	−4	−4	−4	−4	−4	−4	
4	−4	−3	−2	−1	0	1	2	X − 5	Gain of memory
5	−4	−2	0	2	4	6	8	2X − 6	
6	−4	−1	2	5	8	11	14	3X − 7	
7	−4	0	4	8	12	16	20	4X − 8	
8	−4	1	6	11	16	21	26	5X − 9	
9	−4	2	8	14˙	20	26	32	6X − 10	
N	−4	N − 7	2N − 10	3N − 13	4N − 16	5N − 19	6N − 22	(N − 3)X − (N + 1)	

No. of steps (memory locations) used in the subroutine

Figure 4.30

three steps; it will be used twice. Look down column 2 and across row 3 and find that the loss is four memory locations. This problem was solved previously in Table 4.1.

4.6 TIMING LOOPS

A timing loop is a programing routine to idle the processor for a fixed period of time. A constant is loaded into a register, which is decremented until it reaches 0. The loop constant or loop variable is chosen so that the loop terminates in a known period of time. The delay loop is not extremely precise, and if precise timing is required the programmable timer module (PTM) should be used. The total delay using a microprocessor timing loop is determined as follows:

$$\text{Total delay, } T_D = (N \times T_L) + T_P$$

where N = number of times through the loop

T_L = execution time of the loop instructions

T_P = execution time for those instructions not in the timing loop

Example 10. Output, via port A, line Ø, a positive pulse of 1-ms duration;

Execution Time in Cycles	Instruction	Mnemonic	Block Diagram	Remarks

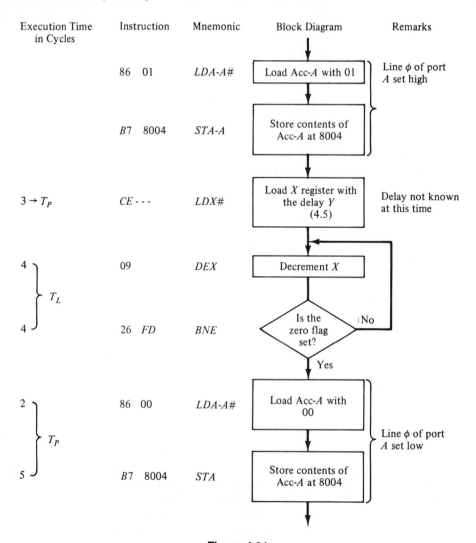

Figure 4.31

assume the PIA port is already set up and that line ∅ is already low. See Figure 4.31.

Assume the clock is operating at 614.4 kHz; then the time of 1 cycle $= \frac{1}{614,400} = 1.6 \ \mu s$.

$$\therefore \quad T_D = 1000 = N(8 \times 1.6) + 10 \times 1.6 \qquad (4.5)$$

$$\therefore \quad N = \frac{1000 - 16}{8 \times 1.6} = 76.875 \simeq 77 \text{ times through the loop}$$

Convert 77_{10} to hexadecimal:

$$77_{10} = 4D$$

The contents of the X register will therefore be $004D = Y$.

4.7 TUTORIAL EXAMPLES

Example 1.   A motor start/stop station is connected via the data lines 0 and 1 of port A of a PIA; line 2 is used to control the solid-state relay that will operate the motor magnetic (see Figure 4.32). Lines 0 and 1 are input data lines. Line 2 is an output data line; the remainder are not used and are set as output lines. The PIA is already set up.

Design a program to interrogate the push buttons via the *bit* instruction. Figure 4.1 will be utilized and partitioned into smaller blocks as shown in Figure 4.33; the program instructions are included. Assume that the program is stored at starting address 0128. Verify the branch displacements, and notice that the two loops conform to Section 4.3. This is the pretest or interrogate section of a complete program.

Example 2.   Change tutorial example 1 to achieve the control circuit of Figure 4.34.

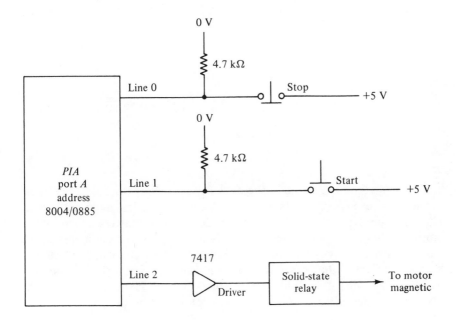

Figure 4.32

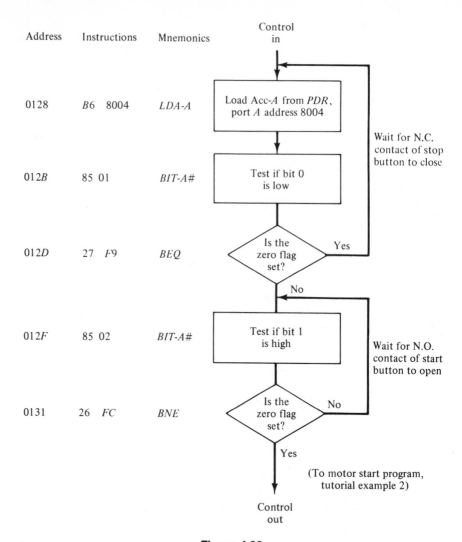

Address	Instructions	Mnemonics
0128	B6 8004	LDA-A
012B	85 01	BIT-A#
012D	27 F9	BEQ
012F	85 02	BIT-A#
0131	26 FC	BNE

Figure 4.33

The method presented in Figure 4.1 of pretesting the control buttons is used because it is essential that motor-control systems be intact before allowing the motor to start. For example, assume that the push-button station is defective; when the microprocessor is activated the motor could start immediately, without depressing the start button, or once started it may not be under the control of the stop button, because it is defective. This could be disaster, to plant personnel, and the plant. Thus, safety is an absolute necessity. The push-button interrogation program could be

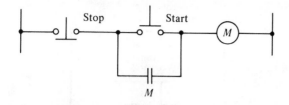

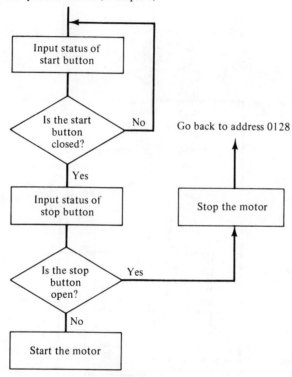

From prestest section (Example 1)

Figure 4.34

made more elaborate, such as initiating an alarm if a push button is not in its quiescent position (see Figure 4.35).

Example 3. A process in a plant uses a countdown loop in its operation. Expand the block diagram of Figure 4.36 to achieve this; use a repeat-until structure.

Assume the process is to flash a lamp three times; the time on and off is Y seconds, approximately. Use a subroutine for the timing loop. As-

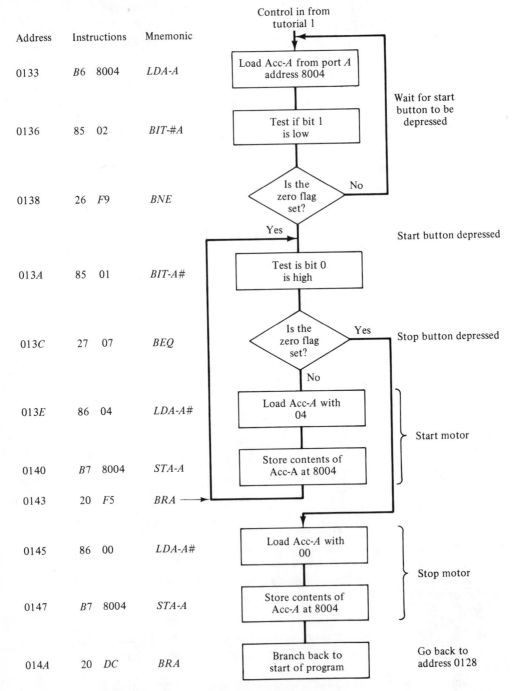

Address	Instructions	Mnemonic
0133	B6 8004	LDA-A
0136	85 02	BIT-#A
0138	26 F9	BNE
013A	85 01	BIT-A#
013C	27 07	BEQ
013E	86 04	LDA-A#
0140	B7 8004	STA-A
0143	20 F5	BRA
0145	86 00	LDA-A#
0147	B7 8004	STA-A
014A	20 DC	BRA

Control in from tutorial 1

Load Acc-*A* from port *A* address 8004

Test if bit 1 is low

Is the zero flag set?   No

Wait for start button to be depressed

Yes   Start button depressed

Test is bit 0 is high

Is the zero flag set?   Yes   Stop button depressed

No

Load Acc-*A* with 04

Store contents of Acc-*A* at 8004

Start motor

Load Acc-*A* with 00

Store contents of Acc-*A* at 8004

Stop motor

Branch back to start of program

Go back to address 0128

Figure 4.35

Block diagram

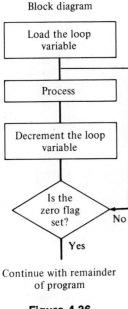

Figure 4.36

Continue with remainder
of program

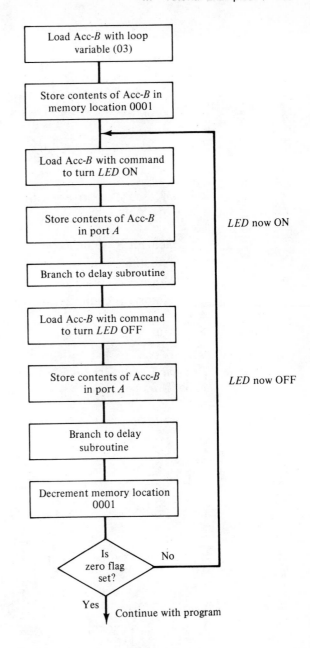

LED now ON

LED now OFF

Continue with program

Figure 4.37

sume the lamp is connected to line 5 of PIA port A. The port is already initialized. See Figure 4.37.

Example 4. Draw the flow chart to replace the relay circuit shown in Fig-

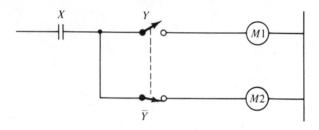

Figure 4.38

ure 4.38. Assume port A is already initialized with lines 0 to 3 input lines and 4 to 7 output lines.

> Contact X is connected to line 0.
> Pressure switch Y is connected to line 1.
> Motor 1 is connected to line 4.
> Motor 2 is connected to line 5.

See Figure 4.39.

Example 5. In a manufacturing process, objects are conveyed from operation A to operation B via a conveyor belt; a detector sets its output high

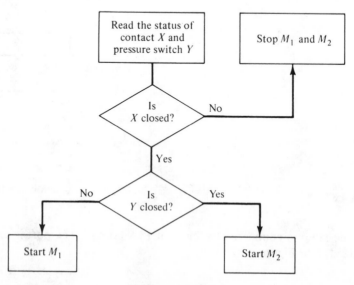

Figure 4.39

when the object is in front of it. Draw a flow chart to detect these positive pulses, and then output a negative pulse each time 10 objects have passed the detector. Assume port A is initialized with lines 0 to 3 input lines and 4 to 7 output lines.

> Address of port A, 8004/8005.
> Signal from detector is connected to line 0.
> Signal for 10 objects counted is on line 4.
> Store the count at address 0001.

See Figure 4.40.

Example 6. Draw a flow chart to measure the time that a contact X is closed. Measure the time to 0.1-s accuracy. The total count does not exceed 2.5 s. Use timer 1 of a programmable timer module to achieve accurate time intervals.

Assume that the timer is initialized and can be controlled via terminal G_1 of the timer module, and the output is via terminal O_1 of the same module. Assume that all signals to and from the processor are via port A, which is already initialized; the lower four lines are input lines, the upper four lines are output, and interrupt line C_{A1} is initialized for positive transition interrupt.

Contact X is connected to line 0. Timer output terminal O_1 is connected via interrupt line C_{A1}. The timer input terminal is connected to the PIA via line 4. The count is to be accumulated in memory location 0001. Assume that the interrupt vector has already been stored in a previous program. See Figure 4.41.

Problems

4-1. Define the term "basic structure."

4-2. What is the difference between an open structure and a closed structure?

4-3. What is meant by the term loop variable?

4-4. Define active test and passive test.

4-5. Reference tutorial example 2; draw the flow chart to emit a 1200-Hz pulse train to a speaker if the push buttons are found to be out of their quiescent position or defective during the pretest. Also, flash an LED to indicate which push button is defective. Use lines 5 to 7 of port A as the alarm output lines: line 5, 1200-Hz pulse train; line 6, start button defective LED; line 7, stop button defective LED.

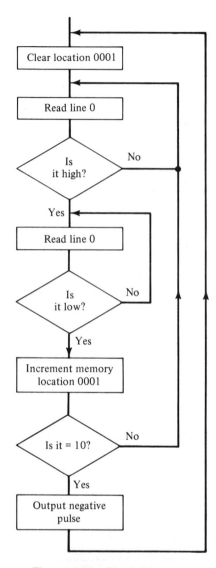

Figure 4.40 Block Diagram

4-6. Write out the program in 6800 code for tutorial example 3. Start the program at 0080; use branch to subroutine in lieu of jump to subroutine. Locate subroutine at 0060.

4-7. Write out the program in 6800 code for tutorial example 3. Use a mask operation on the contents of the POR of port A so that the remaining I/O lines of port A are not altered from their original state.

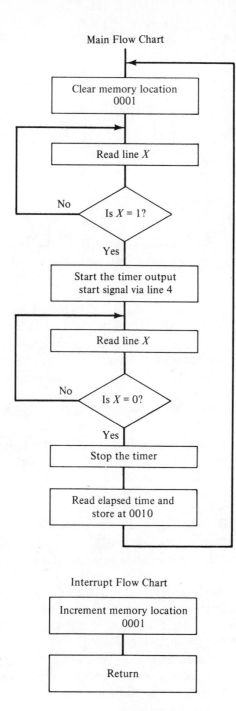

Figure 4.41 Block Diagrams

4-8. Two pulses X and Y are being supplied by a process to a 6800. The process to be performed depends upon the state of these two signals, as shown. Draw a flow chart to perform the four different operations dependent upon the state of X and Y.

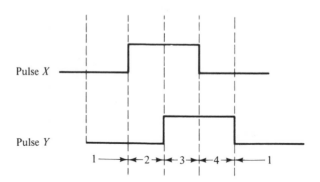

4-9. Expand the flow chart of Figure 4.39 to include all the steps so that a program may be written from the final flow chart.

4-10. Draw the flow chart to output the negative pulse in tutorial example 5. The negative pulse must have a duration of approximately 2 s. No positive pulses from the object detector shall be missed during the 2-s timing loop.

4-11. Draw a flow chart and write the program to debounce a contact. The contact bounces for a maximum of 1 ms.

4-12. Manufactured objects are being automatically loaded into racks in crates ready for the next operation. There are 4 objects to a rack, 12 racks to a crate. Each time a rack is loaded, the crate is repositioned to allow the next empty rack to be filled. When the crate is filled, it is automatically ejected and an empty crate replaces it. The cycle is continuous. Draw a flow chart to execute this operation.

4-13. Determine the time required to execute the following timing loop; the 6800 processor is operating at 1.843 MHz. (See figure on page 171.)

4-14. A 6800 microprocessor is operating at 614.4 kHz. A subroutine timing loop using the decrement X register instruction (09) is being used to generate a *total* delay of 20 ms. Calculate the contents of the X register to generate this delay.

4-15. A subroutine is used in a program to read data from an external source; the data are required to be passed back to the main program. Draw a flow chart to illustrate this.

4-16. Draw and define the operation of the do-while structure and the repeat-until structure.

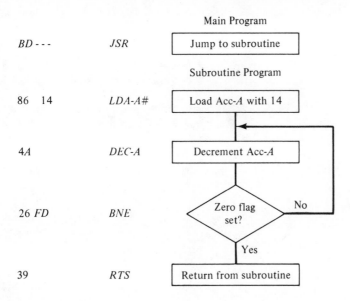

4-17. Define exclusive masking. What is the mask to clear bit 3 of the hexadecimal number DC? What logic operation will be used to execute this task?

4-18. What is meant by the terms memory map and memory space?

4-19. What is the principal advantage of memory-mapped I/O?

4-20. Define operand and data source.

4-21. What is a register-to-register transfer? List the instructions in the 6800 instruction set that fit in this category.

4-22. What is the difference between a logic flow chart and a machine-dependent flow chart?

List of Additional Reading:

The practical guide to structured systems design. Meillir Page-Jones, Yourdon Press, New York, N.Y. 10036.

IEEE Transactions on software engineering, Vol SE-3, No 1, January 1977.

Principles of program design, M. A. Jackson, Academic Press, New York, N.Y. 1975.

Composite/structured design, G. Myers, Van Nostrand, 1978.

Structured design, W. Stevens et al., IBM systems journal Vol 13, No 2, May 1974, pp 115-139.

Structured programming, B. Dahl, E. Dijkstra, Academic Press, London 1972.

Top down structured programming, C. McGowan, J. Kelly, Petrocelli/Charter, New York, N.Y.

Chapter 5

The MC 6800
Microprocessor

This chapter will describe the MPU, its control lines and its output lines; its support chips (the peripherals to the MPU) have already been discussed in Chapter 3. The control and output lines can be divided into four categories: data bus, address bus, peripheral control bus, and the MPU supervisory control lines.

5.1 THE 6800 MPU AND ITS CONTROL

The data bus consists of eight lines, each carrying some binary number (1 or 0); it is bidirectional and is used to convey data *to* and *from* the MPU, *to* and *from* memory and the I/O devices. Between seven and ten loads (RAM, ROM, ACIA, PIA, etc.) can be connected to the data bus before extra driving power (buffer) will be required.

The address bus consists of 16 lines; it is unidirectional. It carries a 16-bit code word, the address of the device being accessed, be it memory, PIA, and so on. The control bus consists of 5 lines, 4 output and 1 input. It carries a mixed set of signals that are used by the MPU to control and regulate the operation of the system peripherals. The signals contained on the bus are as follows:

1. VMA (valid memory address): this is an output line that is normally held low; it goes high when the address (that is the coding

of 1's and 0's) on the 16-line address bus is valid. When the binary state of the address lines are changed to call a different address, there is a time interval in which there is a mix of old address and new address signals. To inform the peripherals that the new address is on the bus and it is valid, the VMA line will go high; it is supplied to each peripheral via its control select line (CSX[1]).

2. Phase 2 clock (ϕ_2): this is an input line to the MPU and is used as the timing medium during data movements along the data bus into and out of the MPU.

3. Read/write line ($R/\overline{W}$) is an output line; it is used to inform the peripherals of the operating mode of the MPU. When the MPU is in the write mode, data will be transferred from the MPU to the peripheral; the $R/\overline{W}$ line will be low. Similarly, when the MPU is in the read mode and data are being transferred from the peripheral to the MPU, the $R/\overline{W}$ line will be high. The $R/\overline{W}$ line's signal level determines the direction of data flow. When the line is not in use, it is placed into a high-impedance state by the MPU; that is, it is isolated from the MPU and is floating. The potential on this line at this time is approximately 2 to 2.5 V.

4. The interrupt request line ($\overline{IRQ}$) is an input line that the processor sets high via an internal pull-up resistor. The line is pulled low by the peripherals PIA, ACIA, PTM, and so on, when these devices are requesting additional service from the MPU.

5. Reset: this is an input line that is normally held high by external circuitry; it is pulled low when external reset is initiated. This same line can be used to reset the peripheral equipment to which the MPU is connected, that is, PIAs, PTMs, and so on.

Normally, reset is used after power failure or on initial start-up, but it can be used at any time to put the MPU back to its initialized condition. The line *must* be held low for a minimum of eight phase 1 (ϕ_1) clock cycles. This ensures that the internal controls and the like have had time to take their starting conditions. For example, while the reset line is being held low, the VMA line and bus available (BA) line will be held low, the $R/\overline{W}$ line is set high, the data bus lines are in a high-impedance state. The address bus will have FFFE (1111 1111 1111 1110) on its lines. The programmable registers, Acc-A, Acc-B, and so on, retain their contents;

---

[1] X indicates a number, since there is usually more than 1 chip select line (e.g., CS2 indicates chip select line 2, which will be active when it is high). This line will be held low until the chip is activated; then it will be set high. Similarly, with $\overline{CS0}$, select line zero and active low, this line is normally high.

however, bit 4 of the condition codes register, the interrupt mask bit, will be set. This prevents the processor from being interrupted during its initialization program, which will now be described. When the reset line goes high, the processor goes into an initialization sequence:

1. The processor will not recognize an interrupt via the $\overline{\text{IRQ}}$ line, since the interrupt mask bit is set. It will, however, recognize an interrupt via the nonmaskable interrupt line (yet to be described).

2. The address on the address bus is FFFE and the processor will automatically load the data that are residing at this address (ROM, EPROM) into the program counter as its high byte; it then automatically increments the address to FFFF, and the contents of this location are now loaded into the program counter as the low byte. Thus, the program counter now contains the starting address, both page (high byte) and line (low byte), of the program that the processor is to execute. The processor now starts to execute that program.

In the period immediately after the power has been switched on to the microprocessor system, its output lines will be in an indeterminate state. Therefore, if the system contains some RAM memory that is being powered by batteries, the information contained within could be damaged, since the state of the VMA line in this time interval is not known. To eliminate this possibility, the VMA line must be forced low until at least eight cycles of the phase 1 (ϕ_1) clock have elapsed. The MPU has nine supervisory lines, eight input and one output; two are shared with the control bus, reset and ϕ_2. The remainder will now be discussed:

1. Bus available (BA): this is an output line; it is low when the MPU is using the address or data buses to execute a program. In the intervals between the MPU fetching instructions and storing data, this line goes high; the buses are not being used. However, the line goes high automatically when the processor is halted and all internal activity has ceased. This line is used by those peripherals that use the address bus, data bus, and R/$\overline{\text{W}}$ line exclusive of the MPU. The technique is known as *cycle stealing;* its use requires special knowledge and special application techniques, and will not be discussed here. The line is *not connected* when its use is not required.

2. $\overline{\text{Halt}}$ is an input line; it is normally held high by a 4.7-kΩ pull-up resistor tied to the 5 V supply. This line is pulled low by external circuitry when the processor is to be put into an idle mode; that is all program activity has stopped. During this time the address bus, data bus, and R/$\overline{\text{W}}$ line are all placed in a high-impedance state, while the VMA line is forced low and the BA line is set

high. Should an interrupt occur, either via the $\overline{\text{IRQ}}$ line or the nonmaskable interrupt line ($\overline{\text{NMI}}$) while the MPU is in the halted state, it will be latched into the MPU and serviced as soon as the MPU is taken out of the halt mode.

3. Three-state control (TSC): this is an input line and is used by peripherals that want exclusive use of the address bus, data bus, and R/$\overline{\text{W}}$ line. The line is normally low and will be set high when access to the buses is required. The MPU responds by setting the BA line low and putting itself into a semihalted state.[2] The TSC line is used in conjunction with the BA line, and when not required it *must* be tied low via a 10-kΩ resistor. This line requires the use of special knowledge if it is to be used effectively and efficiently.

4. Data bus enable (DBE): this is an input line; it is normally tied to the phase 2 clock line. It is used by the MPU to switch the data bus from a high-impedance state (DBE low) to the active, or data transfer, state (DBE high).

5. Nonmaskable input line ($\overline{\text{NMI}}$): this is an input line; it is normally pulled high by the 5 V supply via a 3.3-kΩ resistor. It is connected to those peripherals that demand high-priority attention. Its effect cannot be masked off under program control, and if the line goes low, the processor will act in a manner similar to an $\overline{\text{IRQ}}$ interrupt.

6. Phase 1 (ϕ_1) clock: an input line that controls the timing of the address placed upon the address bus.

To improve the performance of the 6800 and its peripherals, all the address lines from the MPU should be tied to the 5 V line through 10-kΩ resistors. These resistors are called *pull-up* resistors and assist in the prevention of extraneous signals accumulating on these lines when they are floating or in the high-impedance state.

5.2 THE 6800 SYSTEM CLOCK

The 6800 requires a good-quality, 5 V, two-phase clock whose outputs ϕ_1 and ϕ_2 are *nonoverlapping*. The Motorola 6800 Microprocessor Applications Manual contains a detailed description of how to build such a clock (section 4.1, page 4.1). Because wave shape, rise time, and the like are so important for satisfactory operation of a microprocessor system, no discussion will be given here. The Motorola manual should be followed.

---

[2] Semihalted state means that the microprocessor can only be held in this state for a maximum time of 4.5 microseconds.

The ϕ_1 clock, which is only applied to the MPU, is responsible for the internal timing of the chip. ϕ_2 is applied to the MPU to time the data movements; also, it is applied to the enable terminal or one of the chip select lines of the peripheral chips to activate these devices when data transfers are impending. It is usually used in conjunction with the VMA signal (i.e., VMA $\cdot$ ϕ_2).

Care should be taken with the distribution of the ϕ_2 clock lead around the system since it may pick up extraneous signals that could affect the MPU's operation. To reduce and/or eliminate this possibility, this lead should be isolated; the method used by the authors is to *AND* the VMA and ϕ_2 signals close to the MPU and distribute the combined signal to the peripherals via their enable terminal or one of their chip select terminals.

5.3 STACK AND STACK POINTER (SP)

The stack is a successive block of RAM memory locations that is used as a temporary storage area for data and the contents of the programmable registers when control is being temporarily transferred from one memory location to another. For example, when an interrupt occurs, the processor leaves the main program and executes an alternative program at a specified address. When this program is completed, the processor will return to the main program and continue execution where it left off.

To facilitate this transfer of control, the status of the MPU, that is, the contents of the condition codes register, Acc-A, Acc-B, and the index register must be saved, as well as the contents of the program counter, which contains the address of the *next* instruction to be executed. These data are retrieved upon return to the main program, and the processor continues program execution with all registers exactly as they were when the program location was originally changed.

The stack pointer is a 16-bit register that contains the *address* of the next "empty" (the word empty does not mean "contains nothing"; it means it contains irrelevant data or information) location of the stack. The original position of the stack pointer is set under program control during the early stages of the initialization program after the reset line goes high.

Data are stored on the stack in byte-sized pieces at the address contained in the stack pointer, which is then automatically *decremented* by 1. Thus, the stack pointer always points to the *next* lower empty memory location. Conversely, when data are taken out of the stack, the stack pointer is *first incremented by 1*; then the data resident at that address are copied. The pointer is now pointing to an "empty" memory location.

The stack works on the principle of the last data on to the stack are the first data to come off the stack (LIFO; last in, first out).

A further use of the stack is as a temporary storage area for data that are being manipulated. For example, Acc-A is required for a data transfer to a peripheral; however, the accumulator contains information that will be required in succeeding steps of the program and therefore it must be saved. It could be saved virtually anywhere in the RAM memory via a STA-A instruction, but since a stack is in existence, why not use it via the PSH instruction, since this is a 1-byte instruction, as compared to the 2- or 3-byte STA-A instruction? Both time and memory are saved.

Example 1.   Draw a flow chart (see Figure 5.1) and write out the machine language instructions to temporarily save the contents of Acc-A while it is being used to transfer data from memory location 0378 to PIA port A at address 8004.

Care must be exercised when using the stack as it is very easy to get the stack pointer out of position with the relevant data; for example, save (or store) the contents of Acc-A and Acc-B during some program operation, and then return the data to their original location (see Figure 5.2).

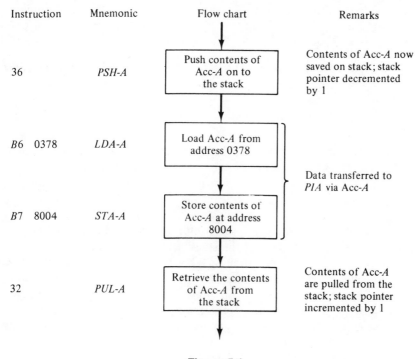

Figure 5.1

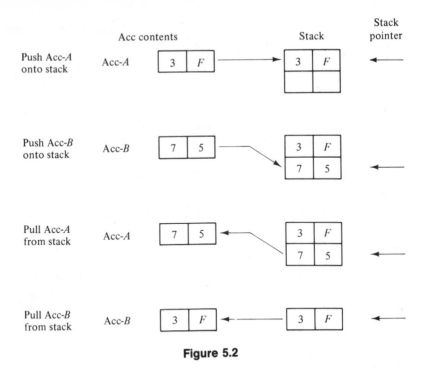

Figure 5.2

In Figure 5.2 the contents have been interchanged, which is clearly incorrect. Therefore, the rules for operating a stack are as follows:

1. Pull the data from the stack in the reverse order in which they were pushed onto the stack.
2. Pull as many pieces of data from the stack as were pushed onto the stack. Keep the pointer in the correct relative position. An example illustrating this will be presented in the tutorial examples.

When the processor saves the contents of a register on the stack, it simply copies the contents onto the stack; it does *not* destroy the contents of the register in the storing process. Whatever was in the register before the save will be in the register after the save. Similarly, when information is pulled from the stack, the information is copied into the recipient register, destroying whatever was in the register originally, but leaving the contents of the stack in their original condition. The stack pointer now points at an "empty" location, empty because the data have been retrieved. Thus, that which remains is redundant.

Interrupts are used by peripherals that demand some special service. They basically fall into two categories, maskable and nonmaskable. The nonmaskable input will be considered first.

5.4.1 Nonmaskable Interrupt (NMI)

This type of interrupt is used exclusively in high-priority applications; that is, if damage to property can result and/or it could be hazardous to human life if the problem is not acted upon immediately, instant action is required. Therefore, the processor monitoring this process must be ever on the alert to such happenings.

The $\overline{\text{NMI}}$ line is an active low line. The line is normally high, and when it goes low the processor will start its $\overline{\text{NMI}}$ interrupt sequence, which is shown in Figure 5.3 in block format (assume that the stack pointer is pointing at address A078, the same as in the JBUG monitor program of the MEK 6800 D2 kits).

The interrupt program being executed *must* include certain features other than that required by the peripheral initiating the interrupt. For example, the last instruction in the interrupt program *must be* return from interrupt (RTI). This instruction informs the processor that the interrupt program is complete. The processor then goes through a return from interrupt sequence as shown in block diagram form in Figure 5.4.

The state of the interrupt mask bit (bit 4 of the condition codes register) is the same as it was prior to storage in the stack; that is, if it was set when the contents were stored, it will be set on return to the main program.

Furthermore if the interrupt was generated by a PIA, a bit will be automatically set in the control register of the originating port. For example, assume that the interrupt to the port came in on line C_{A1}; then bit 7 of control register A will be set. Therefore, before the interrupt program currently being executed is completed, this bit in control register A must be cleared. When it is cleared depends upon the programmer; as long as it is set, no other interrupt can come in on that line (the C_{A1} line in this case). Failure to clear it will cause the processor to continue to execute the interrupt program; it is now in a never-ending loop.

Example 2.   Assume that an interrupt has been originated from PIA port A via line C_{A1}; write that part of the interrupt program that clears bit 7 of control register A and the final instruction. Bit 7 is to be cleared at the beginning of the program. Port A is located at address 8004/8005. See Fig-

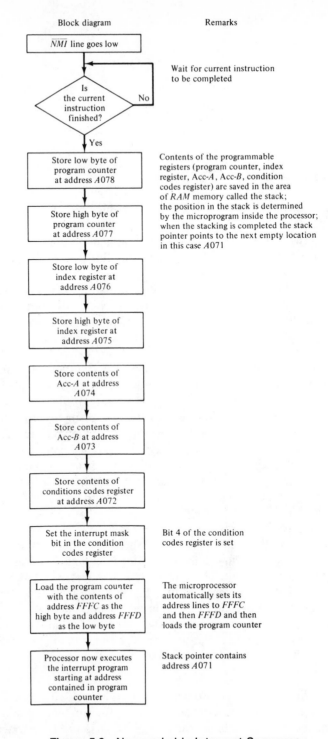

Block diagram   Remarks

$\overline{NMI}$ line goes low

Is the current instruction finished?   No   Wait for current instruction to be completed

Yes

Store low byte of program counter at address $A078$

Store high byte of program counter at address $A077$

Store low byte of index register at address $A076$

Store high byte of index register at address $A075$

Store contents of Acc-A at address $A074$

Store contents of Acc-B at address $A073$

Store contents of conditions codes register at address $A072$

Contents of the programmable registers (program counter, index register, Acc-A, Acc-B, condition codes register) are saved in the area of RAM memory called the stack; the position in the stack is determined by the microprogram inside the processor; when the stacking is completed the stack pointer points to the next empty location in this case $A071$

Set the interrupt mask bit in the condition codes register

Bit 4 of the condition codes register is set

Load the program counter with the contents of address $FFFC$ as the high byte and address $FFFD$ as the low byte

The microprocessor automatically sets its address lines to $FFFC$ and then $FFFD$ and then loads the program counter

Processor now executes the interrupt program starting at address contained in program counter

Stack pointer contains address $A071$

Figure 5.3 Nonmaskable Interrupt Sequence

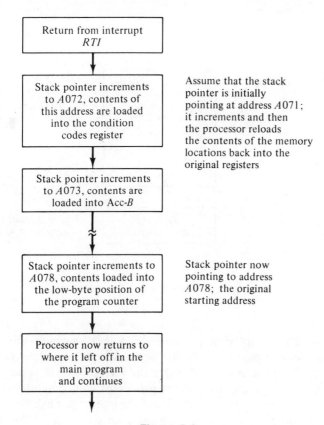

Figure 5.4

ure 5.5. It will be noticed that the interrupt mask bit was set during the NMI sequence (Figure 5.3) after the contents of the registers have been saved. This prevents the processor from recognizing interrupts on the IRQ line that may come in while the processor is servicing the current interrupt. However, the mask bit may be cleared at any time under program control via the CLI instruction, OE; it is cleared automatically when the RTI instruction is executed.

5.4.2 Interrupt Request Line (IRQ)

This line is the most widely used of the two interrupt lines. It can be masked off under program control via the set interrupt mask bit instruction (SEI, op-code OF). When this instruction is executed, bit 4 of the condition codes register will be set and *no* interrupts coming in on the IRQ line will be recognized. This instruction is used when the processor does

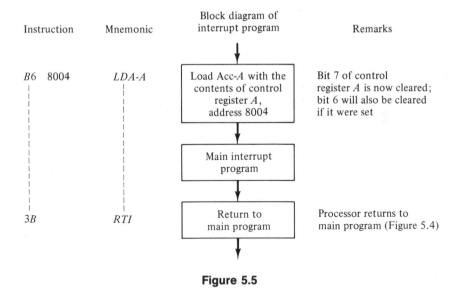

Instruction	Mnemonic	Block diagram of interrupt program	Remarks
B6 8004	*LDA-A*	Load Acc-*A* with the contents of control register *A*, address 8004	Bit 7 of control register *A* is now cleared; bit 6 will also be cleared if it were set
		Main interrupt program	
3B	*RTI*	Return to main program	Processor returns to main program (Figure 5.4)

Figure 5.5

not wish to be interrupted in its current task (e.g., a timing loop). However, should an interrupt occur while the mask is set, it will be recognized immediately the mask is cleared, since the interrupt latch is set.

When the $\overline{\text{IRQ}}$ line goes low, the processor will complete its current instruction and then it will check to see if the interrupt mask bit is set; if it is *not* set, the processor will enter the interrupt sequence. A flow chart of the $\overline{\text{IRQ}}$ sequence is shown in Figure 5.6. If it is set, the processor continues with the main program. The processor returns to the main program in the same manner as in the $\overline{\text{NMI}}$ interrupt.

5.4.3 Software Interrupt (SWI)

This interrupt is program initiated via the op-code 3F. When this instruction is encountered in a program, the processor will save the contents of the programmable registers on the stack, as shown in the block diagram of Figure 5.7. The interrupt program and the return from interrupt (RTI) instruction behave in exactly the same manner as an $\overline{\text{IRQ}}$ or $\overline{\text{NMI}}$ interrupt.

5.4.4 Wait Instruction (WAI)

The wait instruction is used in conjunction with the $\overline{\text{IRQ}}$ and/or the $\overline{\text{NMI}}$ interrupt line. It is used when the processor is to stop its normal fetch/execute cycle and wait for an external event or events to occur and at the

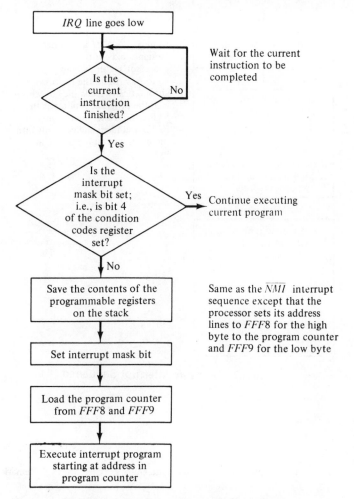

Figure 5.6 Interrupt Request Sequence

same time to release control of the address bus, data bus, and R/$\overline{\text{W}}$ line; virtually the processor is in a quiescent state. It can be stimulated into action via an interrupt line making a high to low transition. When this occurs, the processor immediately picks up the starting address of the interrupt program and begins to execute that program, since the contents of the programmable registers *have already* been saved on the stack.

The block diagram of Figure 5.8 illustrates the sequence of events that occur when a WAI (3E) instruction is encountered in the program. A typical program containing the WAI instruction is shown in Figure 5.9. When the processor has serviced the interrupt, the RTI instruction at the

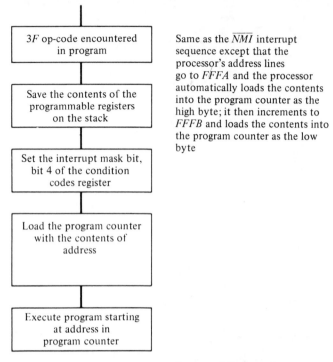

Same as the $\overline{NMI}$ interrupt sequence except that the processor's address lines go to $FFFA$ and the processor automatically loads the contents into the program counter as the high byte; it then increments to $FFFB$ and loads the contents into the program counter as the low byte

Figure 5.7 Software Interrupt Sequence

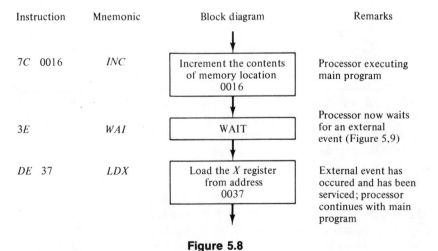

Figure 5.8

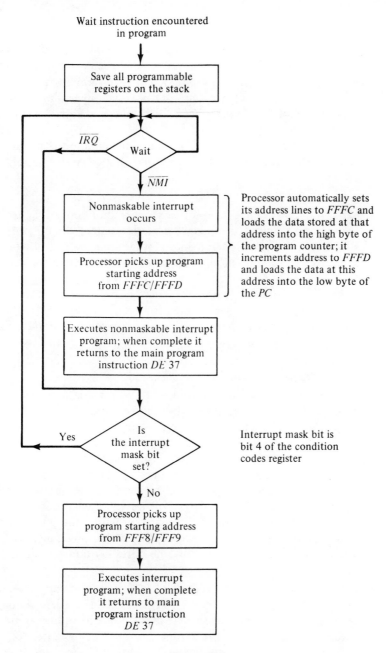

Wait instruction encountered
in program

Save all programmable
registers on the stack

$\overline{IRQ}$

Wait

$\overline{NMI}$

Nonmaskable interrupt
occurs

Processor picks up program
starting address
from *FFFC/FFFD*

Processor automatically sets
its address lines to *FFFC* and
loads the data stored at that
address into the high byte of
the program counter; it
increments address to *FFFD*
and loads the data at this
address into the low byte of
the *PC*

Executes nonmaskable interrupt
program; when complete it
returns to the main program
instruction *DE* 37

Is
the interrupt
mask bit
set?

Yes

Interrupt mask bit is
bit 4 of the condition
codes register

No

Processor picks up
program starting address
from *FFF8/FFF9*

Executes interrupt
program; when complete
it returns to main
program instruction
DE 37

Figure 5.9

end of the interrupt program will cause the processor to return to the *next*
instruction in the main program, typically DE 37 (see Figure 5.8).

5.5 BUILDING A 6800 SYSTEM

This section will discuss the technique of designing, building, and initializing an MC 6800 controlled microprocessor system; the design will include the peripheral chips that support the MPU. Every system must contain a number of essentials such as a clock, I/O ports serial and/or parallel, memory in the form of RAM and ROM or EPROM, and some logic gates for signal control.

The basic design is not difficult if some consideration is given initially to salient features of the system; the following questions should be considered:

1. What function is the processor to serve (motor control, data collection and manipulation, etc.)? This determines the quantity of RAM memory required for data storage; motor control usually requires a larger amount of ROM or EPROM for program storage than it does RAM, since the amount of data read in and stored is minimal. The data-collection scheme requires a larger section for data storage (RAM) than it does for program storage (ROM or EPROM). This of course varies from application to application.

2. When the memory requirements are determined, the memory space required for permanent (ROM, EPROM) memory and temporary memory (RAM) can be assessed.

3. How many I/O lines will be required? Will the data output be in serial or parallel form? If the system is a motor-control scheme, the I/O ports will undoubtedly be PIAs and usually a minimum of two wires is required to start and stop a motor; thus the actual number of ports required can be easily found. If the system is a data-collection system and 8-bit data are being collected, then one port per collection point will be required; but if 16-bit data are being collected, two ports will be required (one complete 6820) and the addressing of the registers within the port will require attention if the data are to be read with a single instruction. Thus the I/O requirements must be carefully considered at the outset.

4. How much stack will be required? Is this an interrupt actuated system? How many interrupts will be used? The stack requires seven empty locations for storage of the contents of the programmable registers for each interrupt currently being serviced. An estimate of the stack requirements is essential to prevent its contents from overflowing into some other area of memory usage.

5. How will the memory space be assigned? What location will be assigned to stack? What location will be assigned to I/O? This as-

pect of memory-space assignment is a crucial point as address decoding can become a costly feature, and it must be remembered that the initialization programs, which are held in ROM or EPROM, must be at high addresses in the memory space, since the MPU automatically sets its lines high on reset and interrupts.

6. Will the halt line be utilized?
7. Will tristate control be required?
8. Will the $\overline{\text{NMI}}$ line be used?
9. Will the bus available line be used?

To answer some or all of these questions, the end use of the system should be known and its application completely understood.

A design of a minimum system, that is, a system built with the chips shown in Table 5.1, will now be presented. This system requires a +5 V power supply capable of delivering 2.5 A at +5 V. Although the basic system does not require this much power by itself, the power requirements can become large when interfacing devices such as photocoupled solid-state ac switches are connected to the output lines of the PIAs and the like.

Some of the output/input pins of the MPU will not be required in this system; they are itemized in Table 5.2. If they are connected to some potential, it will be noted.

Table 5.1

Chip No.	Type	Remarks
2716	EPROM	Storage 2048 bytes
MCM 6810	RAM	Storage 128 bytes
MCM 6820	PIA	I/O port parallel
MCM 6850	ACIA	I/O port serial
MC 6871B	clock	614.4-kHz clock
N7410, N7400	NAND gates	3 input and 2 input NAND gates

Table 5.2

Pin No.	Function	Connected to	Remarks
2	$\overline{\text{HALT}}$	+5 V via 4.7 kΩ	Must be tied high
6	$\overline{\text{NMI}}$	+5 V via 4.7 kΩ	Must be tied high
7	BA	Not connected	Bus available
35	Not used	Ground	
38	Not used	Ground	
39	TSC		Must be tied low via 10 kΩ

The chips should be laid out in such a manner that the conductors are as short as possible. The clock should be located as close as is physically possible to the MPU; this will avoid "ringing" on the clock lines. Furthermore, the power supply lines between these two devices should be kept short; this reduces the noise that can be picked up and transmitted to the MPU via these lines. Phase 2 (ϕ_2) of the clock is required as part of the control for several of the peripheral chips; this clock line must be buffered before it is distributed to different parts of the system to reduce the possibility of picking up unwanted noise. Finally, the power supply lines + 5 V and ground must be big enough, no less than 50 to 100 mils, to provide a low-impedance path for the current; this will avoid low voltages at crucial points in the system.

The whole can be wired via a printed circuit card, which is costly in small quantities, or it can be wired using wire-wrap sockets and wire-wrapping techniques. *Caution:* Keep the runs between terminations as short as possible; this will help to prevent the wires behaving like antennas.

The systems used by the authors have all been built on SK-10 universal socket component boards and wired with no. 22 gauge wire. Good power distribution practice was followed. No special care was taken with layout other than that detailed previously.

The circuit used for the minimum system is shown in Figure 5.10a. The memory space was apportioned as shown in Figure 5.10b.

Address bits A_{11}, A_{12}, and A_{13} are not used in this minimum system; thus they are available whenever an extension to the system is required. Bits A_{14} and A_{15} are used as the peripheral chip select bits.

The RAM is placed in memory so that it can be reached by the low-order 8 bits of the 16-bit address bus. Thus the RAM can be reached by the direct addressing mode, which can only access page 00. The RAM chip has 128 bytes of storage capacity, which was considered big enough for process applications, although the memory map (Figure 5.10b) could contain much more if required.

The EPROM chip is placed at the highest end of the memory space because the 6800's address lines all go high on initialization, and the MPU automatically loads the program counter with the contents of the permanent memory connected to those lines.

The PIA and ACIA are placed somewhere in between the RAM and the EPROM. It will be noticed that only the low-order address bits, A_2, A_1, and A_0, are brought into the PIA; these are the address bits needed to reach all the internal registers in the chip. The ACIA is similarly connected except that address bits A_3 and A_0 are used to access the internal registers. A program to initialize the MPU and get the processor going will now be written; it is very basic since all that is being attempted at this time

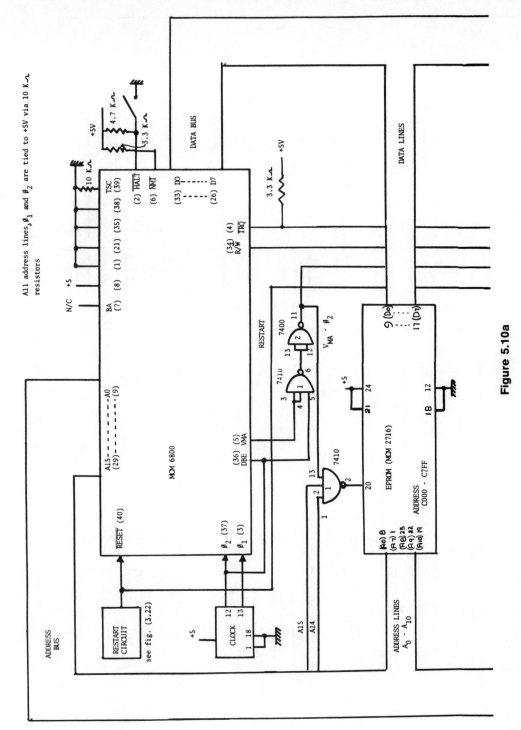

All address lines, θ_1 and θ_2 are tied to +5V via 10 K Ω resistors

Figure 5.10a

189

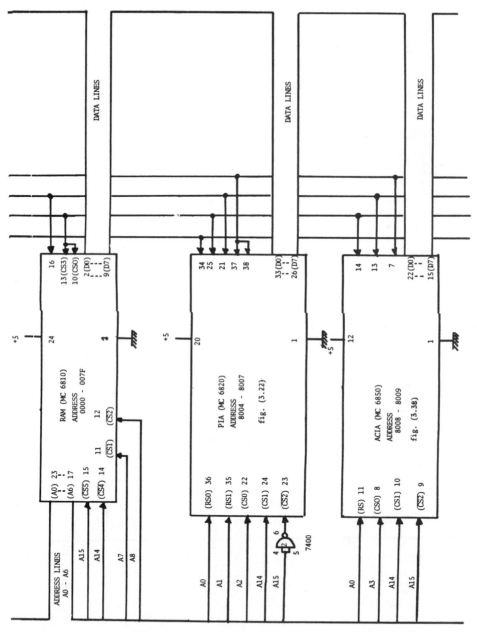

Figure 5.10a *(cont.)*

Address lines A_{15} to A_0

	15	14	13	12	11	10	9	8	7	6	5	4	3	2	1	0	
RAM	0	0	—	—	—	—	—	0	0	0	0	0	0	0	0	0	0000
	0	0	—	—	—	—	—	0	0	0	0	0	0	0	0	1	
	0	0	—	—	—	—	—	0	0	0	0	0	0	0	1	0	
	0	0	—	—	—	—	—	0	0	1	1	1	1	1	1	1	007F
PIA	1	0	—	—	—	—	—	—	—	—	—	—	1	0	0		8004
	1	0	—	—	—	—	—	—	—	—	—	—	1	1	0		
	1	0	—	—	—	—	—	—	—	—	—	—	1	1	1		8007
ACIA	1	0	—	—	—	—	—	—	—	—	—	1	0	0	0		8008
	1	0	—	—	—	—	—	—	—	—	—	1	0	0	1		8009
EPROM	1	1	—	—	—	0	0	0	0	0	0	0	0	0	0	0	C000
	1	1	—	—	—	0	0	0	0	0	0	0	0	0	0	1	
	1	1	—	—	—	0	0	0	0	0	0	0	0	0	1	0	
	1	1	—	—	—	1	1	1	1	1	1	1	1	1	1	1	C7FF

Figure 5.10b

is to see if the microprocessor that has been built works. The program will initialize a PIA port and flash a light-emitting diode that is connected to one of its output lines via a driver chip. The wiring diagram will follow the instruction program. A block diagram of the basic steps required is given in Figure 5.11; this will be followed by the detailed block diagram and the program.

The stack pointer will be initialized at address 007F (top of the RAM memory), although it will not be required in the program, since no interrupts, subroutines, and the like, will be entertained.

The layout of the EPROM memory at the top end *must* conform to the rigid format set out by the manufacture. Table 5.3 illustrates this. In this initial program the nonmaskable interrupt, interrupt request, and software interrupt will *not* be used; therefore, no addresses will be placed in memory locations FFF8 to FFFD inclusive.

Figure 5.12 illustrates the initialization program as well as the operating program. The main program will start at C000. When the EPROM is "blasted," the instructions indicated are set permanently into memory at their associated addresses.

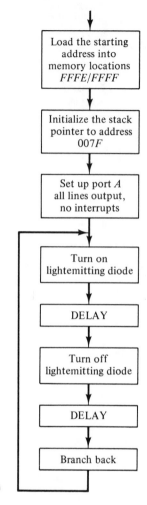

Figure 5.11 Block Diagram

Table 5.3 Table of Contents of the Upper 8 Bytes of EPROM Memory

Memory Location	Contents	
FFFF	Main program starting address	Low byte
FFFE		High byte
FFFD	Nonmaskable interrupt starting address	Low byte
FFFC		High byte
FFFB	Software interrupt starting address	Low byte
FFFA		High byte
FFF9	Interrupt request line starting address	Low byte
FFF8		High byte

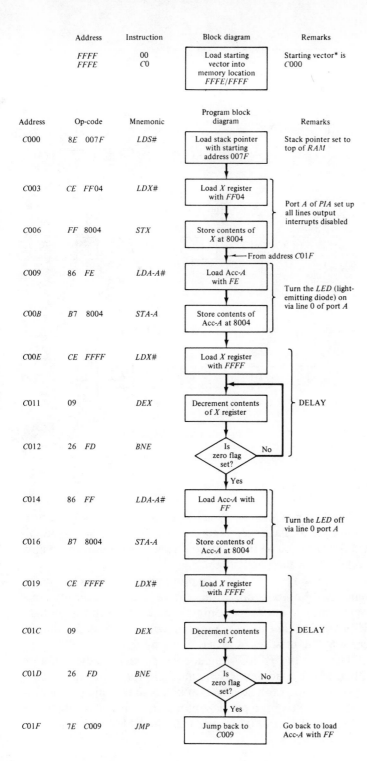

Address	Instruction	Block diagram	Remarks
FFFF	00	Load starting vector into memory location *FFFE/FFFF*	Starting vector* is *C*000
FFFE	*C*0		

Address	Op-code		Mnemonic	Program block diagram	Remarks
*C*000	8*E*	007*F*	LDS#	Load stack pointer with starting address 007*F*	Stack pointer set to top of *RAM*
*C*003	*CE*	*FF*04	LDX#	Load *X* register with *FF*04	Port *A* of *PIA* set up all lines output interrupts disabled
*C*006	*FF*	8004	STX	Store contents of *X* at 8004	
				← From address *C*01*F*	
*C*009	86	*FE*	LDA-A#	Load Acc-*A* with *FE*	Turn the *LED* (light-emitting diode) on via line 0 of port *A*
*C*00*B*	*B*7	8004	STA-A	Store contents of Acc-*A* at 8004	
*C*00*E*	*CE*	*FFFF*	LDX#	Load *X* register with *FFFF*	DELAY
*C*011	09		DEX	Decrement contents of *X* register	
*C*012	26	*FD*	BNE	Is zero flag set? — No	
				↓ Yes	
*C*014	86	*FF*	LDA-A#	Load Acc-*A* with *FF*	Turn the *LED* off via line 0 port *A*
*C*016	*B*7	8004	STA-A	Store contents of Acc-*A* at 8004	
*C*019	*CE*	*FFFF*	LDX#	Load *X* register with *FFFF*	DELAY
*C*01*C*	09		DEX	Decrement contents of *X*	
*C*01*D*	26	*FD*	BNE	Is zero flag set? — No	
				↓ Yes	
*C*01*F*	7*E*	*C*009	JMP	Jump back to *C*009	Go back to load Acc-*A* with *FF*

Figure 5.12

193

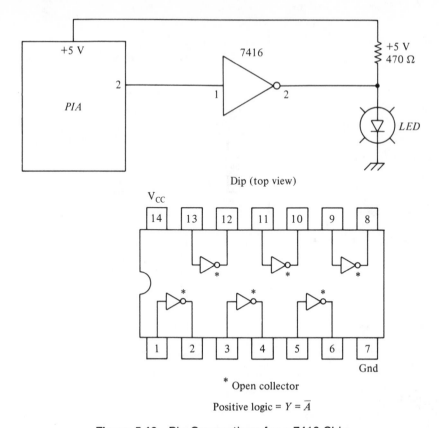

Dip (top view)

* Open collector

Positive logic = $Y = \overline{A}$

Figure 5.13 Pin Connections for a 7416 Chip

To wire in the LED to the system, output line 0 of port A (pin 2) will be used, since line 0 will be alternately set to 1 and then to 0 via the program, causing the LED to flash (see Figure 5.13). There are six open collector drivers in the chip; a 0 on the input (pin 1) will cause current to flow through the LED, which then emits light.

Now that the system is operating, the program in the EPROM can be changed to something more elaborate; this can be accomplished by erasing the program already in existence. Subject the chip to ultraviolet light for the specified period of time and the program will be erased. It can now be reprogrammed to the new set of program instructions.

Before ''blasting'' the EPROM again, the *complete* new program should be written, tested, and debugged, since all programs usually contain errors (bugs). The role of the evaluation kits is to test and debug programs before committing them to permanent memory. However, to assist the reader in setting up the EPROM for use with the NMI, SWI, and IRQ, a sample program will be offered.

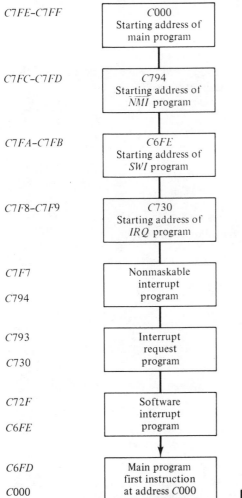

C7FE–C7FF	*C*000 Starting address of main program
C7FC–C7FD	*C*794 Starting address of *NMI* program
C7FA–C7FB	*C*6FE Starting address of *SWI* program
C7F8–C7F9	*C*730 Starting address of *IRQ* program
C7F7 *C*794	Nonmaskable interrupt program
*C*793 *C*730	Interrupt request program
C72F *C6FE*	Software interrupt program
C6FD *C*000	Main program first instruction at address *C*000

Figure 5.14

A careful analysis of Figure 5.10b shows that address lines A_{13}, A_{12}, and A_{11} are not connected, so FFFF becomes C7FF and therefore Table 5.3 can be amended. A corresponding memory map (a layout of how the EPROM memory is assigned) is given in Figure 5.14.

5.6 TUTORIAL EXAMPLES

Example 1. The stack pointer for a 6800 microprocessor is initialized at address A078 and the following has occurred:

1, jump to subroutine (JSR)

3, push data onto stack (PSH)

What is the address contained in the stack pointer?

Assume that the PC contains 0137, and the first, second, and third pushes contained 05, 27, and 2F, respectively.

Address	Content of Stack	Remarks
A078	37	Program counter low byte
A077	01	Program counter high byte
A076	05	Push
A075	27	Push
A074	2F	Push
A073		

Address contained in stack pointer

The stack then experiences two pulls; where is the stack pointer now?

Address	Contents of Stack
A078	37
A077	01
A076	05
A075	

Address contained in stack pointer

Q: What happened if a return from subroutine instruction is now executed?

A: The microprocessor would return to address 0501, use the data stored at this address as an op-code, and execute these data. In other words, the processor has returned to the wrong place in the memory space.

Example 2. Show the connections to add another PIA to the microprocessor system of Figure 5.10a; the new PIA is to be located at address 8000 to 8003 inclusive. See Figure 5.15. The truth table for the logic is as follows:

A_2	A_3	C_{S0}
0	0	1
0	1	0
1	0	0
1	1	0

$$\therefore \quad C_{S0} = \overline{A_2} \cdot \overline{A_3}$$

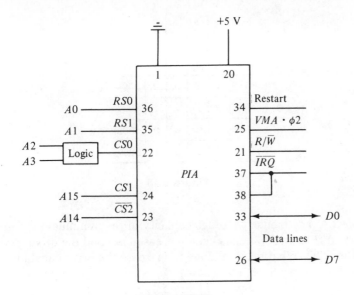

Figure 5.15

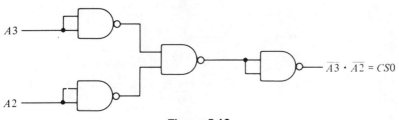

Figure 5.16

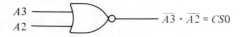

Figure 5.17

The derivation of the logic circuit using two-input NAND gates is shown in Figure 5.16, and that using two-input NOR gates is shown in Figure 5.17.

The logic to generate C_{S0} is required; otherwise, when the ACIA is addressed, port A's control register or its data direction/peripheral data register will also be addressed, which will cause confusion on the data bus. Inspection of the memory map will confirm this. See Figure 5.18.

	3	2	1	0	(see Figure 5.10b)
PIA-A (new)	0	0	0	0	8000
	0	0	1	1	8003
PIA-B (old)	×	1	0	0	8004
	×	1	1	1	8007
ACIA	1	0	0	0	8008
	1	0	0	1	8009

Figure 5.18

Example 3. The memory storage capacity of the minimum microprocessor system (Figure 5.10a) has been increased beyond the driving capability of the 6800. What must be done to increase the drive capability of the data lines, address lines, and so on?

The solution is shown in Figure 5.19.

Example 4. The following figure is a part of a microprocessor program:

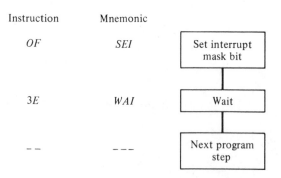

Instruction	Mnemonic	
OF	SEI	Set interrupt mask bit
3E	WAI	Wait
- -	- - -	Next program step

Q: What external stimulus will cause the processor to execute the next program step?

A: A nonmaskable interrupt.

Example 5. The condition codes register contains 17; an interrupt occurs and the processor enters an interrupt routine.

Q: Which interrupt line caused the interrupt?

A: The nonmaskable interrupt line. The interrupt mask bit is set in the

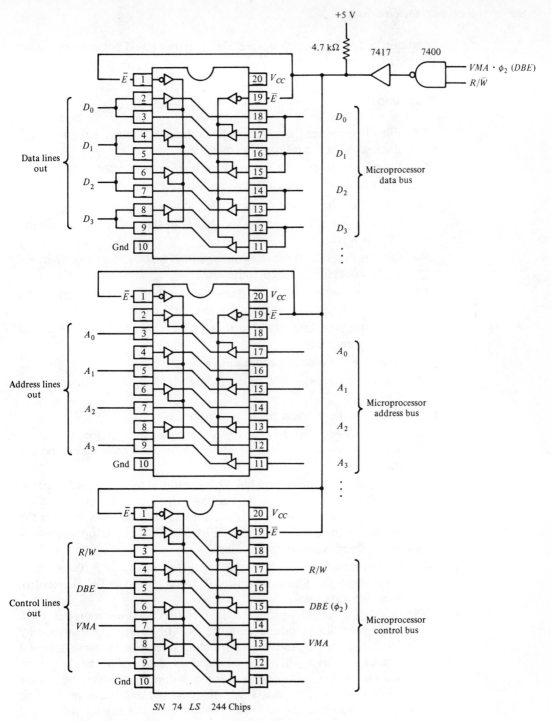

SN 74 LS 244 Chips

Octal buffer line drivers with three state outputs
Only four data and address lines for simplicity

Figure 5.19

199

condition codes register; therefore, the interrupt could not have come via the $\overline{\text{IRQ}}$ line.

Problems

5-1. What determines the order in which the programmable registers are stacked when an interrupt occurs?

5-2. Why is the $\overline{\text{NMI}}$ terminal used for high-priority interrupts?

5-3. What is the function of the ϕ_1 and ϕ_2 clocks?

5-4. Why must the ϕ_1 and ϕ_2 clocks be nonoverlapping?

5-5. What causes a microprocessor that is in the wait state to restart? Give two reasons.

5-6. What is the state of the $R/\overline{W}$ line, address bus and data bus when the microprocessor is in the wait state?

5-7. Refer to Figure 5.6. Why does the microprocessor complete the current instruction before going to an interrupt sequence?

5-8. Why must the interrupt flag in a PIA be cleared before returning from an interrupt?

5-9. Why use the indexed addressing mode?

5-10. Why is it necessary to initialize the stack pointer? When is this done?

5-11. What are the advantages of a subroutine?

5-12. Why is it necessary to keep the stack balanced? That is, why must the same amount of data be pulled from stack as is stored on the stack?

5-13. A microprocessor system has 32,768 address locations, RAM occupies 0000 to 08FF, EPROM occupies the top 2048 bytes, and I/O devices occupy 256 memory locations starting at 1000. What is the top and bottom address of the EPROM? What is the top address of the I/O? Draw the memory map.

5-14. Fit a timer module to the minimum system of Figure 5.10a, starting at address 8010. Show the address decoding logic and *all* the control lines and data lines.

5-15. Develop a control system to operate two 6800 microprocesors in the master/slave mode. The master will be controlling a system by dedicated control and will, at random intervals, transmit selected data to the slave. The slave will read the data and at the same time will inform the master controller it has done so. Before the master sends new data, it must verify that the old data have been read. Should the slave fail to read the data sent, the master will sound

an alarm and then continue with its main program. It must *not* stop
because the slave has stopped responding.

5-16. Design a logic circuit to hold the VMA line, to battery backed
RAM, low for at least 8 cycles of the ϕ_1 clock after the microprocessor, to which the RAM is attached, is switched on.

5-17. Fit the 2K of RAM shown below to the minimum system of Figure
5.10a starting address 0080. Show all address decoding, address
lines, data lines, and control lines.

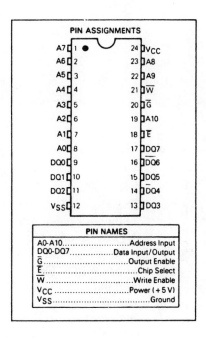

Chapter 6

The MC 6805E2 Microprocessor

The MC 6805E2 MPU chip is a 40-pin device; it is virtually a complete minimum system in itself, since it contains two 8-bit I/O ports similar to the PIAs used in the 6800, a timer, and some on-board RAM memory. The processor (Figure 6.1) has seven control lines and a 13-bit address bus, part of which, the lower eight lines, A_7 to A_0, are time shared to form an 8-bit data bus, D_7 to D_0; they are illustrated as B_7 to B_0. The time sharing or multiplexing is a feature used to reduce the number of output pins required on the MPU chip; this feature also requires the use of strobe signals to inform the external circuits whether data or address information is actually present on the address/data bus.

6.1 THE 6805 MPU AND ITS CONTROL

The control bus consists of seven lines, three input and four output. It carries a mixed set of signals, some of which, the output lines, are used by the MPU to control the operation of the peripherals. The remainder, three lines, are the reset line, the external interrupt control line $\overline{IRQ}$, and the external clock line to the on-board timer. A description of each of the signals carried by the seven control lines will now be given. Two of the output lines, address strobe and data strobe, are nonoverlapping square waves and are specifically associated with the eight multiplexed address/data lines; these will be considered first.

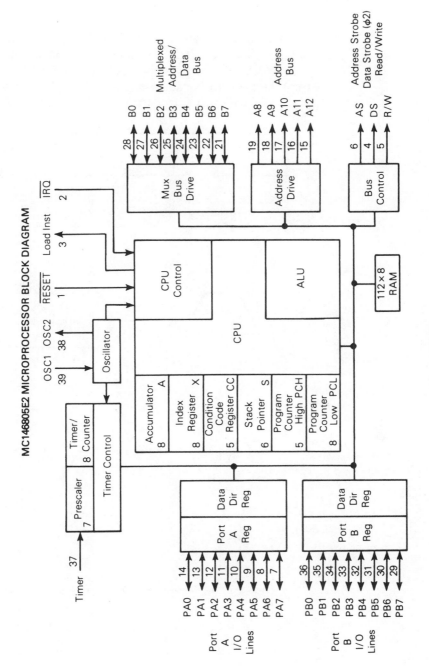

MC146805E2 MICROPROCESSOR BLOCK DIAGRAM

Figure 3.21

1. Data strobe (DS): an output line used by the MPU to inform the peripherals that the information on the address/data bus is valid data (i.e., a write or store cycle is in progress), or that the MPU is ready to receive data via the data lines as a read cycle is in progress (see Figure 6.2). The line is normally low and goes high when a data transfer cycle is pending, the actual transfer of data from the MPU to the addressed peripheral, or vice versa, takes place when the data strobe (DS) makes a *negative transition.*

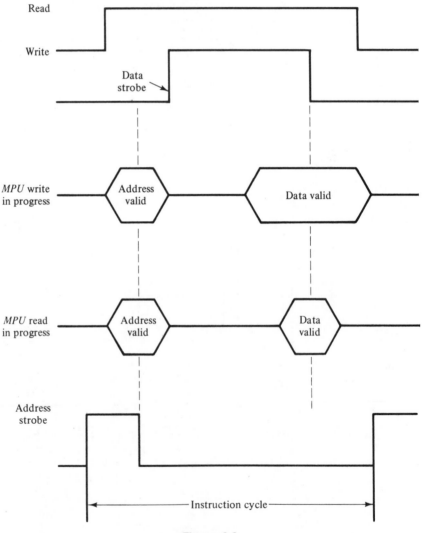

Figure 6.2

Thus the transfer action via the data bus takes place on the negative edge of the data strobe.

2. **Address strobe (AS):** an output line used by the MPU to inform the peripherals that the information on the address/data bus is valid address information. The line is normally low and goes high at the start of a new fetch/execute cycle (see Figure 6.2). Thus the information on the bus is considered to be a valid address when the address strobe makes a *negative transition*. The peripherals require complete address information at the time that the actual data transfer takes place, but at that time the address, on lines b_7 to b_0, has been replaced. Therefore, an address latch *must* be fitted to these lines to retain the address. The latch is actuated when the strobe make its negative transition and will be reset to a new address when the *next* valid address is on the bus. There are memory elements on the market that have a latch internal to the memory chip; thus external latching is not required. Typically, these chips are MCM 65516.

3. **Read/write line (R/$\overline{\text{W}}$):** an output line used by the MPU to inform the selected peripherals whether its operating mode will be read or write when the *next* data strobe is received. The MPU is in the read mode when the R/$\overline{\text{W}}$ line is higher (see Figure 6.2) and in the write mode when the line is low. The line is always either high or low; it does not have a high impedance or floating state.

4. **Load instruction (LI):** an output line used by the MPU to inform interested peripherals that the MPU is in the process of fetching the *next op-code*. The line is normally low and goes high when a fetch op-code operation is in progress; normally the line is not connected since its principal use is for program debugging. The line will remain low during an external interrupt or timer interrupt.

5. **External interrupt request line ($\overline{\text{IRQ}}$):** an input line; it is normally high and will be pulled low by a peripheral or peripherals when the MPU is called upon to service some additional external request.

6. **Timer external clock:** an input line used to supply an external clock, when required, to the timer. This signal line finds its principal application in the field of counting external events and in determining external frequencies.

7. **Reset line:** an input line; it is normally high and goes low to reset the MPU and whatever other devices may be connected to it. Its normal function is to reinitialize the MPU by forcing the address bus to the reset vector location, that is, 1FFE, 1FFF.

The reset line must be held low for a period of not less than five cycles of the input oscillator clock. When the line is low, the following events will occur automatically within the MPU.

1. The address bus is forced to address 1FFE, which contains the high byte of the restart program address.
2. The interrupt request bit (bit 7) of the timer control register is cleared, meaning no interrupt will be forthcoming from the timer.
3. The timer interrupt mask bit (bit 6) of the timer control register is set; thus no interrupt can be initiated by the timer until this bit is cleared.
4. All the data direction bits in the I/O ports are cleared to 0; thus all' lines are input lines.
5. The stack pointer is reset to address 007F, the top address of the stack.
6. The MPU interrupt mask bit (bit 3 of the condition codes register) is set, meaning the MPU will not recognize an interrupt until this bit is cleared.
7. If the processor is in the halted or stopped condition (stop instruction 8E has been executed), the reset line going low will return the processor to the restart condition.
8. If the processor is in the wait condition (wait instruction 8F has been executed), a low on the reset line will force the processor to its restart condition.
9. If the processor is servicing an interrupt request at the time the reset line goes low, it will terminate the program at its current location and return the processor to its restart condition.

When the reset line makes a positive transition (i.e., the line goes high after being held low for the prescribed time), the processor will immediately load the contents of address 1FFE into the high-byte position of the program counter. It will then automatically step to the *next* higher memory location 1FFF and load its contents into the low-byte position of the program counter. Thus the program counter is loaded with the starting address of the restart program, and program execution starts immediately at that address.

The 6805E2 is equipped with two reset modes:

1. Reset via pin 1 external reset.
2. Power-on reset or internal reset.

The MPU will automatically reset itself when power is applied; this is achieved as follows: when the power supply to the processor is turned on, the dc voltage rises to its operating level of $+5$ V; the MPU clock oscillator will start up and a circuit internal to the MPU chip will detect the first clock cycle. When this cycle is detected, a delay of 9600 clock cycles takes place, after which the processor will automatically load the restart vector and begin execution of the restart program.

During the built-in delay, the processor is initializing itself in the same manner as it would if an external reset had occurred (i.e., the nine initialization steps). The clock controlling the 6805 can be from dc values or very low frequencies up to 5 MHz.

6.2 MEMORY SPACE

The memory space (Figure 6.3) has a total capacity of 8192_{10} bytes, of which 80_{10} bytes are committed to the MPU for internal usage: 70_{10} bytes of RAM and 10_{10} bytes of permanent memory. The remainder, 8112_{10} bytes, are available for program and data storage. Part of the total memory capacity is built into the chip, that is, 128_{10} bytes of RAM of which 64_{10} bytes are reserved for the stack, 4 bytes are used for I/O port data and data direction, and 2 bytes are for the timer data and control, a total of 70_{10} bytes; the remainder of the RAM, 58_{10} bytes is available for other uses, such as scratch pad and data storage. The 10_{10} bytes of permanent memory are reserved for the restart and interrupt vectors; they are located at the top end of the memory map, 1FF6 to 1FFF. The remaining 8064_{10} bytes are available for external memory devices such as RAM, EPROM, ROM, PIAs, and ACIAs.

6.3 ADDRESS BUS

The address bus consists of 13 lines, of which 5 are permanent address lines A_{12} to A_8; the remainder B_7 to B_0 are time shared with the transmission of data, thus forming an 8-bit address/data bus.

6.3.1 Five-Bit Section

The 5-bit high address bus is unidirectional and carries the high address code word (e.g. 1F, 03 etc.). It continually has addressed information on its lines. That is, it does *not* go into a high-impedance state at any time during the period that the processor is in operation.

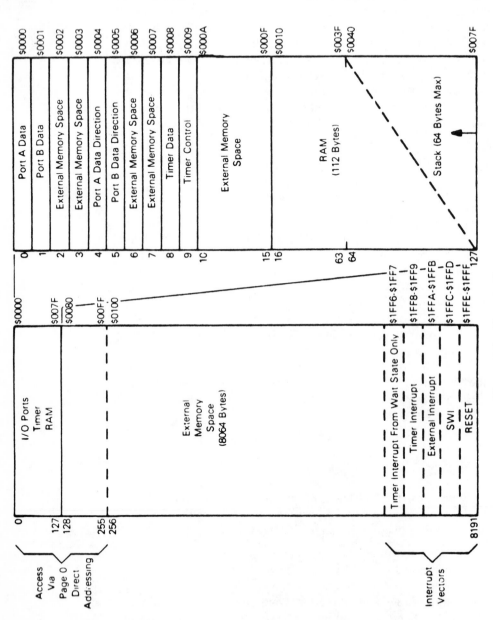

Figure 6.3 Memory Space

6.3.2 Address/Data Bus

The lower 8-bit address bus is time shared with the data transfers; thus the bus is a unidirectional output bus when carrying address information and a bidirectional bus when carrying data (Figure 6.4). When an instruction is being executed, the MPU will place address information on the bus at the same time it will emit a positive strobe pulse on the address strobe line; some logic circuitry will latch or retain the contents of the bus for later use. The MPU will then alter the mode of the bus to a data carrier and, if a write operation is being executed, will place data on these lines; at the same time it will emit a positive strobe on the data strobe line. The selected peripheral now knows that valid data are at its input terminals. Should the processor be executing a read operation, the MPU will emit a positive strobe, which informs the selected peripheral that the address/data bus and the MPU are ready to receive data; the peripheral now places the data on the bus. The MPU subsequently loads the data into its internal registers. The latch *must* be a negative edge triggered device

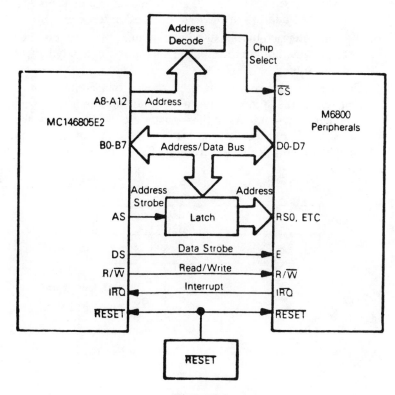

Figure 6.4

since the address is considered to be valid when the address strobe makes a negative transition.

The timing diagram of Figure 6.2 illustrates the importance of the address and data strobes.

6.4 REGISTERS

The 6805 has five registers; their titles and functions are as follows:

1. Accumulator A (Acc-A): the processor has one 8-bit register that can be used for data manipulation and arithmetic calculations. This register is usually the workhorse of the processor.
2. Index or X register: an 8-bit device that can be used for data manipulations, as a temporary storage register, or its contents can be used as the base address when indexed mode addressing is used.
3. Program counter: a 16-bit register, although only 13 bits are used. The top 3 bits, 15, 14, and 13, are permanent zeros. This register contains the address of the *next* instruction to be executed.
4. Stack pointer: a 6-bit register that contains the address of the *next* "empty" location on the stack. This register can only be used as a stack pointer.
5. Condition codes register: a 5-bit device that contains the flags that reflect the status of the MPU following the last instruction executed. The contents of this register are used as part of the decision-making process of the MPU. The format of the register is shown in Figure 6.5. A description of the function of each bit will follow.

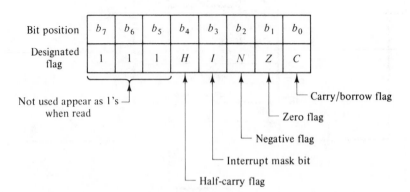

Figure 6.5

Bit 0 is the carry/borrow flag and will be set when the contents of the selected register or memory exceeds 255_{10} (i.e., a carry out of the bit 7 position) or when a borrow occurs (i.e., the result of the preceding subtraction contained in a selected register or memory is negative). This flag is also affected by bit test, rotate, shift, and the logic operation negate.

Bit 1 is the zero flag and will set when the selected register or memory is loaded with 00 or when the result of the *last* arithmetic or data manipulation or logic operation caused the result to be 00. This flag is affected by shift, and rotate.

Bit 2 is the N flag or negative flag; this bit will be set when bit 7 of the selected register or memory is 1, indicating a negative number in signed numbers arithmetic. The result may also be due to an arithmetic or logic operation, a shift, or a rotate.

Bit 3 is the interrupt mask bit, and when set it indicates that the interrupts via the $\overline{\text{IRQ}}$ line (external interrupt) and from the timer are inhibited; until this bit is cleared, the processor will *not* acknowledge an interrupt.

Bit 4 is the half-carry bit; it is set when a carry occurs between bits 3 and 4 of the selected register or memory. It finds its principal use with binary-coded-decimal addition.

6.5 ADDRESSING MODES

There are 10 different addressing modes in the 6805 instruction set; this gives the programmer the opportunity to write minimum-sized programs and yet achieve maximal results. Indexed addressing with and without offset permit addressing via 1-, 2-, and 3-byte instructions; this type of addressing permits the use of tables and the like anywhere in the memory space, as well as the very short form of page 00 addressing (1-byte instruction). Direct or page 00 addressing and extended addressing modes are also present; thus any address can be reached with a single 2- or 3-byte instruction. Bit test and branch instructions are included, as well as bit set and bit clear instructions; these instructional modes can save many programming steps and therefore optimize program size and performance.

1. Immediate addressing (IMM): the operand or data are the second byte of the instruction; it is used to access constants required during program execution.

2. Inherent addressing (INHERENT) can be used with accumulator or indexed addressing, and all the information required for instruction execution is contained in the op-code.

3. Direct addressing (DIR) or page 00 addressing: this addressing mode requires a 2-byte instruction; the second byte contains the address

of the data to be operated on or transferred to. The 128 bytes of on-board RAM that contain the I/O ports and timer can be directly accessed by this very efficient addressing mode, as well as a further 128 bytes of external memory.

4. Extended addressing (EXT): this is a 3-byte instruction and any address within the memory space can be reached by this mode.

5. Indexed addressing, no offset (IND NO OFFSET): this is a page 00 addressing mode and the 8-bit index register contains the address to be accessed (i.e., the index register is being used as a pointer). All the instructions using this addressing mode are 1-byte instructions. For example, the X register contains 36; the instruction load Acc-A will utilize the contents of this address; the op-code for LDA–(IND NO OFFSET) is F6 (see Figure 6.6). This type of addressing is frequently used for accessing a table of values or the I/O ports.

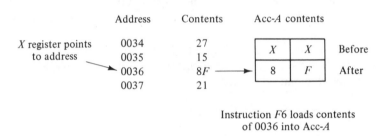

Instruction *F*6 loads contents
of 0036 into Acc-*A*

Figure 6.6

6. Indexed addressing 8-bit offset (IND 8 BIT OFFSET): this is a 2-byte instruction; the second byte contains a *positive* 8-bit number that is automatically added to the contents of the index register to form the address being accessed when the instruction is executed. The operand can be located anywhere in the 511 memory locations of the first 2 pages. There is 1 memory location that is *not* accessible by this mode, and that is 01FF (the top address on page 01). This is because the largest number that can be contained in the X register is FF and the largest 8-bit offset is also FF; therefore,

$$FF + FF = 01FE \tag{6.1}$$

is the largest address that can be accessed by this mode. An example of the use of this addressing mode will now be offered. Store the contents of Acc-A at 2F relative to the current contents of the X register; the X register contains E7. The address to be accessed will therefore be

$$E7 + 2F = 0116 \tag{6.2}$$

The op-code for STA–(IND 8 BIT OFFSET) is E7; the complete instruction will be E7 2F. Thus the contents of Acc-A will be stored on page 01, location 16. The original contents of the X register are *NOT* altered by this operand, as the addition of E7 + 2F takes place in a temporary register.

   7. Indexed addressing 16-bit offset (IND 16 BIT OFFSET): the offset is a 16-bit positive number that is automatically added to the contents of the 8-bit index register to form the address being accessed at the time that the instruction is executed. Any memory location in the whole memory map can be accessed by this mode of addressing (e.g., OR the contents of a specified memory location with the contents of Acc-A, use the 16-bit offset indexed mode of addressing). The index register contains E9 and the offset is 1E46; thus the selected address would be

$$1E46 + E9 = 1F2F \qquad (6.3)$$

The op-code for ORA–(IND 16 BIT OFFSET) is DA; the complete instruction would be DA 1E46. The contents of the X register are *not* altered by this operand. The result of the OR operation is contained in Acc-A.

   Note: Only forward displacements can be made with any form of indexed addressing.

   8. Relative addressing (REL): this mode of addressing is only used to transfer control of the program to a point other than the next sequential instruction. The transfer can be forward or backward of the relative position of the program counter. The 2's complement number system is always used for relative addressing; therefore, the span is limited to 126_{10} bytes backward and 129_{10} bytes forward of the *branch instruction op-code memory location* or 128_{10} backward and 127_{10} forward from the PC current position.

   9. Bit set/clear: this instruction is used to set or clear *individual* bits in memory locations on page 00 only. When this mode of addressing is used, direct addressing is automatically implied. The instruction has 2 bytes; the first byte contains the op-code and the bit number to be cleared or set. The second byte contains the address of the bit to be operated on. For example, the op-code to set a bit is $10 + (2 \times n)$, where *n* is the bit number to be set. Table 6.1 illustrates the op-codes for the 8 bits of a byte. A complete instruction to set bit 6 in memory location 0093 would be 1C93.

   Similarly, to clear a bit the op-code is $11 + (2 \times n)$. Table 6.2 illustrates the op-codes for the 8 bits of a byte. A complete instruction to clear bit 3 in memory location 0004 would be 1704.

   Note: Only 1 bit can be set or cleared per instruction; also, the bit must be located in the first 256 bytes of memory (i.e., page 00).

   10. Bit test and branch: this instruction is used to test the status of a specified bit at a selected address; when this addressing mode is used, rel-

Table 6.1

Bit No. to Be Set (n)	2 × n	Op-code 10 + (2 × n)
7	2 × 7 = E	10 + E = 1E
6	2 × 6 = C	10 + C = 1C ←
5	2 × 5 = A	10 + A = 1A
4	2 × 4 = 8	10 + 8 = 18
3	2 × 3 = 6	10 + 6 = 16
2	2 × 2 = 4	10 + 4 = 14
1	2 × 1 = 2	10 + 2 = 12
0	2 × 1 = 0	10 + 0 = 10

Table 6.2

Bit No. to Be Cleared (n)	2 × n	Op-code 11 + (2 × n)
7	2 × 7 = E	11 + E = 1F
6	2 × 6 = C	11 + C = 1D
5	2 × 5 = A	11 + A = 1B
4	2 × 4 = 8	11 + 8 = 19
3	2 × 3 = 6	11 + 6 = 17 ←
2	2 × 2 = 4	11 + 4 = 15
1	2 × 1 = 2	11 + 2 = 13
0	2 × 0 = 0	11 + 0 = 11

Table 6.3

Bit No. to Be Set (n)	2 × n	Op-code 0 + (2 × n)
7	2 × 7 = E	0 + E = 0E
6	2 × 6 = C	0 + C = 0C
5	2 × 5 = A	0 + A = 0A
4	2 × 4 = 8	0 + 8 = 08
3	2 × 3 = 6	0 + 6 = 06 ←
2	2 × 2 = 4	0 + 4 = 04
1	2 × 1 = 2	0 + 2 = 02
0	2 × 0 = 0	0 + 0 = 00

ative addressing is implied. An instruction has 3 bytes; the first byte contains the op-code and the bit number to be tested. The second byte contains the address on page 00 of the bit to be tested, while the third byte contains the displacement from the program counter's current position, op-code address + 3. Table 6.3 illustrates the op-code for the 8 bits of a byte for a branch if and only if (iff) the selected bit is set, BRSET, op-code 0 + (2 × n).

A complete instruction to test bit 3 of memory location 0092 and branch to 00A6 if set is given in Table 6.4. This instruction is located at address 0083.

Table 6.4

Address	Op-code/Operand	Remarks
0083	06	Op-code to branch if bit 3 is set
0084	92	Address of bit to be tested
0085	20	Displacement to be added to the program counter's current position, which is 0086

Similarly, to branch if and only iff (iff) the selected bit is clear, Table 6.5 presents the op-codes for the 8 bits of a byte; the op-code for BRCLR is $01 + (2 \times n)$. Thus a complete instruction to test bit 5 of memory location 00AF and branch to 0086 if it is clear is as shown in Table 6.6. The instruction op-code is located at address 0091.

Table 6.5

Bit No. to Be Tested (n)	$2 \times n$	Op-code $01 + (2 \times n)$
7	E	$01 + E = 0F$
6	C	$01 + C = 0D$
5	A	$01 + A = 0B \leftarrow$
4	8	$01 + 8 = 09$
3	6	$01 + 6 = 07$
2	4	$01 + 4 = 05$
1	2	$01 + 2 = 03$
0	0	$01 + 0 = 01$

Table 6.6

Address	Op-code/Operand	Remarks
0091	0B	Op-code to branch if bit 5 is clear
0092	AF	Address of bit to be tested
0093	F2	Displacement to be added to the program counter's current position, which is 0094

Note: Only 1 bit can be tested per instruction; also the bit must be located in the first 256 bytes of memory.

The programs that will be written using the 6805 instruction set will use the same symbols as in the 6800 programs, that is, # for immediate addressing and (N,X) for indexed addressing. Since there are three indexed addressing modes, the number of bytes in the instruction will indicate which indexed mode is being used, since

Index no-offset requires 1 byte
Index 8-bit offset requires 2 bytes
Index 16-bit offset requires 3 bytes

The examples that accompany this chapter will illustrate these and the other addressing modes.

6.6   THE 6805 INSTRUCTION SET

The instruction set of the MC 6805E2 microprocessor is contained in Appendix D and can be divided into 5 specific instruction groups:

1. Data transfer instructions
2. Arithmetic logic instructions
3. Control transfer instructions
4. Processor control instructions
5. Bit manipulation instructions

6.7   CLOCK

The clock for the 6805E2 can be either crystal controlled or an external clock. The MPU has an internal oscillator that can be interfaced with the output from a crystal via pins 38 and 39, as shown in Figure 6.7, or a good-quality external clock via pins 39 and 20 (ground) (Figure 6.8). Pin 38 is not connected.

The clock should be mounted as close as possible to the MPU chip to avoid ringing of the clock lines, noise pickup, and the like. This MPU-controlled microprocessor system does not require that clock lines run to the peripherals, as is the case with the 6800 system, since data-transfer timing is achieved via the data strobe. However, good-quality clock signals must still be supplied to the MPU for trouble-free operation.

6.8   TIMER

A timer is built into the MPU and consists of an 8-bit programmable counter as well as a 7-segment software-controlled prescaler or divide by

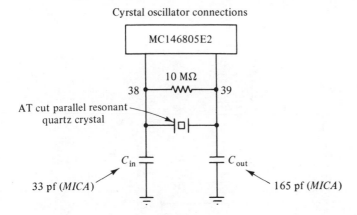

Cyrstal oscillator connections

Figure 6.7

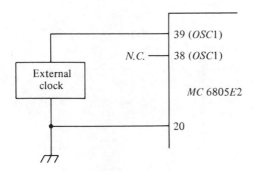

Figure 6.8

system. The prescaler (Figure 6.9) is the recipient of the actuating clock pulses, and when the number of input pulses equals the divide by number, the prescaler will output a pulse, which will decrement the count in the counter by 1. The prescaler and counter are actuated on the negative edge of the internal or external clock.

Thus the actuating clock, whether it be external via pin 37 or internal from the local oscillator or both, is capable of being prescaled; this feature extends the operating range of the timer. The internal clock is normally in phase with the address strobe; however, the address strobe line will be held low during a wait instruction, but the timer remains active. Thus the internal clock, in this case, is still operating at the same frequency and in the same phase as it would be if the MPU were executing program instructions. The timer has three different operating modes, which are selectable under program control via bits 5 and 4 of its control register. Access to the control register is through address 0009.

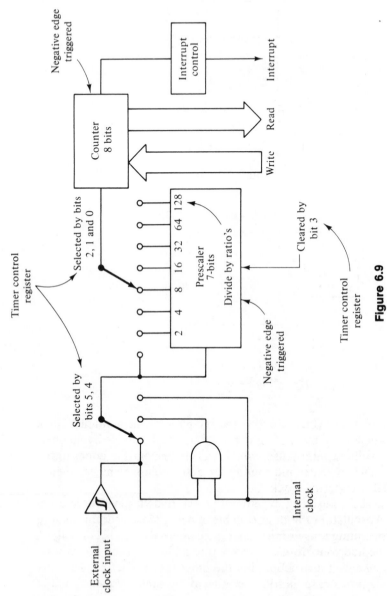

Figure 6.9

The functions of each bit in the control register are shown in Tables 6.7 and 6.8.

7	6	5	4	3	2	1	0	**Bit no.** (6.4)

Bits 2, 1, and 0 control the output of the prescaler, as shown.

Table 6.7

Bit No.

2	1	0	Divide the Actuating Clock by:
0	0	0	1
0	0	1	2
0	1	0	4
0	1	1	8
1	0	0	16
1	0	1	32
1	1	0	64
1	1	1	128

Table 6.8

Bit No.	Binary State	Function	Remarks
3	1	Prescaler reset bit	Contents of the prescaler are automatically set to zero, and will remain at zero as long as this bit remains set
3	0		Prescaler accumulates actuating clock pulse counts
4	1	External clock enable bit	External clock line, pin 37 is enabled; the external clock is available if selected
4	0		External clock line disabled
5	1	External or internal clock select bit	External clock selected as the input to the prescaler
5	0		Internal clock selected as the input to the prescaler
6	1	Timer interrupt mask bit	Timer interrupt of the MPU is inhibited; this bit is automatically set on external reset, power-on reset, stop instruction 8E, or under program control

Table 6.8 (*cont.*)

Bit No.	Binary State	Function	Remarks
6	0		Timer can interrupt MPU on a time out (the count in the counter reaches 00); this bit can also be cleared under program control
7	1	Timer interrupt request bit	Indicates that the timer has experienced a time out and is requesting an MPU interrupt, subject to bit 6 being cleared; this bit can also be set under program control
7	0		No interrupt requested by the timer; this bit is automatically cleared by external reset, power-on reset, stop instruction 8E, or under program control

The number to be decremented is loaded into the counter via address 0008; for example, load the counter with 3F (see Figure 6.10).

Instruction

Op-code	Data/Address	Mnemonic	Block diagram

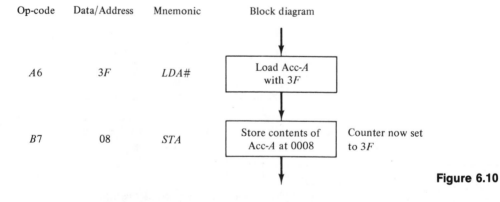

| *A*6 | *3F* | *LDA#* | Load Acc-*A* with 3*F* |
| *B*7 | 08 | *STA* | Store contents of Acc-*A* at 0008 — Counter now set to 3*F* |

Figure 6.10

Example 1. Set up the timer control register to the following specifications:

1. Use the internal clock; disable external clock.
2. Set prescaler to ÷ 32.
3. Enable interrupts; clear interrupt mask bit.
4. Clear interrupt request bit.

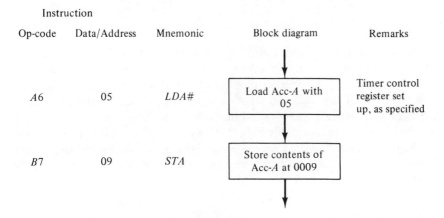

7	6	5	4	3	2	1	0	Bit no.
0	0	0	0	0	1	0	1	Binary state

0	5	Hexadecimal number

Instruction

Op-code	Data/Address	Mnemonic	Block diagram	Remarks
A6	05	LDA#	Load Acc-A with 05	Timer control register set up, as specified
B7	09	STA	Store contents of Acc-A at 0009	

Figure 6.11

The control register must contain the format shown in Figure 6.11. The timer begins to operate immediately; an actuating clock is present at the prescaler. The three operating modes of the timer are summarized in Table 6.9.

Table 6.9

Bit No.

5	4	*Remarks*
0	0	Timer operates from the internal clock; external clock is disabled; timer will continue to operate while the processor is in the wait state
0	1	Internal clock and the external clock are ANDed together to form the input to the prescaler; this mode is used for measurement of external pulse widths
1	1	Internal clock is disabled; the external clock is the input to the prescaler; this mode can be used to count external events or frequencies
1	0	Timer is disabled

The timer will continue to decrement its contents after they reach 00; that is, the contents of the timer counter *after* decrementing the original count to 00 and decrementing 1 more count will be FF. It will then count down from this point (i.e., 256_{10} counts to the next 00). Therefore, the original count, if it is to be repeated in the next cycle, must be reloaded at the start of each new cycle. This timer is not similar to the 6840 programmable timer module in this respect.

6.9 I/O PORTS

Each I/O port has eight input/output lines, a data direction register (DDR), and a peripheral data register (PDR). The I/O lines can be individually programmed to be either input lines or output lines, depending upon the contents of the data direction register. If a bit in the data direction register is cleared, the associated line is an input line. For example, bit 2 of the data direction register is 0; therefore, line 2 is an input line. Similarly, if the data direction bits are high, the associated lines are output lines. Access to the data direction register for port A is via address 0004 and for port B it is 0005.

Example 2. Initialize port A with lines 7, 5, 3, and 1 as output lines and the remainder as input lines. The contents of the data direction register must be as shown in Figure 6.12. Data to and from the peripherals are by way of the peripheral data registers, which are located at address 0000 for port A and 0001 for port B.

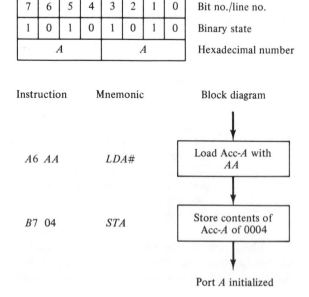

7	6	5	4	3	2	1	0	Bit no./line no.
1	0	1	0	1	0	1	0	Binary state
		A				*A*		Hexadecimal number

Instruction	Mnemonic	Block diagram
A6 AA	LDA#	Load Acc-A with AA
B7 04	STA	Store contents of Acc-A of 0004

Port *A* initialized

Figure 6.12

Instruction	Mnemonic	Block diagram	Remarks

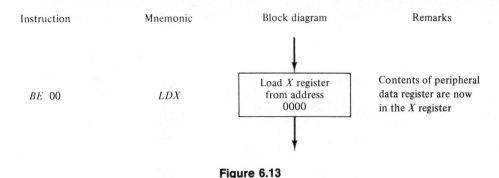

BE 00   *LDX*

Load *X* register from address 0000

Contents of peripheral data register are now in the *X* register

Figure 6.13

Example 3. Load the contents of peripheral data register port A into the index register (see Figure 6.13). When the processor reads the contents of a peripheral data register, it will read the last value written in if the lines are outputs, and the current status of the lines if they are inputs, since only the outputs are latched. When an external reset or power-on reset occurs, all lines of both ports are configured as input lines.

6.10 STACK AND STACK POINTER

The stack is located between addresses 0040 and 007F, 64 bytes, of the on-board RAM, and it operates in the same manner as the 6800 stack; that is, the stack pointer is automatically decremented when a byte is stored or saved on the stack. Thus the stack pointer always points to the next "empty" stack location. Similarly, when a byte is pulled, or copied, from the stack, the stack pointer is incremented first and then the byte is pulled. The stack pointer is automatically loaded with the top address (007F) during an external reset or a power-on reset or when the program instruction, reset stack pointer (RSP op-code 9C), is executed.

Care must be exercised when using the stack in the 6805 not only because it obeys the rule of last in, first out, but also because if the contents of the stack exceed 64 bytes, the stack pointer will automatically "wrap around" and point to the top address again. For example, a subroutine call is initiated, and the contents of the program counter must be saved in the stack for the return to the main program. Assume that the stack pointer is pointing at the last location on the stack (0040) and the program counter contains 0267; when the subroutine call is initiated, events occur as shown in Figure 6.14. The byte that was stored at 007F, that is, 51, has been replaced by 02, the high byte of the program counter; thus the previously stored information is lost. The stack pointer now contains address 007E; the stack is now out of position.

Address in the Stack Pointer	Contents of That Address		Remarks
	Before the subroutine call	After the subroutine call	
0040	73	67	Low byte of program counter
0041	27	27	
0042	01	01	
007*E*	21	21	
007*F*	51	02	High byte of program counter

Stack pointer "wraps around"

Figure 6.14

When an interrupt occurs, the program registers of the 6805E2 are automatically stored on the stack in the order shown in Figure 6.15 (assume that the stack pointer contains 007F).

6.11 INTERRUPTS

The processor has four interrupt vectors, as well as a restart vector stored in the top 10_{10} memory locations; the memory at these locations must be permanent (i.e., ROM, EPROM, etc.). The locations are assigned as shown in Figure 6.16. There is a priority among these vectors, as follows:

1. Top priority: external reset as it can restart or reinitialize the MPU at *any* time.

2. Stop: an instruction that places the processor in a low-power consumption mode; the internal oscillator is turned off. Thus *all* internal processing and the timer are stopped. The data strobe and address strobe lines are forced low, the read/write line is set high and the address/data bus goes to the data input state. The 5-bit section of the address bus contains the address of what would have been the next instruction in the executing program. The routine followed by the processor at the inception of a stop instruction is shown in Figure 6.17. The processor remains in the stopped condition until either reset or an external interrupt occurs.

Stack Pointer
Contents   Block diagram   Remarks

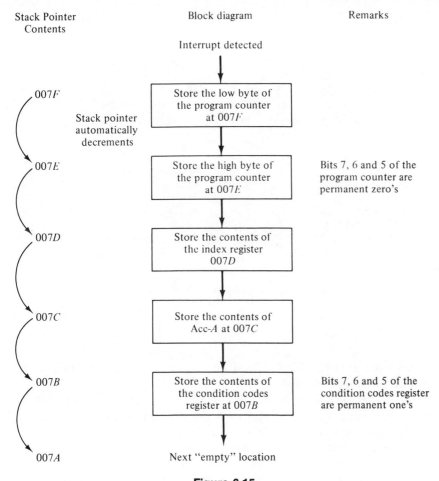

Figure 6.15

3. Wait instruction: places the processor in a low-power consumption mode, slightly higher wattage than the top mode, and places the address and data strobe lines, the read/write line, and so on, in the same condition as the stop instruction. The internal clock is disabled from all the internal circuitry of the MPU except the timer, which continues to count down. The routine followed by the processor when a wait instruction is encountered is shown in Figure 6.18. The processor remains in the wait state until either a reset or external interrupt or timer interrupt occurs.

4. Software interrupt (SWI): instruction op-code 83; when this instruction is encountered in a program, the contents of the program registers are saved on the stack and the software interrupt vector is picked up

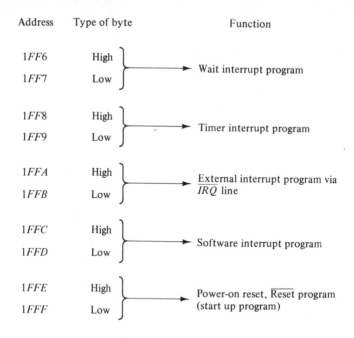

Figure 6.16

from addresses IFFC/IFFD. This instruction is executed regardless of the state of the interrupt mask bit.

5. External interrupt: $\overline{IRQ}$ line makes a negative transition and the interrupt routine is executed (Figure 6.19). The processor will only recognize the interrupt via the $\overline{IRQ}$ line when the interrupt mask bit is clear. When several peripherals are connected to the $\overline{IRQ}$ line, they may be wired in hard-wired OR configuration, as shown in Figure 6.20.

6. Timer interrupt: has the lowest priority, since not only can the interrupt mask bit inhibit timer interrupt but so can the timer's own interrupt mask bit (bit 6 of the timer control register). When the timer experiences a time out, counter decrements from 01 to 00, the internal logic of the timer goes through the routine of Figure 6.21. The processor will only recognize an interrupt from the timer when bit 7 of the timer control register is set, and the interrupt mask bit (bit 3 of the condition codes register) and the timer interrupt mask bit (bit 6 of the timer control register) are cleared.

When the processor is executing the timer interrupt program, bit 7 of its control register must be cleared; otherwise, the processor will continue to execute the interrupt program. Also, if the timer interrupt will not be required again, the timer interrupt mask bit must be set, since the timer is continually counting down, and failure to inhibit the interrupt will cause other timer interrupts at some later point in the main program.

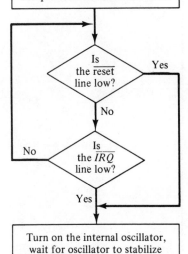

Figure 6.17

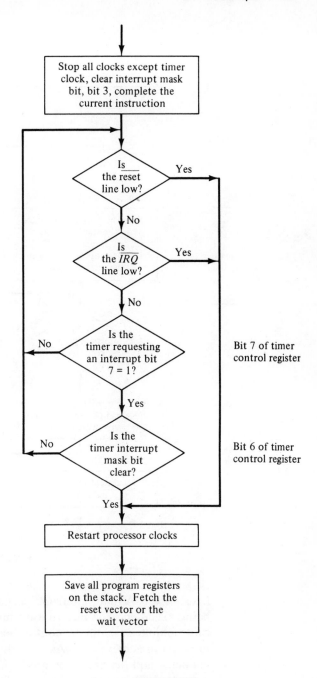

Figure 6.18

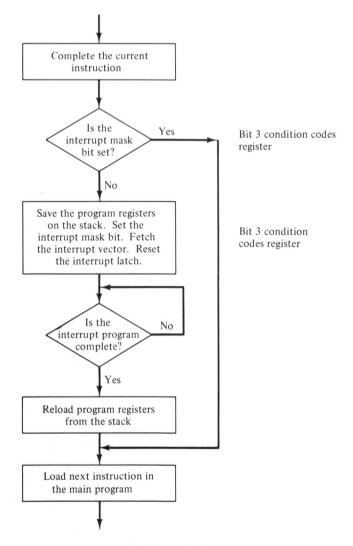

Figure 6.19

Example 4. A timer interrupt has occurred, and the interrupt program is being executed. The interrupt bit must be reset, and further timer interrupts inhibited; set up the control register of the timer during the interrupt program to achieve this. Assume the timer was operating from the internal clock and the prescaler was ÷ 32. See Figures 6.22 and 6.23.

An RTI instruction (op-code 80) *must* be *last* instruction of *any* interrupt program; this instruction will return the processor to the main program.

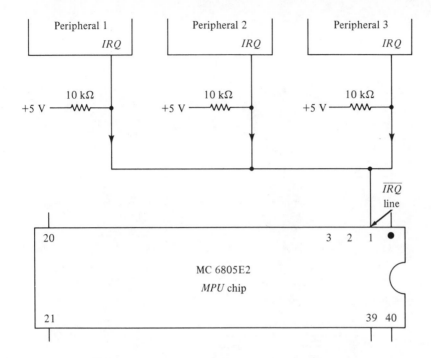

Figure 6.20

6.12   MINIMUM MICROPROCESSOR SYSTEM WITH THE MC 6805E2 CHIP

The memory map of Figure 6.24 and wiring diagram of Figure 6.25 illustrate a suggested minimum system with the MC 6805 microprocessor chip. This system contains all requirements of a microprocessor system (RAM, PIA, ACIA, etc.). The wiring diagram is suggested as a guide; other more elaborate systems can be designed when an understanding of the basic device and its operation are understood. The circuit was wired on SK-10 boards with no. 22 gauge wire in exactly the same manner as that used to wire the 6800. The program "blasted" into the EPROM was simple and initially typed $ (dollar sign) onto a CRT screen via the ACIA, and then went into a CRT echo program. The program is given in Table 6.10.

The program loops between 3D and 2A, as new characters are typed in. Initially, when the processor starts, it goes to the highest address to find its starting vector, 1C 10; thus this vector must be included in the original program.

Address	Instruction	
1FFF	10	Start up vector, address
IFFE	1C	of start of program

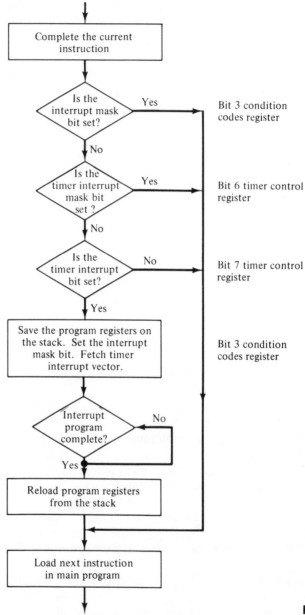

Figure 6.21

The remainder of the memory locations in the EPROM are left in their "unblasted" state, FF.

A program to flash an LED is offered in the tutorials. The bit-rate generator is a crystal-controlled oscillator and is the clock source for the

Timer control register

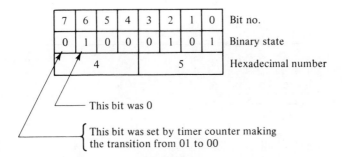

Figure 6.22

Instruction	Mnemonic	Block diagram	

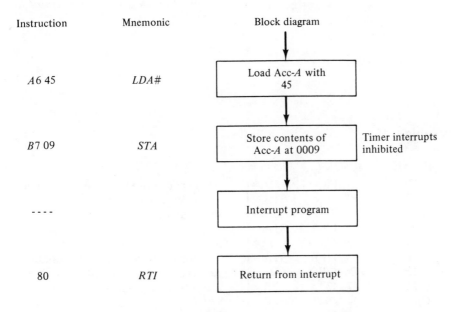

A6 45	LDA#	Load Acc-A with 45	
B7 09	STA	Store contents of Acc-A at 0009	Timer interrupts inhibited
- - - -		Interrupt program	
80	RTI	Return from interrupt	

Figure 6.23

6805E3 microprocessor; it also provides a multiple frequency of 4800 Hz for the ACIA, which is operating at the 300-baud rate, and the external clock for the timer when required (tutorial example 3 and Problem 6.15). The device is programmable and in the minimum system used here it is set to ×64 with the processor connected to F1 (614.4 kHz). Other output frequencies may also be selected.

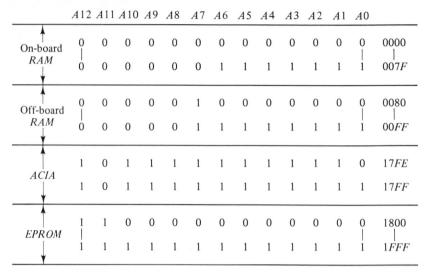

Figure 6.24

Table 6-10

Address	Instruction	Mnemonic	Remarks
1C 10	A6 F0	LDA# ⎱	Set up port A of PIA, upper 4 lines out-
1C 12	B7 04	STA ⎰	put, lower 4 lines input
1C 14	A6 03	LDA# ⎱	Reset ACIA
1C 16	C7 17FE	STA ⎰	
1C 19	A6 49	LDA# ⎱	Set up ACIA, ÷ 16, 7 bits even parity,
1C 1B	C7 17FE	STA ⎰	1 stop bit, transmit interrupt disabled
1C 1E	→C6 17FE	LDA ⎫	Read status register of ACIA, wait for
1C 21	47 47	ASR ⎬	transmit data register to empty
1C 23	└24 F9	BCC ⎭	
1C 25	A6 24	LDA# ⎱	Load Acc-A with $ sign
1C 27	C7 17FF	STA ⎰	Sent to CRT
1C 2A	→C6 17FE	LDA ⎫	Read status register of ACIA, wait for
1C 2D	46	ROR ⎬	arrival of new data
1C 2E	←24 FA	BCC ⎭	
1C 30	C6 17FF	LDA	Load new data into Acc-A
1C 33	→CE 17FE	LDX ⎫	Wait for transmit data register to
1C 36	56 56	ROR ⎬	empty, use the X register
1C 38	└24 F9	BCC ⎭	
1C 3A	C7 17FF	STA	Transmit data back to CRT
1C 3D	└20 EB	BRA	Go back for next character

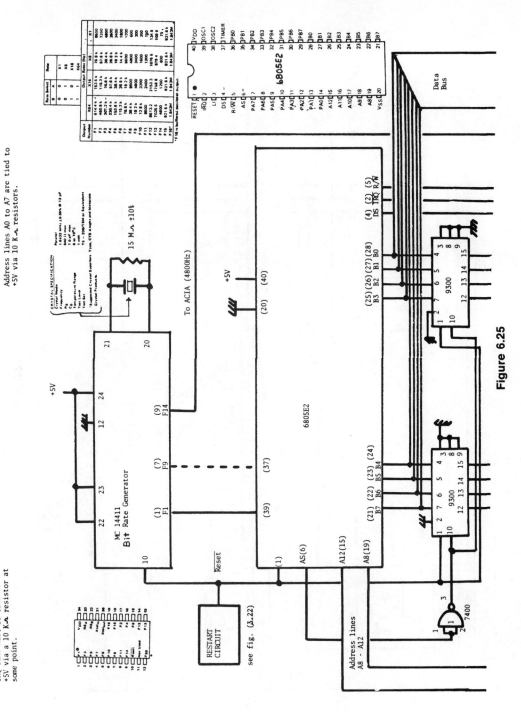

Figure 6.25

233

DATA BUS

ADDRESS BUS

Figure 6.25 (cont.)

6.13 TUTORIAL EXAMPLES

Example 1.   Write the initial start-up program to initialize port B all lines output; then flash an LED using a microprocessor delay loop.

A delay loop via the 6805 microprocessor is slightly more difficult than via the 6800 since the X register in the 6805 is only 8 bits wide. The flow chart of Figure 6.26 illustrates a suggested method. Assume the required delay is 3FF3. Acc-A will initially be loaded with 3F and the X register will be loaded with F3. The complete program to start up the microprocessor and flash the LED is given in Table 6.11; assume the LED is connected via the appropriate interface to line B_0 (pin 36) of the 6805 chip.

The start-up vector must be included at IFFE/IFFF:

> IFFF   10   Start-up vector
> IFFE   1C

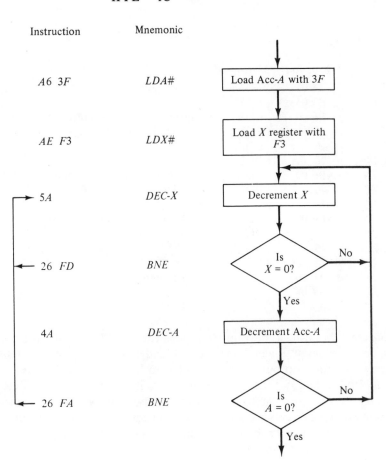

Figure 6.26

Table 6.11

Address	Instruction	Mnemonic	Remarks
1C 10	A6 FF	LDA#	Set up port B all lines output
1C 12	B7 05	STA	
1C 14	→A6 01	LDA#	Set output line B_0 to 1, LED now ON
1C 16	B7 01	STA	
1C 18	Delay shown in Figure 6.26		
1C 22	A6 00	LDA#	Reset output line B_0 to 0, LED now OFF
1C 24	B7 01	STA	
1C 26	Delay		
1C 30	20 E2	BRA	Go back to 1C 14

Remainder of EPROM contains FF.

Table 6.12

Address	Instruction	Mnemonic	Remarks
1C 10	A6 0F	LDA#	Set up port A
1C 12	B7 04	STA	
1C 14	B6 00	LDA	Interrogate the start button; branch back if
1C 16	46	ROR	start button is closed
1C 17	46	ROR	
1C 18	25 FA	BCS	
1C 1A	→B6 00	LDA	Interrogate the stop button; branch back
1C 1C	46	ROR	if stop button is open
1C 1D	24 FB	BCC	
IC 1F	→B6 00	LDA	Wait for start button to close
1C 21	46	ROR	
1C 22	46	ROR	
1C 23	24 FA	BCC	
1C 25	A6 20	LDA#	Start motor
1C 27	B7 04	STA	

Example 2. Write a start-up program (Table 6.12) to initialize port A upper 4 lines input, lower 4 lines output, and control the motor as detailed next. A motor stop button is connected to line A_0 and a start button is connected to line A_1; the motor starter, via its interface, is connected to line A_4. Write a program to interrogate the start/stop buttons and start the motor when the start button is activated. The program to stop the motor is left to the reader as Problem 6.2. The start/stop buttons are wired in the manner shown in Figure 4.1.

Note: The reader should continue the program in Problem 6.2.

IFFF 10 ⎫
1FFE 1C ⎭   Start-up vector

Example 3.   Write a start-up program (Table 6.13) to initialize port B (line B_0 is an output line and the remainder are input lines) and initialize the timer section to achieve a delay similar to that of tutorial example 1. Use the interrupt from the timer to signal the program to change the state of the output to the LED.

Table 6.13

Address	Instruction	Mnemonic	Remarks
1F 10	9B	SEI	Set interrupt mask bit
1F 11	A6 01	LDA# ⎫	Set up port B, line B_0 output, remainder
1F 13	B7 05	STA ⎭	input
1F 15	A6 7F	LDA# ⎫	Set timer, external clock ÷ 128, interrupt
1F 17	B7 09	STA ⎭	masked, interrupt request bit cleared,
			prescaler reset and held at zero
1F 19	A6 72	LDA# ⎫	Load delay count into timer
1F 1B	B7 08	STA ⎭	
1F 1D	A6 01	LDA# ⎫	Set output line B_0 to 1 LED now ON
1F 1F	B7 01	STA ⎭	
1F 21	AE 37	LDX# ⎫	Start timer
1F 23	BF 09	STX ⎭	
1F 25	9A	CLI	Clear interrupt mask bit
1F 26	8F	WAIT	Wait for timer interrupt
1F 27	20 F8	BR	Branch back to 1F 21
		Interrupt Program	
1F 40	A8 01	EOR	Exclusive OR contents of Acc-A
1F 42	B7 01	STA	Change output state of line B_0
IF 44	AE 7F	LDX# ⎫	Stop timer and reset interrupt flag and
1F 46	BF 09	STX ⎭	mask interrupt bit
1F 48	AE 72	LDX# ⎫	Reload counter in timer
1F 4A	BF 08	STX ⎭	
1F 4C	80	RTI	Return from interrupt

    This example illustrates the use of the wait instruction; the processor waits for the timer interrupt to set the processor into action again. When the interrupt occurs, the processor automatically goes to address 1FF9/8 and picks the starting address of the timer interrupt program. Thus the top end of the EPROM must be programmed as follows:

1FFF 10 ⎫
1FFE 1F ⎭   Address of start-up program

1FF9 40 ⎫
1FF8 1F ⎭   Address vector of the interrupt program

The contents of all memory locations not programmed in the EPROM remain at FF. The clear interrupt mask bit instruction CLI (address 1F25) could have been placed before the instruction start timer (address 1F21/23) if desired. The SEI instruction at the beginning of the program is used only as insurance against unwanted interrupts. The timer counter must be reloaded with the count required for the next cycle at the start of each new cycle (address 1F48/4A).

Example 4.   The timer in a 6805E2 MPU operated system is being used to count the number of objects passing a detector. When the count reaches 07, start motor 01 and reset the timer counter; operate the motor until its motion is arrested by a limit switch X. The program repeats continually (see Figure 6.27).

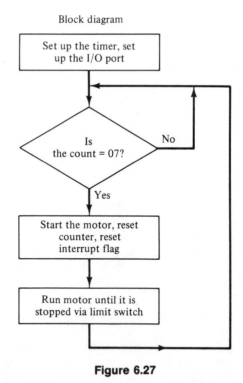

Block diagram

Figure 6.27

The timer interrupt vector is to be set for address 1F50. Assume the following:

1. Motor 01 operates for *less* time than it takes for 07 objects to pass the detector.

2. The output from the detector is a positive pulse.

3. Port A is used as the I/O device: line 0, input line from detector; line 1, output line to motor; remainder are programmed as output lines.

See Table 6.14.

Table 6.14

Address	Instruction	Mnemonic	Remarks
1F 10	9B	SEI	Set interrupt mask bit
1F 11	A6 07	LDA#⎱	Load counter with desired count
1F 13	B7 08	STA ⎰	
1F 15	A6 78	LDA#⎱	See figure for format of timer control
1F 17	B7 09	STA ⎰	register

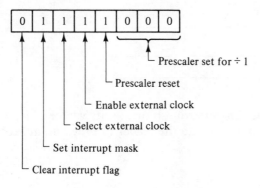

1F 19	A6 01	LDA#⎱	Line 0 set high, motor *not* running
1F 1E	B7 00	STA ⎰	
1F 1D	A6 FE	LDA#⎱	Set up port A
1F 1F	B7 04	STA ⎰	
1F 21	9A	CLI	Clear interrupt mask bit
1F 22	A6 30	LDA#⎱	Start timer, clear interrupt mask bit
1F 24	B7 09	STA ⎰	
1F 26	⌐→9D	NOP	Wait for counter to reach 00
1F 27	⌐20 FD	BRA	

This example illustrates the use of the NOP instruction to wait for the interrupt (i.e., the count to reach 07). Thus it is imperative that the timer be reset in the interrupt program as the RTI instruction will return the processor to the wait loop in which there can be no restart timer command. Furthermore, the timer must be reloaded with its count 07 at the start of

Interrupt program block diagram

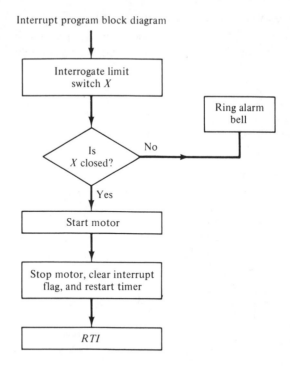

each new cycle. The interrupt program, its vector address as well as the start-up address, are left to the reader to program, in Problem 6.14.

Problems

6-1. Rewrite tutorial example 14 of Chapter 3 in 6805 op-code. There are no control registers on the I/O ports of the 6805; therefore interrupts cannot be entertained via the I/O ports.

6-2. Complete tutorial example 2 of Chapter 6; that is, write the program to stop the motor when the stop button is activated.

6-3. Write a program in 6805 code that will solve the problems outlined in tutorial examples 3, 5, and 6 of Chapter 4.

6-4. Write a program in 6805 code to solve Problems 4.5, 4.7 and 4.12.

6-5. Write a program using a 6805 microprocessor to generate a 20-ms delay.

6-6. Flash an LED via port A of a 6805 using a subroutine to generate the delay.

6-7. Write a program to start four motors consecutively via port B at 20-s intervals; the signal to start comes from a contact X via port A.

6-8. Show the connections to add a PIA (chip 6821) to the minimum system (Figure 6.25) at address locations 17FA to 17FD inclusive. Show all address decoding necessary to operate this amended minimum system.

6-9. Write a program to convert hexadecimal numbers to decimal numbers; the largest to be converted is $63 \equiv 99_{10}$.

6-10. Write a program to display the decimal digits converted in Problem 6.9 on a two-section, seven-segment LED display.

6-11. The control register of the internal timer in 6805E2 chip contains 3F; decode these data, listing what each bit or group of bits means to the operation of the timer.

6-12. Design the interface for tutorial example 3; a high will cause the LED to be lit.

6-13. Determine the flashing rate of the LED in tutorial example 1. The 6805 microprocessor is operating at an input clock frequency of 614.4 kHz.

6-14. Complete the program in tutorial example 4.

6-15. Determine the flashing rate of the LED used in tutorial example 3. The external clock of the timer is connected to pin 7 (F9) of the bit rate generator used in the minimum system (Figure 6.25).

Chapter 7

Programming Examples

The programs contained in this text are written in the MPU's language, that is binary numbers or hexadecimal op-codes; and since there is no easy way to remember each op-code, mnemonics are included to assist the reader in following the programs offered.

Some programming of a microprocessor has been previously discussed as well as the method of structuring a program; therefore, this chapter will contain worked examples of an industrial nature to further illustrate the material covered in the previous chapters.

All the programs pertain to process-control applications, as well as applying the CRT or teletype to the processor. The examples will also include CRT echo programs, program editing via the CRT, instructions not previously used, and techniques not previously illustrated.

Example 1 illustrates the conversion of a 6-bit hexadecimal number to a decimal number by using the DAA instruction, as well as the processor flashing an LED. The example further illustrates a method of holding the output lines of port B of a PIA in a high condition during and after initialization.

Example 2 illustrates a method of displaying a 6-digit decimal or hexadecimal number. Multiplexing is employed to minimize the data-handling system.

Example 3 demonstrates a method of using the X register as a pointer in a table of contents. The table used in Example 2 (7.1) has been previously set up at a known location with the codes to enable the decimal numbers to be displayed. This example augments Example 2, as the number to be displayed now is not known, and the display code for that number must be found.

Example 4 illustrates the method of echoing the transmission back to a CRT; the transmission is via a 20-mA loop, which is illustrated in Figure 7.9.

Example 5 illustrates the transmission of a message to the CRT, ALARM-MOTOR #1. This type of message is usually coded and stored in memory (Table 7.10) as a subroutine. The message is only printed *once*.

Example 6 is a motor sequence control problem, with four motors. Two motors, M_1 and M_2, operate in a cyclic manner; M_3 and M_4 operate in a similar manner upon activation of a contact X. Start process and stop process are under operator control. Further safety aspects are illustrated.

Example 7 illustrates the use of the memory cell as an alternative to resetting the stack, as in Example 6.

In Example 8 a table moves back and forth between two limit switches. This problem has been solved via logic gates in tutorial example 3 of Chapter 2. The problem is repeated using the microprocessor to illustrate how a process that is currently being controlled by logic gates can be changed to microprocessor control. The example will also illustrate the interrogation of the arresting limit switch before reversal is allowed.

Example 9 is an illustration of scanning selected switches for positional status and the continual updating of the visual display representing the status, that is, an alarm annuciator system.

Example 10 illustrates the use of the X register to chain a process through its parts. The order in which the parts occur may be required to vary from product to product. This example illustrates the versatility of indexed mode addressing.

In Example 11 lines C_{A1} and C_{A2} are being used as interrupt lines. When an interrupt occurs, which line was responsible? This example will illustrate interrupt polling.

7.1 WORKED EXAMPLES

Example 1. Write a program to read the hexadecimal data contained on the six input lines when button X is depressed; convert the number to decimal and save the result; the previous result must also be saved. The processor will be doing other things in between conversions; in this case it will flash an LED (see Figure 7.1).

Initialize port A, all lines inputs, positive transition interrupt on line C_{A1}. Initialize port B, all lines outputs; do not allow the lines to go low during initialization. Program starts at 0080 when the system reset line

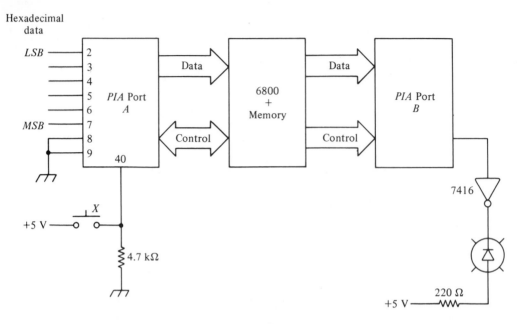

Figure 7.1 System Block Diagram

goes high. The previous data must be stored at address 0001; the new data must be stored at 0002. Thus, when a new piece of data is converted, the previous data are shifted from 0002 to address 0001; the new data are then stored at 0002. Use a subroutine for the LED flashing delay.

Refer to Figure 7.2 and Tables 7.1 to 7.3.

The processor is flashing an LED as something to do while waiting for the interrupt. It could be doing any useful task.

This example illustrates the use of the decimal adjust instruction; it will *only* adjust the contents of Acc-A. Also, correct operation of this instruction is dependent upon the state of the half-carry and carry bits in the condition codes register. These bits were cleared with an AND instruction (84 1E) after the contents of the condition codes register had been transferred to Acc-A via the TPA instruction (07). When the bits were cleared, the contents of Acc-A were transferred back to the condition codes register via a TAP instruction (06). The DAA instruction will *not* adjust hexadecimal numbers correctly unless the half-carry and carry bits are correct for the number present in the Acc-A. For example, assume Acc-A contains the data 09. From row 1 of Table 7.4, which is a copy of the DAA instruction (19), the following can be seen:

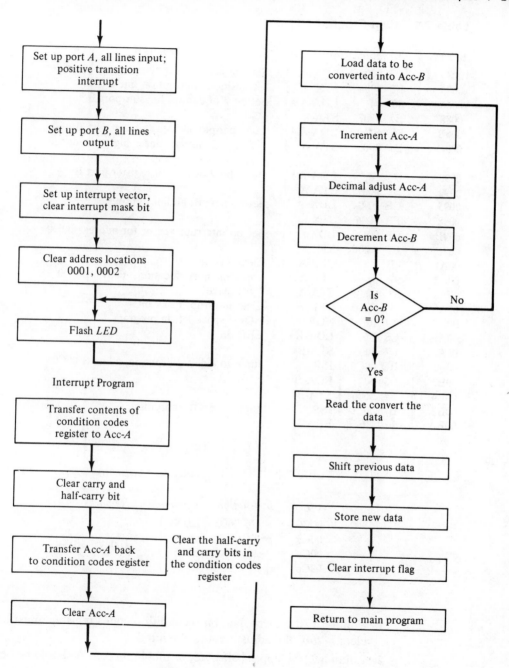

Figure 7.2 Program Block Diagram

Table 7.1 Main Program

Address	Instruction	Mnemonic	Remarks
0080	CE 0007	LDX#	Set up port A, all lines input, positive
0083	FF 8004	STX	transition interrupt on line C_{A1}
0086	86 04	LDA-A#	Set bit 2 of control register port B
0088	B7 8007	STA-A	
008B	86 FF	LDA-A#	Load peripheral data register port B with
008D	B7 8006	STA-A	FF, all output lines high-impedance state
0090	86 00	LDA-A#	Reset bit 2 of control register port B
0092	B7 8007	STA-A	
0095	CE FF04	LDX #	Set up port B, all lines output
0098	FF 8006	STX	
009B	CE 00D0	LDX#	Set up interrupt vector for address 00D0
009E	FF A000	STX	
00A1	4F	CLR-A	Clear Acc-A
00A2	97 01	STA-A	Clear memory locations 0001 and 0002
00A4	97 02	STA-A	inclusive
00A6	0E	CLI	Clear interrupt mask bit
00A7	010101	NOP	NOPs; space left in program
00AA	C6 FF	LDA-B#	LED on
00AC	F7 8006	STA-B	
00AF	BD 00C0	JSR	Jump to delay subroutine, address 00C0
00B2	C6 00	LDA-B#	LED off
00B4	F7 8006	STA-B	
00B7	BD 00C0	JSR	Jump to delay subroutine
00BA	20 EE	BRA	

Table 7.2 Delay Subroutine Program

Address	Instruction	Mnemonic	Remarks
00C0	CE 0600	LDX#	
00C3	09	DEX	Delay
00C4	26 FD	BNE	
00C6	39	RTS	

1. Column 1, the carry flag, bit 0 condition codes register, must be clear *before* the adjustment is executed.
2. Column 2, the upper half-byte, or nibble, in Acc-A should be between 0 and 9, it is in this case.
3. Column 3, the half-carry flag, bit 5 condition codes register, must be clear.

Table 7.3 Interrupt Program

Address	Instruction	Mnemonic	Remarks
00D0	07	TPA	Transfer contents of condition codes register to Acc-A
00D1	84 1E	AND-A#	Clear half-carry and carry bit
00D3	06	TAP	Transfer contents of Acc-A back to condition codes register
00D4	4F	CLR-A	Clear Acc-A
00D5	F6 8004	LDA-B	Load Acc-B with the hexadecimal data
00D8	→4C	INC-A	Increment contents of Acc-A
00D9	19	DAA	Decimal adjust contents of Acc-A
00DA	5A	DEC-B	Decrement contents of Acc-B
00DB	└26 FB	BNE	
00DD	D6 02	LDA-B	Load Acc-B with previous data, address 0002
00DF	97 02	STA-A	Store new data, in Acc-A, at address 0002
00E1	D7 01	STA-B	Store previous data, in Acc-B, at address 0001
00E3	F6 8004	LDA-B	Clear interrupt flag in port A
00E6	3B	RTI	Return to main program

Table 7.4
DAA

Decimal Adjust ACCA

Operation: Adds hexadecimal numbers 00, 06, 60, or 66 to ACCA, and may also set the carry bit, as indicated in the following table:

State of C-bit before DAA (Col. 1)	Upper Half-byte (bits 4-7) (Col. 2)	Initial Half-carry H-bit (Col.3)	Lower to ACCA (bits 0-3) (Col. 4)	Number Added after by DAA (Col. 5)	State of C-bit DAA (Col. 6)
0	0-9	0	0-9	00	0
0	0-8	0	A-F	06	0
0	0-9	1	0-3	06	0
0	A-F	0	0-9	60	1
0	9-F	0	A-F	66	1
0	A-F	1	0-3	66	1
1	0-2	0	0-9	60	1
1	0-2	0	A-F	66	1
1	0-3	1	0-3	66	1

Note: Columns (1) through (4) of the above table represent all possible cases which can result from any of the operations ABA, ADD, or ADC, with initial carry either set or clear, applied to two binary-coded-decimal operands. The table shows hexadecimal values.

Description: If the contents of ACCA and the state of the carry-borrow bit C and the half-carry bit H are all the result of applying any of the operations ABA, ADD, or ADC to binary-coded-decimal operands, with or without an initial carry, the DAA operation will function as follows.

Subject to the above condition, the DAA operation will adjust the contents of ACCA and the C bit to represent the correct binary-coded-decimal sum and the correct state of the carry.

4. Column 4, the lower half-byte of Acc-A should be between 0 and 9; it is in this case.

5. Column 5, the number to be added to the contents of Acc-A to form the decimal number is 00.

6. Column 6 indicates the state of the carry bit after the adjustment.

Therefore, Acc-A contains 09 and 00 is added to its contents; thus the decimal adjusted number is 09, which is correct. Now assume that Acc-A contains 16 in hexadecimal notation, the upper half-byte is 1, and the lower half-byte is 6. The decimal equivalent of 16 is 22_{10}; that is, 06 must be added to the contents of Acc-A. Clearly, row 2 or 3 of the table meets this condition. However, row 3 will not fit the contents of Acc-A as column 4 is not correct; similarly, row 2 will not fit for the same reason. Therefore, the result obtained by executing a DAA instruction will be incorrect; similarly, if the contents of Acc-A are 32 and a DAA instruction is executed, the result will be incorrect. There are many other examples.

Therefore, the program executed in the interrupt, from address 00D0 to address 00DD, must be executed to obtain a correct conversion each time; the number to be converted is placed in Acc-B and the result appears in Acc-A.

The largest number that can be converted is 63, which is equivalent to 99_{10}; larger numbers can be converted but *not* with this program.

The initialization of port B must be explained. When a PIA is reset, port B's output lines automatically go to the high-impedance state. The voltage on these lines is approximately 2.5 V; therefore, if logic gates are tied to these lines, they can and do respond to this and behave as though the lines were hi (1). When the port is initialized, the lines that are programmed as outputs immediately go low, and if the output lines are active high, the equipment connected will start when the port is reset. On the other hand, if the output lines are active low, the equipment will start when the port is initialized. However, the output lines *can be kept high* during initialization by the method used in this example, program steps 0086 to 0098 inclusive. This method should be carefully studied and applied where applicable. The reader will notice in the examples that follow that the output lines from the port are always active low, where false start-up could be a problem.

Example 2. Write a program to display a number; that is, display 6 digits using seven-segment LED displays (Litronix DL 704 or Monsanto MAN 72 or MAN 74). See Figure 2.41.

Refer to Figure 7.3. The display buffer is port A of a PIA chip, while the position buffer is port B of the same chip. Assume the address of port

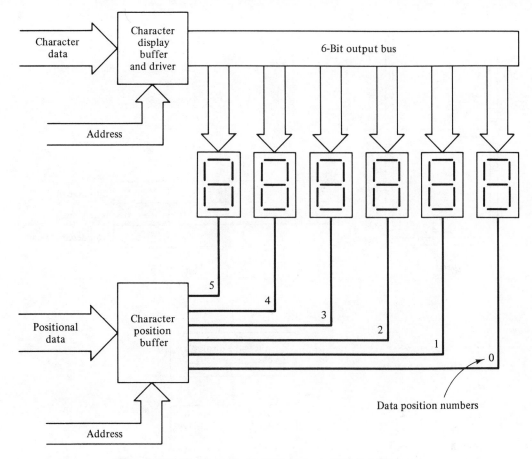

Figure 7.3 System Block Diagram and Wiring Diagram

A is 8020/21 and port B is 8022/23. It can be seen from the block diagram that 8 characters could ultimately be displayed.

The type of displays listed have the bit pattern of Figure 7.4. To illuminate a bit position, a 0 must be placed in that position (e.g., display 7, that is, bits 0, 1, and 2 must be illuminated); therefore the coding of the output of PIA port A will be as in Figure 7.5.

The display has been illustrated in detail in Figure 2.41; the digit is enabled by setting the data position line high. When this line is high, the code that is on the 6-bit bus is applied to the digit, causing those elements that have 0 applied to them to glow, in the case illustrated (Figure 7.5), bits 0, 1, and 2. The program block diagram is shown in Figure 7.6 and the programs in Tables 7.5 and 7.6. Characters are displayed from left to right; that is, line 5 goes high, followed by lines 4, 3, 2, 1, 0.

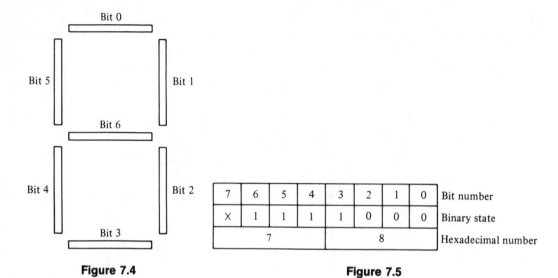

Figure 7.4

7	6	5	4	3	2	1	0	Bit number
X	1	1	1	1	0	0	0	Binary state
7				8				Hexadecimal number

Figure 7.5

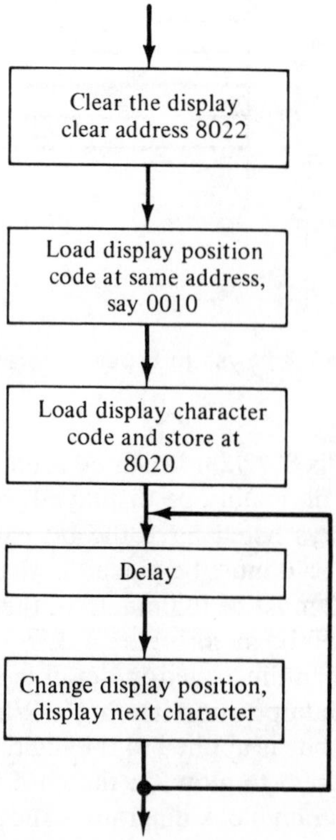

Figure 7.6 Program Block Diagram

Table 7.5 Main Program

Address	Instruction	Mnemonic	Remarks
0100			Reader should initialize port A address 8020/21 all lines output,
⋮			and port B address 8022/23 all lines output. Keep port A lines
⋮			*high* during initialization; port B lines should all go low after
0120			initialization (normally executed by J bug in D2 kit)
0121	86 20	LDA-A# ⎱	Store digit display position at ad-
0123	97 10	STA-A ⎰	dress 0010
0125	CE 0020	LDX#	Load X register with address of dis-
			play data storage area
0128	7F 8022	CLR	Clear display, all display enable
			lines low
012B	A6 00	LDA-A(O,X) ⎱	Load Acc-A with character to be
			displayed
012D	B7 8020	STA-A ⎰	Store character in port A
0130	96 10	LDA-A ⎱	Load Acc-A with position of char-
0132	B7 8022	STA-A ⎰	acter to be displayed, store at
			8022; character displayed
0135	FF 0013	STX	Save contents of X register at ad-
			dress 0013/0014
0138	BD 0150	JMP	Jump to delay subroutine
013B	FE 0013	LDX	Retrieve the contents of X register
			from 0013/0014
013E	08	INX	Increment contents of X register
013F	74 0010	LSR	Shift the digit in positional memory
			location 0010
0142	24 E7	BCC	Branch if carry flag is clear, next
			character
0144	7E 0121	JMP	Jump back to start and repeat

Characters to be displayed are stored initially at address 0020, and they must be stored in the code required to activate the elements of the display. For example, assume the characters to be displayed are 1, 2, 3, 4, 5, 6; the required codes are shown in Figure 7.7. Thus the codes stored will be as follows:

$$\begin{array}{ccc} \textit{Address} & \textit{Code Stored} & (7.1) \\ 0020 & 79 & \\ 0021 & 24 & \\ 0022 & 30 & \\ 0023 & 19 & \\ 0024 & 12 & \\ 0025 & 02 & \end{array}$$

Table 7.6 Delay Subroutine

0150	CE 3000	LDX#	Delay; this delay causes the display to
0153	→09	DEX	appear steady; actually the characters
0154	└26 FD	BNE	are flashing on and off. Various visual
0156	39	RTS	effects can be created by experimenting
			with the delay, the data stored at
			0151/0152

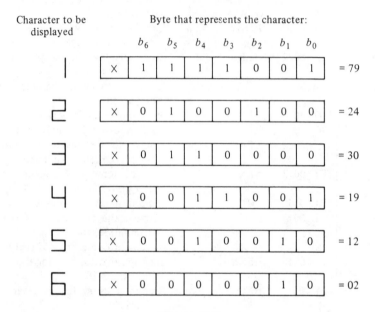

Figure 7.7

The X register is used as a pointer to point to the next character to be displayed. The multiplexing method of display is economical as regards equipment and programming.

This example illustrates multiplexing and a method of displaying certain data. For example, the previous example read a hexadecimal number and converted it to a decimal number; now that number can be displayed.

Example 3. Write a program that uses the X register as a pointer to the contents of a previously erected table. The table contains the codes required to convert decimal numbers into seven-segment LED displays. See Figure 7.8. Assume that the data that are to be encoded are located at address 0002, Example 1. The table is located in EPROM starting at address 7750. See Tables 7.7 and 7.8.

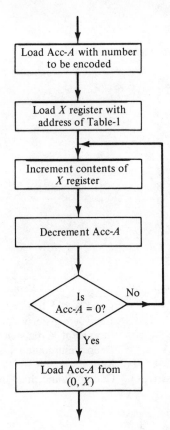

Figure 7.8 Block Diagram

This example illustrates the use of the X register as a pointer to locate a code number. The disadvantage with this program is that, if the character loaded into the Acc-A at program step 0176 is *not* included in the table, the program will drive the contents of the X register beyond the bounds of the table. Problem 7.1 will leave the task of restricting the bounds of the X register to the reader.

The reader should tie Examples 1, 2, and 3 together; only two 7-segment displays will be required as 63_{10} is the largest number that can be converted with six data lines.

Example 4. Write a program to echo the transmissions from a CRT terminal; the transmissions are received from the CRT terminal via the ACIA and then retransmitted by the same chip back to the CRT. The transmissions will be at the 300-baud rate and will use a 20-mA loop (see Figure 7.9). The terminal is set for 300-baud rate, 7-bit code, even parity,

Table 7.7

Address	Contents	Decimal Number
7750	40	0
7751	79	1
7752	24	2
7753	30	3
7754	19	4
7755	12	5
7756	02	6
7757	78	7
7758	00	8
7759	18	9

Table 7.8 Program

Address	Instruction	Mnemonic	Remarks
0176	96 02	LDA-A	Load Acc-A from address 0002
0178	CE 774F	LDX#	Load X register with 7750 − 1 = 774F
017B	08	INX	Increment X register
017C	4A	DEC-A	Decrement contents of Acc-A
017D	26 FC	BNE	Repeat
017F	A6 00	LDA-A(O,X)	Code is found; load into Acc-A

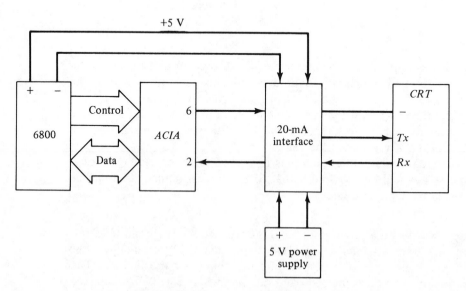

Figure 7.9 System Block Diagram

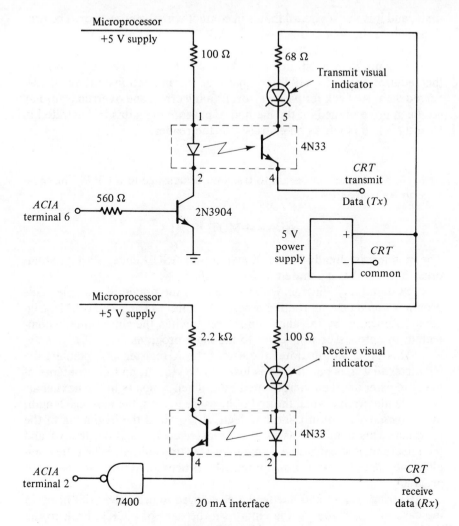

Figure 7.9 (*cont.*)

1 stop bit; therefore, the ACIA must also be set to transmit and receive at this rate and with the code format. The natural question to ask is why have the processor transmit back that which has already been received?

The answer is the operator very often must hold a conversation with the processor; for example, the time of a delay may have to be changed or a different sequence of events must be called up. Thus, the operator types in the code word and the processor responds. When the operator types in the code word, she or he wants to read what has been typed; the terminal does not always print the keyboard output as they can be two separate units, and it is the keyboard that is in contact with the processor. The processor, in turn, is in contact with the CRT.

Refer to Figure 7.10 and Table 7.9.

This program illustrates a method of initializing the ACIA to meet the requirements of a CRT terminal reading and retransmitting the received data. A check for parity error, framing error, and overrun error has *not* been programmed in; the method of programming this is illustrated in Chapter 3 and is left as an exercise for the reader.

Example 5.   Write a program to transmit a message to a CRT. The message to be transmitted is

ALARM-MOTOR #1

The message is already stored in memory in ASCII code, starting at address 0216, and is illustrated in Table 7.10.

System block diagram is the same as in the previous example. This program would be a subroutine program (Figure 7.11 and Table 7.11), triggered into action by the alarm on motor 1; when the subroutine is completed, the processor will return to the main program.

This program illustrates the use of the X register as a pointer; the character that X is pointing to is loaded into Acc-A, and the contents of the X register are then incremented by 1. Then X points to the next character. To determine when the end of message occurs, the message length, including carriage return and line feed, is stored at the beginning of the program. This number was initially transferred to page 00, line 01 and then decremented each time a character was transmitted. When the complete message has been transmitted, the processor returns to the main program.

Carriage return and line feed are required to bring the print head, in the case of a teletype, or the cursor, in the case of a CRT, back to the beginning of a line. The ACIA should be reset and initialized by the reader as an exercise.

Example 6.   A certain process in a plant requires that two motors operate in a cyclic manner; first M_1 will operate for T_1 seconds, and then M_2 will operate for T_1 seconds. The process continues in this way. When a contact X in the plant closes, motors M_1 and M_2 will stop, and motors M_3 and M_4 will start to cycle with a time of T_2 seconds. When contact X opens, the process reverts back to M_1 and M_2, with M_3 and M_4 stopped.

The system block diagram (Figure 7.12), interface diagram (Figure 7.13), and state diagram (Figure 7.14) illustrate the problem.

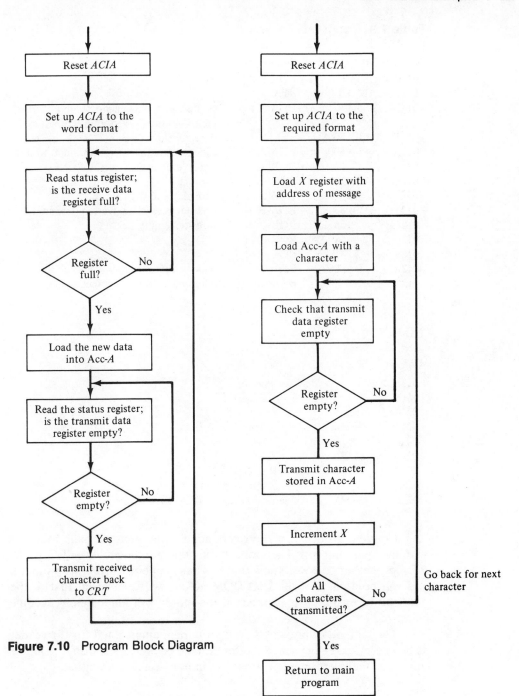

Figure 7.10 Program Block Diagram

Figure 7.11 Program Block Diagram: Subroutine

Table 7.9 Program

Address	Instruction	Mnemonic	Remarks
0110	86 03	LDA-A#	Reset ACIA
0112	B7 8008	STA-A	
0115	86 49	LDA-A#	Set up ACIA transmit/receive format, ÷ 16, 7 bits, even parity, 1 stop bit
0117	B7 8008	STA-A	interrupt disabled
011A	B6 8008	LDA-A	Read status register, for new data, wait for
011D	46	ROR-A	receive data register full flag
011E	24 FA	BCC	
0120	B6 8009	LDA-A	Read received data
0123	F6 8008	LDA-B	Read status register, wait for transmit data
0126	56	ROR-B	register empty flag
0127	56	ROR-B	
0128	24 F9	BCC	
012A	B7 8009	STA-A	Transmit received data back to CRT
012D	20 E8	BRA	Branch back to 011A (repeat)

Table 7.10

Address	Data	Remarks
0215	10	Message length
0216	414C	AL
0218	4152	AR
021A	4D2D	M-
021C	4D4F	MO
021E	544F	TO
0220	5220	R space
0222	2331	#1
0224	0D0A	Carriage return/line feed

Caution: Notice that the overloads for the motor remain in series with the operating coil. Therefore, if they operate the motor will stop; the microprocessor does *not* know that this has happened, with this program, and will hold pin 4 of the CRYDOM active, with the result that, if the overload reset button is depressed, the motor will immediately become energized.

The method of preventing this from happening is left for the reader as an exercise. A suggestion is offered: feed back a signal from the motor to inform the microprocessor that the motor is/is not operating.

1. The control lines to the motors from the interface are active high, while the output lines from the processor to the interface are active low. This is for safety reasons.

Table 7.11 Subroutine Program (starting address 0136)

Address	Instruction	Mnemonic	Remarks
0136	86 03	LDA-A#	Reset ACIA
0138	B7 8008	STA-A	
013B	86 49	LDA-A#	÷ 16, 7 bits, even parity, 1 stop bit
013D	B7 8008	STA-A	interrupts disabled
0140	B6 0215	LDA-A	Load message length into Acc-A
0143	97 01	STA-A	Store message length at address 0001
0145	CE 0216	LDX#	Load X register with starting address of message
0148	A6 00	LDA-A(0,X)	Load Acc-A with character
014A	F6 8008	LDA-B	Read status register and wait for trans-
014D	56	ROR-B	mit data register flag to indicate
014E	56	ROR-B	empty (flag raised = 1)
014F	24 F9	BCC	
0151	B7 8009	STA-A	Transmit character held in Acc-A
0154	08	INX	Increment contents of X register
0155	7A 0001	DEC	Decrement message length counter
0158	26 EE	BNE	Go back for next character
015A	39	RTS	Return to main program

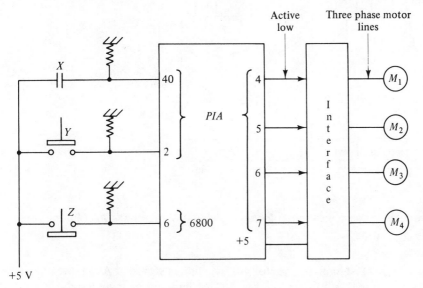

Resistors are part of a Beckman resistor network, part no 641-1, 4.7 kΩ

Y = start process
X = transfer of control contact
Z = emergency stop, or stop

Figure 7.12 System Block Diagram

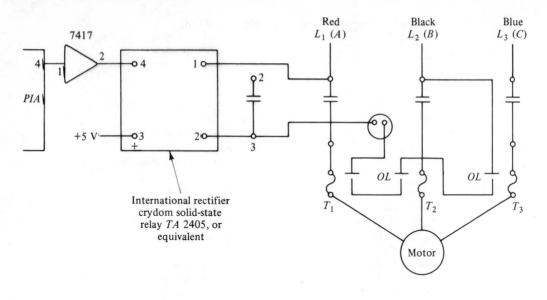

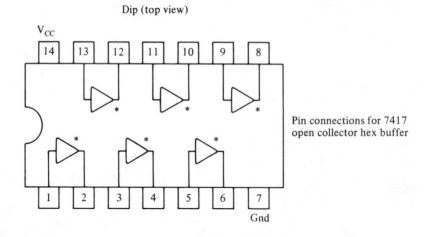

Dip (top view)

Pin connections for 7417
open collector hex buffer

Figure 7.13 Interface Diagram for an Individual Three-phase AC Motor

2. The interrupt vector via $\overline{\text{IRQ}}$ line is stored at A000 (the high byte) and A001 (the low byte). This line will be used for the X contact via line C_{A1} pin 40 of the PIA chip.

3. The nonmaskable interrupt vector, $\overline{\text{NMI}}$, has been previously programmed into EPROM as address 00B0; this line, pin 6 of the MC 6800, will be used as the stop line from push button Z.

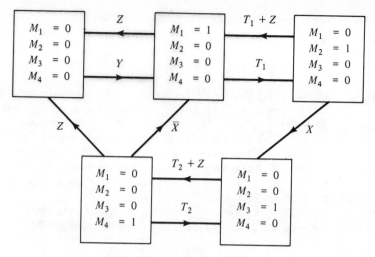

Figure 7.14 State Diagram

Initialize port A of the PIA chip as follows:

I/O line 0 (pin 2) input
Lines 5, 4, 3, and 2 (pins 7, 6, 5, 4, respectively) output
Lines 7, 6, and 1 not used

The PIA port must be initialized such that the output lines *do not* go low during initialization.

When the reset line of the MPU and PIA chips goes low for the predetermined time and then high, the program of Table 7.12 will begin execution at 0010; a block diagram of the (Figure 7.15) will illustrate the example.

The program continues in the loop illustrated until the X contact closes again; then program control will be transferred to address 0070 (Table 7.13).

This problem has illustrated the use of increment stack pointer, positive and negative transition interrupts, resetting the interrupt vector, clearing interrupt flags, and how interrupts can be used to transfer control only when the process is in a specified condition (use of the SEI and CLI instructions). The NOP between instructions 0E and 0F, clear interrupt mask and set interrupt mask, is necessary to give the processor time to check the status of the interrupts; that is, has an interrupt occurred? Failure to include this delay could prevent an interrupt from being detected. The disadvantage with this method of interrupts is the time and program space required to increment the stack pointer (14 memory locations for

Table 7.12

Address	Instruction	Mnemonic	Remarks
0010	0F	SEI	Set interrupt mask bit (bit 4 condition codes register); the bit should be set but this is insurance; also it is required when using $\overline{\text{NMI}}$
0011	86 04	LDA-A#⎫	Set up bit 2 of control register of PIA port A, address 8005; processor can now access peripheral data register
0013	B7 8005	STA-A ⎭	
0016	86 FF	LDA-A#⎫	Load peripheral data register of PIA port A with FF; all output lines are set high
0018	B7 8004	STA-A ⎭	
001B	86 00	LDA-A#⎫	Reset bit 3 of the control register of port A; processor can now access data direction register
001D	B7 8005	STA-A ⎭	
0020	CE FC07	LDX#⎫	Load the illustrated code into PIA port A control register, address 8004
0023	FF 8004	STX ⎭	

PIA, port A	7	6	5	4	3	2	1	0	Bit number
Control register			0				7		Hexadecimal number
	0	0	0	0	0	1	1	1	Binary state

— Interrupt not masked
— Interrupt on positive edge
— Processor can access PDR
— Not used in this program
— Not used in this program
— This bit is set when an interrupt occurs on C_{A1}

I/O lines 7 to 2 inclusive set up as output lines
Lines 1 and 0 set up as input lines
Port A set up to interrupt on positive transition of line C_{A1}

0026	┌→ B6 8004	LDA-A	Load Acc-A from port A
0029	│ 46	ROR-A	Rotate contents of peripheral data register right one place, to check if the Y push button is depressed; carry flag will be set if button is depressed
002A	└─ 24 FA	BCC	Branch back to 0026 if the carry flag is clear; push button Y is open
002C	7E 0037	JMP	Jump unconditionally to address 0037, start motor M_1

Table 7.12 (*cont.*)

Address	Instruction	Mnemonic	Remarks

Interrupt program for X contact opening, i.e., a positive transition interrupt

Address	Instruction	Mnemonic	Remarks
0030	31		Increment stack pointer to reset the stack
0031	31		back to its position before the interrupt
0032	31		occurred
0033	31	INS	
0034	31		
0035	31		
0036	31		
0037	86 07	LDA-A#	Set up control register A, in port A, for a
0039	B7 8005	STA-A	positive transition interrupt
003C	CE 0070	LDX#	Set up new interrupt vector, address 0070
003F	FF A000	STX	
0042	B6 8004	LDA-A	Read PDR port A to clear interrupt flag bit 7 of control register.
0045	86 FB	LDA-A#	M_1 on, M_2 off, M_3 off, M_4 off
0047	B7 8004	STA-A	
004A	CE FF00	LDX#	Delay T_1, decrement contents of X_1
004D	09	DEX	
004E	26 FD	BNE	
0050	86 F7	LDA-A#	M_1 off, M_2 on, M_3 off, M_4 off
0052	B7 8004	STA-A	
0055	CE FF00	LDX#	Delay T_1
0058	09	DEX	
0059	26 FD	BNE	
005B	0E	CLI	Clear interrupt mask bit, wait, then set
005C	01	NOP	interrupt mask again; interrupts are only
005D	0F	SEI	allowed after M_2 has completed its operation
005E	20 E5	BRA	Branch back to address 0045

Table 7.13

Interrupt program for X contact closing, i.e., a negative transition interrupt

Address	Instruction	Mnemonic	Remarks
0070	31		Increment stack pointer to reset the stack
0071	31		back to its position before the interrupt
0072	31		occurred
0073	31	INS	
0074	31		
0075	31		
0076	31		
0077	86 05	LDA-A#	Set up control register A for negative inter-
0079	B7 8005	STA-A	rupt
007C	CE 0030	LDX#	Set up new interrupt vector, address 0030
007F	FF A000	STX	
0082	B6 8004	LDA-A	Read PDR, port A to clear interrupt flag

Table 7.13 (*cont.*)

Address	Instruction	Mnemonic	Remarks
0085	→ 86 EF	LDA-A# ⎤	M_1 off, M_2 off, M_3 on, M_4 off
0087	B7 8004	STA-A ⎦	
008A	CE F000	LDA# ⎤	Delay T_2 seconds
008D	→ 09	DEX	
008E	⌐ 26 FD	BNE ⎦	
0090	86 DF	LDA-A# ⎤	M_1 off, M_2 off, M_3 off, M_4 on
0092	B7 8004	STA-A ⎦	
0095	CE F000	LDX# ⎤	Delay T_2 seconds
0098	→ 09	DEX	
0099	⌐ 26 FD	BNE ⎦	
009B	0E	CLI	Clear interrupt mask bit, wait, then set
009C	01	NOP	interrupt mask again; interrupts are only
009D	0F	SEI	allowed after M_4 has completed its operation
009E	⌐ 20 E5	BRA	Branch back to 0085

Emergency stop interrupt program via Z push button

Address	Instruction	Mnemonic	Remarks
00B0	86 FF	LDA-A# ⎤	Stop all motors
00B2	B7 8004	STA-A ⎦	
00B5	31 ⎤		
00B6	31		Reset stack pointer
00B7	31		
00B8	31 ⎬	INS	
00B9	31		
00BA	31		
00BB	31 ⎦		
00BC	0F	SEI	
00BD	7E 6020	JMP	Jump back to 0020

the main programs). This could have been eliminated with the technique of using a memory cell to remember which program is being executed. This method will be illustrated in Example 7.

The process start button Y, the process transfer control contact X, and the emergency stop button Z were *not* pretested before allowing the program to go to address 0023 (wait for start signal); this was left out deliberately. The reader should program in the pretest for the Y button. The pretest for contact X and button Z will be contained in Example 9. The method of connecting the push buttons to the I/O port should be noted (i.e., the connection of the pull-up resistor). Figure 7.16 demonstrates some of the safety aspects. The line leaves the safety of the control room and enters the plant, where it is vulnerable to damage by shorting to ground or open circuit. If either of these events occurs, the control system via the I/O will *not* be stimulated into action. Similarly, with the stop but-

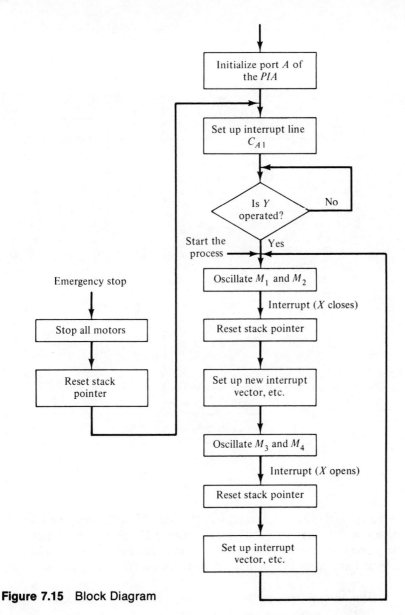

Figure 7.15 Block Diagram

ton Z (Figure 7.17), if the line goes to ground or breaks, an interrupt occurs and the process will stop.

Example 7.   A process in a plant is started when Y is depressed and uses two motors, M_1 and M_2. A contact X will close and energize motor 2;

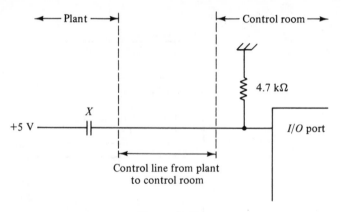

Figure 7.16

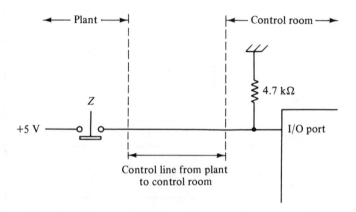

Figure 7.17

when the contact closes again, M_2 will stop and M_1 will start. The process continues in this manner (shown in Figure 7.18) until it is stopped. A memory cell will be used to remember which motor is operating. The system block diagram and state diagram now follow (Figures 7.18 and 7.19).

1. The control lines to the motors are active high; the control lines from the processor are active low.
2. The interrupt vector via $\overline{\text{IRQ}}$ line is 02A0 and is stored at A000/A001.
3. The nonmaskable interrupt vector $\overline{\text{NMI}}$ has been previously programmed as 02C0.

Initialize port A as follows:

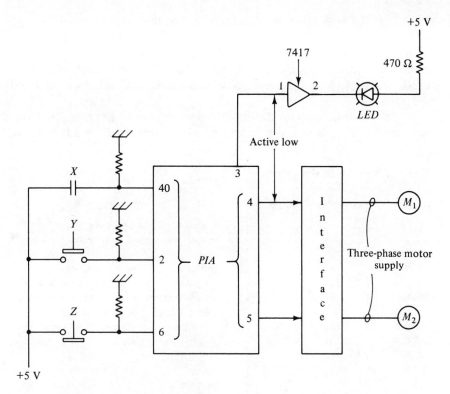

Resistors are part of a Beckman resistor network, part no. 641-1, 4.7 kΩ
For interface for motors, see Figure 7.13.

Figure 7.18 System Block Diagram

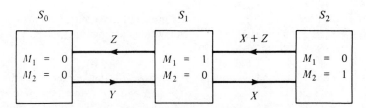

Figure 7.19 System State Diagram

I/O line 0 (pin 2) input
Lines 1, 2, and 3 (pins 3, 4, and 5, respectively) output
Lines 4 to 7 unadopted (not used)

The PIA port must be initialized so that the output lines *do not* go low during initialization; this is left for the reader to program into this example. When the reset line of the MPU and PIA chips makes a positive transition, the accompanying program (Table 7.14) will begin execution at 0230, program block diagrams illustrates the program (Figure 7.20 and 7.21).

The other duties that the processor is performing are many and varied, and for this example the processor will flash an LED. While the pro-

Table 7.14 Main Program

Address	Instruction	Mnemonic	Remarks
0230			Set 1's on all the I/O lines of port A; the output lines *must not* be allowed to go
(Reader to fill in details)			low during initialization; otherwise the motors M_1 and/or M_2 will start
023E	CE FE07	LDX#	Set up port A line 0 input, remainder output, interrupt mask clear, interrupt on
0241	FF 8004	STX	positive edge (verify the code, FE07)
0244	7F 0001	CLR	Clear memory location 0001 (memory cell)
0247	CE 02A0	LDX#	Set up interrupt vector 02A0 at address
024A	FF A000	STX	A000/A001
024D	B6 8004	LDA-A	Wait for the Y push button to close (pre-
0250	46	ROR	test the start button)
0251	25 FA	BCS	
0253	F6 0001	LDA-B	Active test on the contents of the memory
0256	26 38	BEQ	cell; if Z flag is *not* set, branch to 0290 and start M_2; otherwise continue and start M_1
0258	86 FB	LDA-A#	Start motor M_1, stop motor M_2
025A	B7 8004	STA-A	
025D	0F	SEI	Set interrupt mask bit
025E	CE F200	LDX#	Delay, for flashing the LED
0261	09	DEX	
0262	26 FD	BEQ	
0264	86 02	LDA-A#	Exclusive OR the contents of the PDR
0266	B8 8004	EOR-A	port A and the contents of Acc-A; turn
0269	B7 8004	STA-A	the LED *on* but maintain the motor that is running
026C	CE F200	LDX#	Delay
026F	09	DEX	
0270	26 FD	BEQ	
0272	86 02	LDA-A#	Exclusive OR the contents of the PDR
0274	B8 8004	EOR-A	port A and the contents of Acc-A; turn
0277	B7 8004	STA-A	the LED *off* but maintain the motor that is running

Table 7.14 (*cont.*)

Address	Instruction	Mnemonic	Remarks
027A	0E	CLI	Clear interrupt mask bit (bit 4 condition
027B	01	NOP	codes register), wait; set the interrupt
027C	0F	SEI	mask bit again; no interrupts are al-lowed to occur in the processor's other duties
027D	01	NOP ⎫	Fill locations 027D to 028C inclusive with
028C		⎭	01's to be used later
028D	20 C4	BRA	Branch back to 0253, check memory cell again
028F	01	NOP	Not used
0290	86 08	LDA-A# ⎫	Start motor M_2, stop motor M_1
0292	B7 8004	STA-A ⎭	
0295	0F	SEI	Set interrupt mask bit
0296	20 C6	BRA	Branch back to 025E to flash LED

Interrupt Program

02A0	F6 0001	LDA-B ⎫	Active test on contents of memory cell; if
02A3	26 0B	BEQ ⎭	Z flag is *not* set, branch to 02B0 and decrement memory location 0001; otherwise increment the memory location
02A5	7C 0001	INC	Increment memory cell
02A8	F6 8004	LDA-B	Clear interrupt flag in PIA, port A
02AB	3B	RTI	Return from interrupt
02B0	7A 0001	DEC	Decrement memory cell
02B3	20 F3	BRA	Branch back to 02A8

cessor is actively controlling the output to the LED, *no* interrupt will be allowed to occur, other than $\overline{\text{NMI}}$, which cannot be masked off. The interrupt via the C_{A1} line will be a positive transition (i.e., when X closes, line C_{A1} will go high). The memory cell is located at address 0001.

Memory locations 027D to 028C are to be filled with op-code 01 (NOP). These locations have been left to enable the reader to modify the program (Problem 7.3).

In the program, as it exists, when the LED has been turned on and then off and the interrupt mask bit cleared and reset, it is not known if an interrupt occurred in this interval; therefore, the program goes back to location 0253 and rechecks the memory cell and restarts the motor. This can be eliminated with the use of a mask, which can be stored at memory location 0002; that is, if the bit is set, an interrupt has occurred. Therefore, the program must go via location 0253; if the mask is not set, the program will go via 025E.

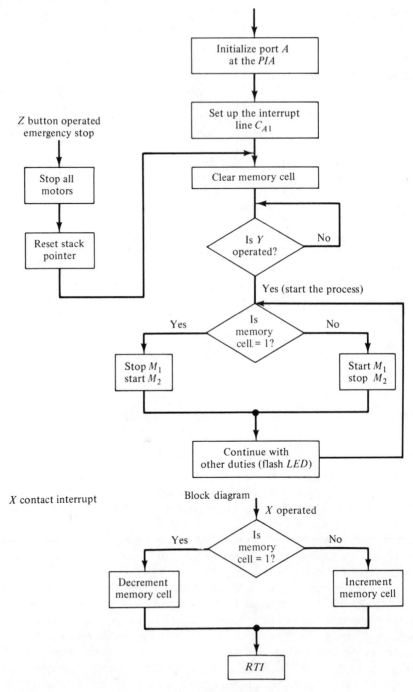

Figure 7.20 Block Diagram

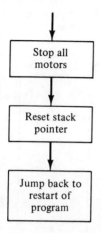

Figure 7.21 Emergency Stop Interrupt

Table 7.15 Nonmaskable Interrupt Program: Stop

Address	Instruction	Mnemonic	Remarks
02C0	86 FF	LDA-A# ⎫	Stop all motors
02C2	B7 8004	STA-A ⎭	
02C5	31 ⎫		Reset stack pointer
02C6	31 ⎪		
02C7	31 ⎪	INS	
02C8	31 ⎬		
02C9	31 ⎪		
02CA	31 ⎪		
02CB	31 ⎭		
02CC	7E 0245	JMP	Jump back to memory location 0245

This example has illustrated the use of the instruction clear a memory location, and the use of Exclusive OR to maintain the original data yet modify that which is to be changed. The program also illustrates how to use the processor for other than the one function, in this case flash an LED, which is simple but effective. Also, spaces were left in the program for further additions. The use of the memory cell has saved the resetting of the stack pointer after each interrupt (except the $\overline{\text{NMI}}$).

Example 8. A machine in a plant moves back and forth along a bed between two limit switches X and Y (e.g., a metal planer or a knife grinder in the log chipping section of a pulp mill). The machine is equipped with the following push buttons: start forward (A), start rev (B), retentive start (C), and stop/reset (Z). The retentive start button is normally used after a stop to ensure that the machine resumes its original direction of travel.

All the numbers are on the *PIA* chip, except (6), which is on
the 6800 chip; i.e., 2 and (6) are joined together

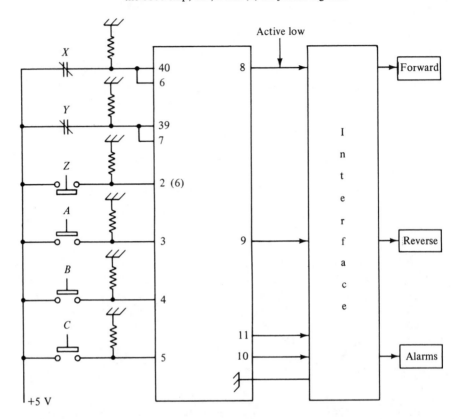

Resistors are part of a Beckman resistor network, part no. 641-1, 4.7 kΩ

Figure 7.22

The motion is arrested by limit switches at each end of its travel, and the
action of a limit switch is communicated to the processor via interrupts;
also, the limit switches are interrogated each time direction of motion is
changed. If the arresting limit switch or its wiring is defective, an alarm
will sound (and the machine will not change direction), warning the operator that a defect in the control scheme has occurred.

Figures 7.22 to 7.26 illustrate the problem.

1. The control lines to the forward/reverse magnetics and alarms are
   active high.

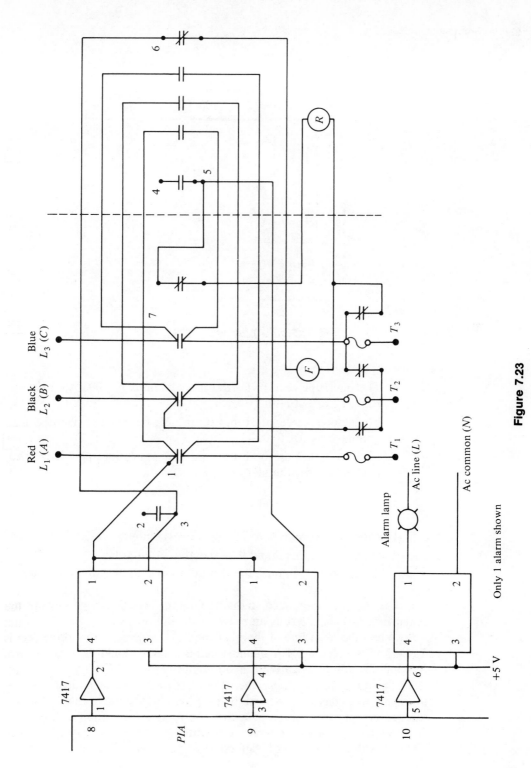

Figure 7.23

Only 1 alarm shown

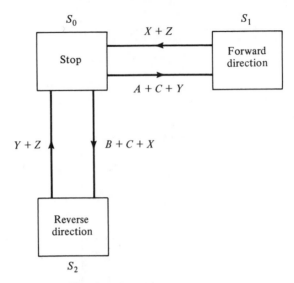

Figure 7.24 State Diagram

2. The interrupt vector via $\overline{\text{IRQ}}$ line is stored at A000/A001; interrupt vector is 01B0.
3. The nonmaskable interrupt vector $\overline{\text{NMI}}$ has been previously programmed as 02C0.
4. Ports A and B are both used in this application; port B is used exclusively for the alarms.

Initialize port A as follows:

I/O lines 0 to 5 (pins 2 to 7, respectively) inputs
Lines 6 and 7 (pins 8 and 9, respectively) outputs
Initialize port B, all lines 0 to 7 (pins 10 to 17, respectively) outputs

The ports must be initialized so that the output lines *do not* go low during initialization. This is left as an exercise for the reader.

When the reset line of the MPU and PIA goes high, the program of Table 7.16 is executed. The memory cell is located at address 0001. Lines C_{A1} and C_{A2} will be negative transition interrupts.

This program illustrates the use of a memory cell as a means of determining what has happened in the past to influence the events of the future. Interrupts are still used for the stop or arrest direction of travel.

Notice the use of the limit interrogation at the end of travel; the philosophy is that, if the limit is defective at the far end of the machine bed,

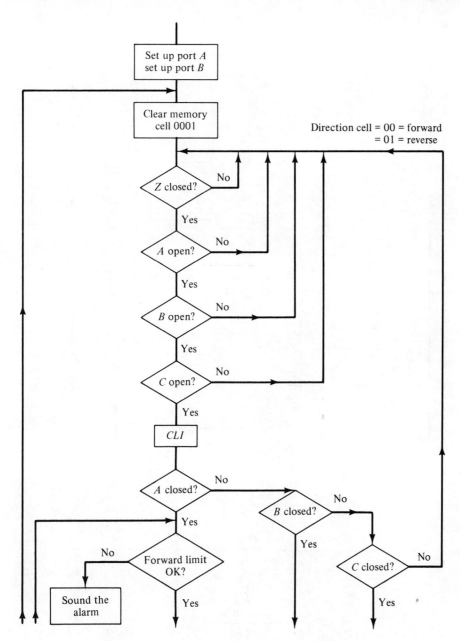

Figure 7.25 Planer Program Block Diagram

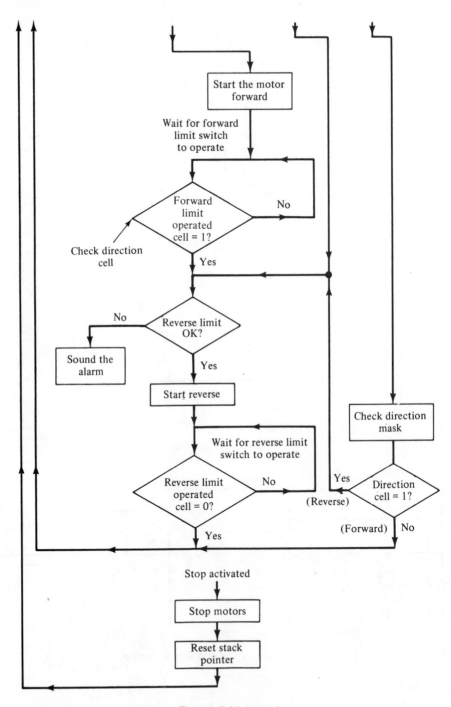

Figure 7.25 (*cont.*)

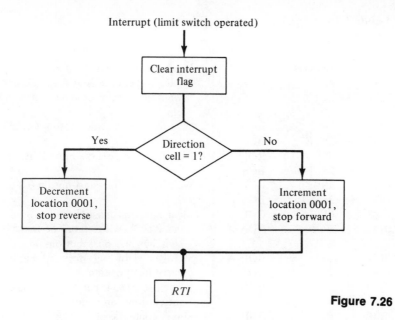

Figure 7.26

why start the machine in that direction as the motion will not be arrested, since the limit is defective and damage will result? Fail-safe wiring methods, as illustrated, should be used. Once the machine has been started the arresting limit switches are no longer interrogated. The reader should incorporate periodic interrogation of the next active limit into the program (see Problem 7.5).

Table 7.16 Main Program

Address	Instruction	Mnemonic	Remarks
0000			Set highs on all I/O lines of port A and port B; the output lines must not be allowed to go low during initialization; otherwise the forward and/or reverse magnetics will close, and alarms will be activated
Reader to fill in details			
011F	CE C00D	LDX#	Set up port A lines 0 to 5 (input) and 6 and
0122	FF 8004	STX	7 (output); interrupt C_{A1} and C_{A2} active on negative edge (verify the code C00D)
0125	CE FF04	LDX#	Set up port B all lines output, interrupts
0128	FF 8006	STX	disabled
012B	7F 0001	CLR	Clear memory location 0001 (memory cell)
012E	CE 01B0	LDX#	Set up interrupt vector 01B0 at address
0131	FF A000	STX	A000/A001

Table 7.16 (cont.)

Address	Instruction	Mnemonic	Remarks
0134	B6 8004	LDA-A ⎫	Load contents of PDR into Acc-A
0137	46	ROR-A ⎬	Rotate contents of Acc-A right one place;
0138	24 FA	BCC ⎭	wait for stop/reset push button (Z) to close
013A	46	ROR-A ⎫	Wait for start forward button (A) to open.
013B	25 F7	BCS ⎭	
013D	46	ROR-A ⎫	Wait for start reverse button (B) to open
013E	25 F4	BCS ⎭	
0140	46	ROR-A ⎫	Wait for retentive start button (C) to open
0141	25 F1	BCS ⎭	
0143	0E	CLI	Clear interrupt mask bit
0144	B6 8004	LDA-A	Load Acc-A with status of the external switches and push buttons
0147	46	ROR-A ⎫	Shift status of start forward button into
0148	46	ROR-A ⎭	carry flag position
0149	24 35	BCC	Branch to 0180 if carry flag is clear; start forward button open
014B	48	ASL-A	Move contents of Acc-A back to their original position
014C	85 10	BIT-A ⎫	Test bit 4; if set, then forward limit switch
014E	27 40	BNE ⎭	is open; branch to 0190 and sound alarm
0150	010101	NOP ⎫	6 NOPS, space in memory
0153	010101	⎭	
0156	86 BF	LDA-A# ⎫	Start motor forward
0158	B7 8004	STA-A ⎭	
015B	D6 01	LDA-B ⎫	Check if contents of memory cell (location
015D	27 FC	BEQ ⎭	0001) are 01; wait for forward limit to
015F	B6 8004	LDA-A ⎫	Test bit 5; if set, the reverse limit switch
0162	85 20	BIT-A ⎭	is open
0164	27 3A	BNE	Branch to 01A0 alarm program 2
0166	010101	NOP ⎫	6 NOPS space in memory (not required)
0169	010101	⎭	
016C	86 7F	LDA-A# ⎫	Start motor reverse
016E	B7 8004	STA-A ⎭	
0171	D6 01	LDA-B ⎫	Check contents of memory cell (location
0173	26 FC	BNE ⎭	0001); wait for contents to become 00
0175	20 D5	BRA	Branch to location 014C
0180	46	ROR-A	Shift status of start reverse button into carry flag position
0181	24 02	BCS	Branch to 0185 if carry flag is clear, start reverse button open
0183	20 DA	BRA	Branch to 015F
0185	46	ROR-A	Shift status of retentive start button into carry flag position

Table 7.16 (*cont.*)

Address	Instruction	Mnemonic	Remarks
0186	24 AC	BCC	Branch to 0134 if carry flag is clear, retentive start button is open
0188	D6 01	LDA-B	Check contents of memory cell; if = 01, start reverse, otherwise start forward
018A	27 D3	BEQ	Branch to 015F (i.e., move in the reverse direction)
018C	20 BE	BRA	Branch to 014C (i.e., move in the forward direction)

Alarm 1 program forward limit switch defective

0190	86 FE	LDA-A# ⎫	Energize alarm 1 audible and visual
0192	B7 8006	STA-A ⎭	
0195	⌐→ 01	NOP ⎫	Wait for stop/reset button to be depressed
0196	└ 20 FD	BRA ⎭	

Alarm 2 program reverse limit switch defective

01A0	86 FD	LDA-A# ⎫	Energize alarm 2 audible and visual
01A2	B7 8006	STA-A ⎭	
01A5	⌐→ 01	NOP ⎫	Wait for stop/reset button to be depressed
01A6	└ 20 FD	BRA ⎭	

Interrupt via C_{A1} or C_{A2} lines (forward or reverse limits)
Forward limit interrupt

01B0	B6 8004	LDA-A	Clear interrupt flag in PIA port A
01B3	96 01	LDA-A ⎫	Check the contents of memory cell (0001)
01B5	⌐ 26 01	BNE ⎭	
01B7	│ 7C	INC	Increment contents of 0001
01B8	└→ 7A	DEC	Decrement contents of 0001
01B9	3B	RTI	

$\overline{NMI}$ program stop-reset button activated

01C0	86 FF	LDA-A# ⎫	Stop motors
01C2	B7 8004	STA-A ⎭	
01C5	313131	INS ⎫	Reset stack pointer to original position
	313131	⎬	
	31	⎭	
01CC	7E 012B	JMP	Jump back to 012B

Example 9.   Write a program to read the status of specified switches and compare the result to the previous scan. If a change in status has occurred, flash the relevant lamp; the operator will acknowledge the change by depressing a push button (see Figure 7.27).

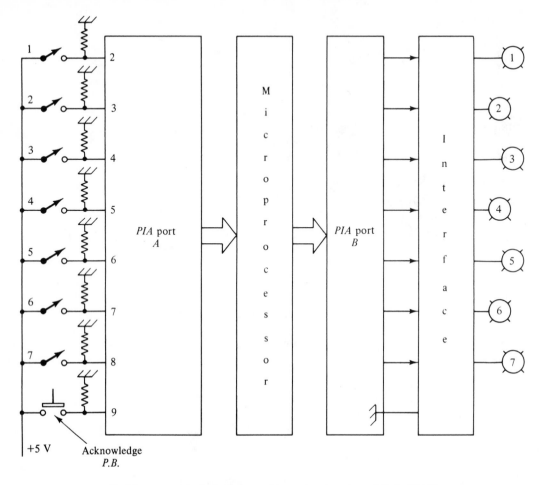

Resistors are part of a Beckman resistor network part no. 641.1, 4.7 kΩ
For interface for lamps, see Figure 7.23.

Figure 7.27 Switches Being Scanned

The I/O ports are set up: port A, all lines input; port B, all lines output. Refer to Figures 7.28 and 7.29 and Table 7.17.

This example illustrates the scanning and display of the status, opened or closed, of seven switches. If the status has changed since the last scan, then flash the lamp that represents that switch. The lamp will remain flashing until the operator depresses the acknowledge button; the lamp will then stop flashing and remain lit or unlit depending upon the status of the representative switch. Thus this program illustrates scanning, setting a mask of the change that has taken place, and saving this mask at

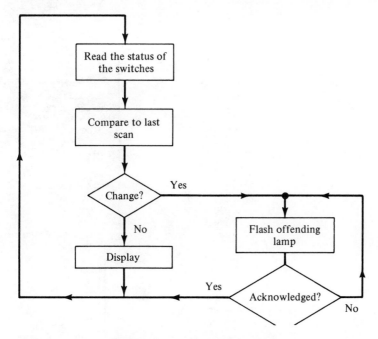

Figure 7.28 Block Diagram

some address, 0011. The mask is created by the operation of exclusive OR the status of the present and previous scans. If the result is 00, no change has taken place; otherwise, the changed bit, representing the offending switch, will be identified.

Example 10.   A process that is divided in five parts, $P_1P_2P_3P_4P_5$, is to be microprocessor controlled. The order in which the parts occur depends upon the process to be performed; for example, when product X is being produced, the process is $P_1P_5P_3P_4P_2$, whereas when product Y is produced, the process is $P_5P_3P_4P_2P_1$, and similarly with other products.

Assume that the program for each part is written and that a space is left, at the end of and between each section, for a program that can be used to chain the sections together in any order. See Figure 7.30.

The whole when completed will be blasted into EPROM memory. Furthermore, assume that RAM scratch pad memory is available at memory locations 0000 to 007F, and that communication with this section is available via a program and CRT already present. Write the program that is to be placed in the vacant spaces (see Figure 7.31).

Five memory locations will be required at the end of each part of the program. The low and high bytes for the order of program execution will

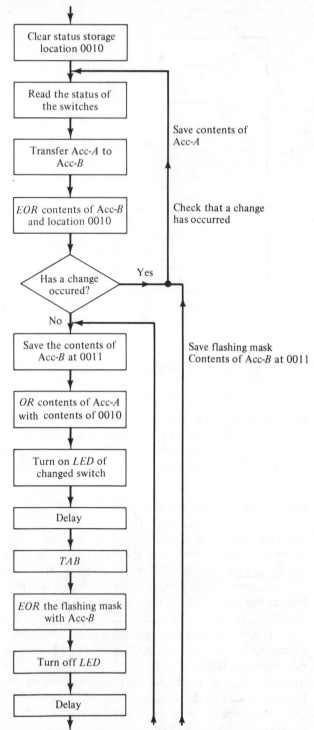

Figure 7.29 Program Block Diagram

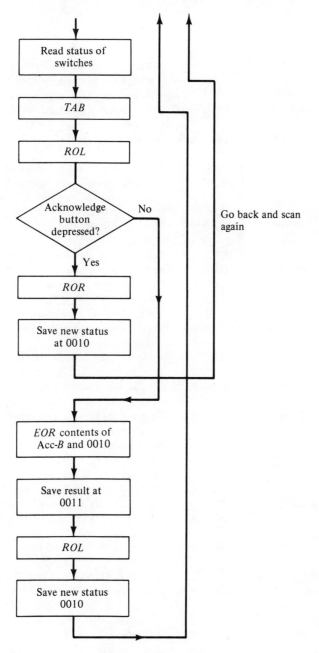

Figure 7.29 (*cont.*)

Table 7.17 Main Program

Address	Instruction	Mnemonic	Remarks
0080			Set up port A all lines input, set up port B
00A0			all lines output; these lines must be held high during and after initialization
00A1	7F 0010	CLR	Clear memory location 0010, storage for status
00A4	B6 8004 ←	LDA-A	Read the status of the switches; status is in Acc-A
00A7	16	TAB	Transfer status from Acc-A to Acc-B
00A8	D8 10	EOR-B	Has any change occurred since last scan; exclusive OR contents of Acc-B and address 0010
00AA	27 04	BEQ	Branch if zero flag is set to 00B0 (no change)
00AC	97 10	STA-A	Save current status of the switches
00AE	20 F4	BRA	Scan switches again
00B0	→D7 11	STA-B	Save the mask, contents of Acc-B, at 0011
00B2	9A 10	ORA-A	OR contents of 0010 with contents of Acc-A
00B4	B7 8006	STA-A	Turn on the lamp whose switch status has changed
00B7	BD 00DB	JSR	Jump to delay subroutine
00BA	16	TAB	Tab
00BB	D8 11	EOR-B	EXOR the flashing mask address 0011 with contents of Acc-B
00BD	F7 8006	STA-B	Turn off the lamp whose switch status has changed
00C0	BD 00DB	JSR	Jump to delay subroutine
00C3	B6 8004	LDA-A	Read status of the switches into Acc-A
00C6	16	TAB	Transfer Acc-A to Acc-B
00C7	49	ROL-A	Check whether acknowledge button has
00C8	→24 08	BCC	been depressed; branch if *not* pressed
00CA	46	ROR-A	Shift contents of Acc-A back to original position
00CB	97 10	STA-A	Save new status at 0010
00CD	010101	NOP	NOPS
00D0	20 D2 —	BRA	Branch back and continue scanning
00D2	↳D8 10	EOR-B	Has any change occurred since last scan; exclusive OR contents of Acc-B and address 0010
00D4	DA 11	ORA-B	Save mask, at 0011
00D6	46	ROR-A	Shift contents of Acc-A back to original position
00D7	97 10	STA-A	Save new status at 0010
00D9	20 D5	BRA	Branch back and repeat (00B0)

Table 7.17 *(cont.)*

Address	Instruction	Mnemonic	Remarks
Subroutine	*Program*		
00DB	CE 2020	LDX#	Delay, to cause lamp to flash
00DE	09	DEX	
00DF	26 FD	BNE	
00E1	39	RTS	

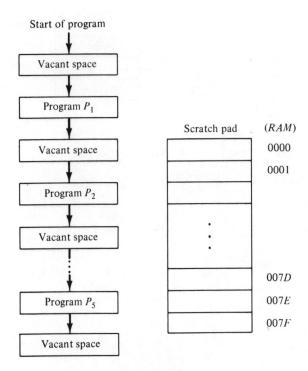

Figure 7.30

be loaded into the RAM section via a CRT. When the process starts, the processor will load the X register from the location specified as shown in Figure 7.31.

The program being executed need not consist of all five parts; how the parts are executed depends upon how the scratch pad is loaded.

Example 11. Interrupt lines C_{A1} and C_{A2} are both being used as interrupt lines. If an interrupt occurs, which line caused the interrupt? Also, the correct interrupt program must be executed for line C_{A1} or C_{A2}. Further-

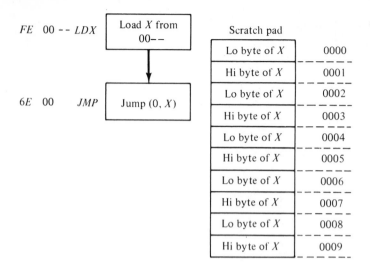

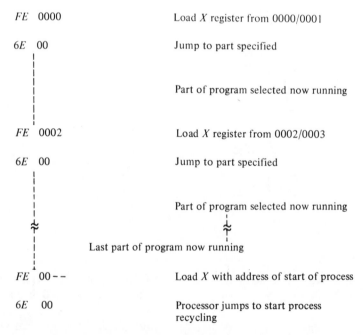

Figure 7.31 Block Diagram of the Program to Be Placed in Each Vacant Space

more, if both interrupts are raised, then a program that deals with this special case must be executed. The interrupt flags must be cleared early in the interrupt program in case another interrupt comes in while executing the original interrupt. Assume the interrupt vector is 0230, and that port A is at address 8004/8005 (see Figure 7.32).

The interrupt programs for interrupt C_{A1} and C_{A2} combined or C_{A1} or C_{A2} individually are located within the range of the branch instruction. All the interrupt programs must end with RTI (return from interrupt), 3B.

If the interrupt programs are so big that the branches from 023C and 023F are outside the range of the branch instruction the following routine may be used. Assume that the subprogram to be executed is located 300_{10} bytes from the branch instruction location (023A); then the BEQ instruction cannot be used by itself and a jump instruction must be used as well, as Figure 7.33 illustrates.

Jump instructions may be used, as they are *nonreturning*-type instructions and store nothing on the stack for later use. If a subroutine call had been used, then return information would have been stored on the stack and the possibility of the stack pointer getting out of position now exists.

This example has illustrated interrupt polling; if four interrupt lines had been used, a similar program would have been written.

Problems

7-1. Modify Example 3 so that if the character contained in Acc-A is *not* in the table the program will cycle *once* and then indicate an error by setting bit 4 of address 0005.

7-2. Two motors, P_1 and P_2, are to operate as main and auxiliary. P_1 is the main pump; if contact X closes, P_2 will cut in and P_1 will remain running. When X opens, P_2 will stop and P_1 will remain running. A block diagram and state diagram should be used to illustrate the problem. The I/O ports must be initialized since the processor is assumed to start from a reset condition. The interrupt line C_{A1}, pin 40, will be used by the switch X to interrupt the processor, as it is doing other things as well as looking after the two motors. The control philosophy to control the motors will be as follows:

1. Initially, when the processor is starting up, *none* of the motors will be operating. Motor P_1 will be started by operator intervention via push button Y.
2. Once the motor is running the processor will flash a lamp to inform the operator that P_1 is operating.
3. Initially, motor 1 will operate, the interrupt will activate a new program, motor 2 will come on, motor 1 will continue operating,

	Instruction		Block diagram
Address	Op-code	Mnemonic	Interrupt detected

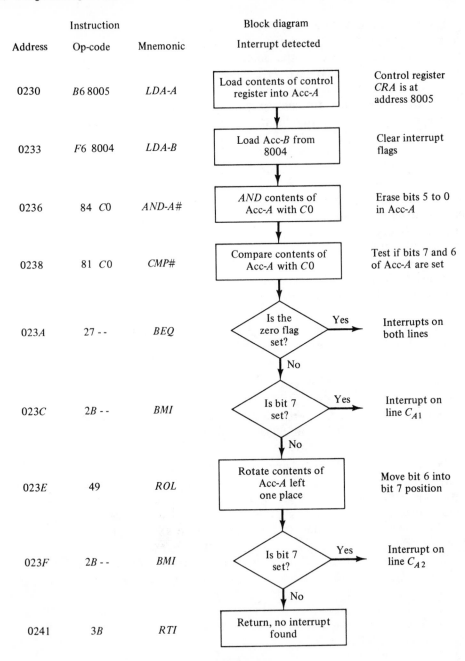

0230	B6 8005	LDA-A	Load contents of control register into Acc-A	Control register CRA is at address 8005
0233	F6 8004	LDA-B	Load Acc-B from 8004	Clear interrupt flags
0236	84 C0	AND-A#	AND contents of Acc-A with C0	Erase bits 5 to 0 in Acc-A
0238	81 C0	CMP#	Compare contents of Acc-A with C0	Test if bits 7 and 6 of Acc-A are set
023A	27 - -	BEQ	Is the zero flag set? — Yes	Interrupts on both lines
023C	2B - -	BMI	Is bit 7 set? — Yes	Interrupt on line C_{A1}
023E	49	ROL	Rotate contents of Acc-A left one place	Move bit 6 into bit 7 position
023F	2B - -	BMI	Is bit 7 set? — Yes	Interrupt on line C_{A2}
0241	3B	RTI	Return, no interrupt found	

Figure 7.32

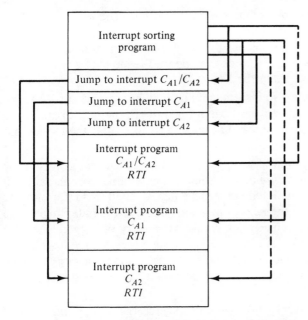

Figure 7.33 Interrupt Program

and a new interrupt vector will be set up. When the interrupt occurs again, motor 2 will stop but motor 1 will continue to operate. The cycle repeats each time the X contact closes.

4. Interrupt vector is located at A000/A001; this vector must be located under program control. Use positive transition interrupts.

a. Write the program for the preceding starting at memory location 0100; interrupt vector when P_1 is operating 0400. Reset the stack pointer at each interrupt.

b. Repeat the problem using a memory cell in lieu of repositioning the stack pointer.

c. If motor P_1 fails in operation or fails to start, then start P_2 and sound an alarm.

d. If motor P_2 fails when called upon to operate, sound an alarm.

7-3. Modify Example 7 so that when an interrupt occurs the interrupt program will set bit 1 of memory location 0005. Thus the LED flashing program will check this bit at this location. If the bit is set, the program will go to memory location 0253; otherwise, it will go to 025E.

7-4. a. A microprocessor, which is monitoring a process, is connected via a 20-mA loop to a CRT. Write a program to transmit the words BEARING OIL LOW to the CRT. Return the cursor to the left-hand margin.

b. Transmit the preceding message and flash the letters LOW, as well sound an audible alarm.

c. Repeat b part and silence the audible alarm, as well as stop flashing the letters LOW when the operator responds by typing in the letters ACK. Echo the letters ACK.

7-5. Modify Example 8 so that the next limit switch to operate is interrogated twice, once before the motion starts in the new direction, as the example illustrates, and again a short time after the motion has started.

7-6. A program is required to process a list that is stored in RAM; the items in the list have been erratically located due to constant addition and deletions. Each entry has a tag attached to it informing the processor where the next item is located. Use the index register to chain through the list. How would each entry be tagged? Assume the items of the list are 1 byte long. Write a program to illustrate the technique used to load this item into Acc-A and then step on to the next item. What the processor does with data once loaded in Acc-A does not matter.

7-7. Design an electronic data switch to switch the output and input of an ACIA from one device to an alternative device. The switch is to be controlled by the same microprocessor that controls the ACIA.

7-8. Expand Example 1 to seven input lines, and if the number on the data lines exceeds 99, sound an alarm.

7-9. A motor is controlled by two switches, A and B. Start the motor when A is closed *and* B is open, *or* when A is open *and* B is closed. Otherwise, the motor is stopped.

7-10. In a manufacturing process, objects on a conveyer belt continually pass through a tunnel where some process is performed. The objects are counted going in and again coming out.

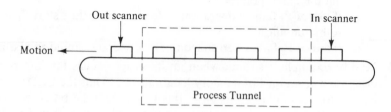

The output from the scanners are positive pulses and normally occur as shown:

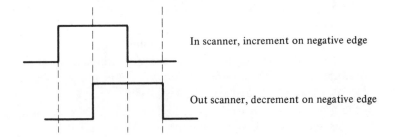

In scanner, increment on negative edge

Out scanner, decrement on negative edge

Use the in scanner pulse to increment a counter and the out scanner pulse to decrement the counter. If the count of objects in the tunnel exceeds four, sound the alarm; something is going wrong.

a. Use the timer module as the counter.

b. Use a memory location as the counter.

7-11. Design the interface and write a program to count the number of objects passing a point. When the number equals a preset demand, output a positive pulse of 20-ms duration via line 0 of port A. Display the accumulative count in decimal form; reset the counter each time the positive pulse is transmitted.

7-12. Add two 16-bit numbers together and store the result.

7-13. Interrupt lines C_{A1}, C_{A2}, C_{B1}, and C_{B2} are all utilized for interrupts. Design a flow chart and then write a program to achieve the following priority.

Top priority $\longrightarrow C_{A2}$
$\phantom{\text{Top priority} \longrightarrow} C_{B2}$
$\phantom{\text{Top priority} \longrightarrow} C_{A1}$
Lowest priority $\longrightarrow C_{B1}$

If more than one line is simultaneously indicating an interrupt, treat the interrupts individually but in the priority shown.

7-14. Incorporate the checks for parity error, framing error, and overrun error into Example 4.

7-15. Increase the range of the program in Example 1 to cover numbers up to and including 999_{10}.

7-16. a. Draw a flow chart and write the program to record how long a contact has been closed. Use the timer module set for 0.1 s intervals. Store the accumulated count in hexadecimal notation in memory locations 0001/0002. The elapsed time will never be greater than 6502.5 s. What is significant about this number?

b. Store the accumulated count in BCD code and do not allow the count to exceed 999.9_{10}.

7-17. Write a program to interrogate the accumulated count in Problem 7.16 and when the count reaches a predetermined number, stop counting; convert the number to decimal if it is in hex code and display on seven-segment LED displays.

7-18. Write a program to search a specified area of RAM for a particular number sequence. If the sequence is found, indicate the fact by turning on an LED and place the address of the start of the number sequence in Acc-A. For example, data are stored in an area of RAM from address 0080 to 00BF inclusive; the data were read in and stored previously. Search this area for the hexadecimal number sequence ABCD. If the sequence cannot be found, turn on an alternative LED. This program is to be used as a subroutine program.

7-19. Write a program to stack specified data into consecutive sections of RAM memory (i.e., to form a queue; first in, first out). Data are read from the queue and processed; the flow of data into and out of the queue occurs in a random manner. If the queue is full, light an LED and accept no new data. If the queue is empty, light an alternative LED and do not try to read data from the queue as it is erroneous. The queue starts at address 0080 and is full when address 00BF is reached. The number of characters in the queue depends upon the number read in minus the number processed. The actual number in the queue varies from 0 when the queue is empty to 64_{10} when the queue is full.

Hint: The position of the head of the queue is always known; it is 0080. The position of the tail end of the queue is *not* known, unless the queue is full; therefore, a register or memory location must remember where the end of the queue is at all times.

7-20. Modify the program of Problem 7.18 to search *only* the characters actually in the queue, *not* the whole 64_{10} bytes.

7-21. Rework the following problems to operate on a 6805E2 controlled microprocessor system:
Examples 3 and 5 to 9
Problems 7.1 to 7.12 and 7.15 to 7.20

Chapter 8

Programmable Controllers

In previous chapters it was shown that motors can be controlled with logic circuits which were derived from the state diagram, or with microprocessors and the use of flow charts. This requires a good understanding of either logic circuits and Boolean algebra, or of programming techniques, using an op-code, assembly language or a high level language. In addition to this the designer has to solve the problems of interfacing, that is how to connect the logic circuit or microprocessor to the actual motor starter or limit switch. This problem has been reduced considerably with the introduction of programmable controllers, which are solid-state devices specifically designed to replace relays and hard-wired logic in their function of motor control. They come complete with interfacing solid-state relays, eliminating interface problems. Programs can be entered or changed without the need for learning a special op-code. The programmer only needs to have a good understanding of relay type logic and its design as developed in Chapter 2, since relay symbols are used. The state diagram can still be used for combinational and sequential circuits. The controller discussed in this and the following chapter is the Modicon 484, equipped with 1K of RAM and the enhanced capabilities level 2. The 470 adapter is used for microprocessor interface. For exact keyboard operation and hardware description the reader is referred to the 484 manual.

A typical programmable controller can be divided into 4 parts as shown in Figure 8.1.

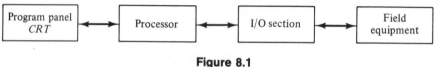

Figure 8.1

8.1 PROCESSOR

The processor is capable of arithmetic and logic functions. It can also store and handle data as well as continuously monitor the status of all its input signals; based on this information it controls specified output signals. The processor is programmed from a program panel (CRT) with a special keyboard (see Figure 8.2).

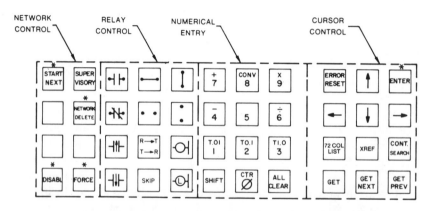

Figure 8.2 P180 Keyboard Layout (Gould Modicon Division)

The basic element of programming is the relay contact, either normally open or normally closed, which when connected in parallel or series forms a ladder diagram (see Figure 8.3). This diagram allows for up to 10 elements in each horizontal rung and up to 7 rungs per network. Within a network, power flow will be allowed only from left to right or vertically (up and down); it is never allowed to flow toward the left.

Once a program has been entered, it can be monitored and modified from the CRT. The former is done by illuminating the "current" path brighter than the remainder of the circuit, and the latter is done by placing the cursor on the device to be altered and then making the change from the keyboard. Contacts can be bypassed and output coils can be turned on or off.

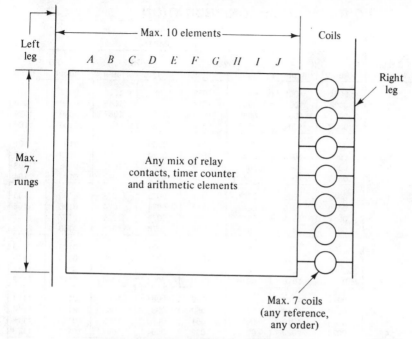

Figure 8.3 (*Courtesy Gould Modicon*)

8.2 INPUT/OUTPUT

The input and output modules are optically coupled solid-state devices. They are used to prevent any transients on the field wiring (110 or 220 V ac) from affecting the internal logic. Each I/O module has its own distinct reference number or address, which is determined by a set of switches at the top of the I/O housing (see Figure 8.4).

All input signals, that is, all push buttons and limit switches, must have a number corresponding to the number of the module to which they are connected. The same holds true for all output signals, which are connected to external relays or starters.

8.3 PROGRAMMING

If a normally open (N.O.) external button is connected to input module 1001, it will be entered on the CRT as follows:

$$\dashv\vdash \atop 1001 \qquad\qquad (8.1)$$

I/O REFERENCE CONFIGURATION

Module Number (Top to Bottom)	Circuit Number	CHANNEL ONE HOUSING NUMBER							
		ONE		TWO		THREE		FOUR	
		Output	Input	Output	Input	Output	Input	Output	Input
1	1	0001	1001	0033	1033	0065	1065	0097	1097
	2	0002	1002	0034	1034	0066	1066	0098	1098
	3	0003	1003	0035	1035	0067	1067	0099	1099
	4	0004	1004	0036	1036	0068	1068	0100	1100
2	1	0005	1005	0037	1037	0069	1069	0101	1101
	2	0006	1006	0038	1038	0070	1070	0102	1102
	3	0007	1007	0039	1039	0071	1071	0103	1103
	4	0008	1008	0040	1040	0072	1072	0104	1104
3	1	0009	1009	0041	1041	0073	1073	0105	1105
	2	0010	1010	0042	1042	0074	1074	0106	1106
	3	0011	1011	0043	1043	0075	1075	0107	1107
	4	0012	1012	0044	1044	0076	1076	0108	1108
4	1	0013	1013	0045	1045	0077	1077	0109	1109
	2	0014	1014	0046	1046	0078	1078	0110	1110
	3	0015	1015	0047	1047	0079	1079	0111	1111
	4	0016	1016	0048	1048	0080	1080	0112	1112
5	1	0017	1017	0049	1049	0081	1081	0113	1113
	2	0018	1018	0050	1050	0082	1082	0114	1114
	3	0019	1019	0051	1051	0083	1083	0115	1115
	4	0020	1020	0052	1052	0084	1084	0116	1116
6	1	0021	1021	0053	1053	0085	1085	0117	1117
	2	0022	1022	0054	1054	0086	1086	0118	1118
	3	0023	1023	0055	1055	0087	1087	0119	1119
	4	0024	1024	0056	1056	0088	1088	0120	1120
7	1	0025	1025	0057	1057	0089	1089	0121	1121
	2	0026	1026	0058	1058	0090	1090	0122	1122
	3	0027	1027	0059	1059	0091	1091	0123	1123
	4	0028	1028	0060	1060	0092	1092	0124	1124
8	1	0029	1029	0061	1061	0093	1093	0125	1125
	2	0030	1030	0062	1062	0094	1094	0126	1126
	3	0031	1031	0063	1063	0095	1095	0127	1127
	4	0032	1032	0064	1064	0096	1096	0128	1128

Figure 8.4   I/O Reference Configuration (*Courtesy Gould Modicon*)

The signal (A) can be inverted ($\overline{A}$) internally without the use of an external normally closed (N.C.) contact or auxiliary relay, as would be the case if relays were used; simply program it in as a closed contact:

$$\frac{\text{─┤↑├─}}{1001}$$

(8.2)

If, on the other hand, an N.C. external button is used, it will *not* be entered on the CRT as closed, since it must produce the opposite effect internally from that of an N.O. button, and inverting the external button as well

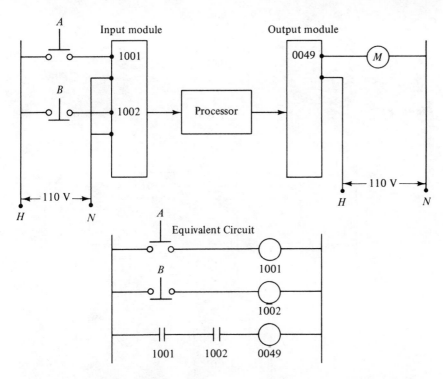

Figure 8.5 Coil 1001 and its contact 1001 represent push button A, while coil
1002 and its contact represent push button B. Press button A and
coil 1001 is energized, closing contact 1001. This in turn ener-
gizes output coil 0049, because contact 1002 is already closed.

as its internal signal constitutes a double inversion. It is for this reason
that all *normally closed external buttons are programmed as normally
open* on the CRT (see Figure 8.5).

Example 1.   Design a stop-start motor control circuit with the start button
(N.O.) connected to I/O module 1001 and the stop button (N.C.) connected to module 1002. The output signal is module 0049, which is connected to a motor starter. This circuit will be programmed on the CRT as
shown in Figure 8.6. It is suggested that the programmer underline all
N.C. external contacts to avoid confusion with N.O. contacts.

8.4 TIMERS AND COUNTERS

Timers or time-delay signals can be used anywhere in the program. They
require two elements, one to store the preset time and one to keep track of

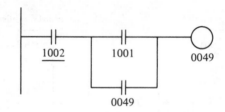

Figure 8.6

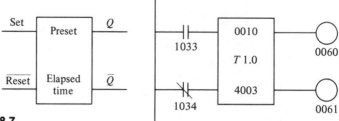

Figure 8.7

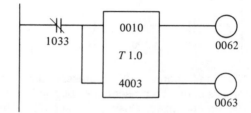

Figure 8.8

the elapsed time. The preset time is a fixed value of four digits (0001 to 0999), which represents up to 999 seconds (timer in seconds). The timer can never exceed this preset value (see Figure 8.7). Coil 0060 will be energized and coil 0061 deenergized if N.O. button 1033 has been closed for a total of 10 s, *provided* that contact 1034 remains closed. The elapsed time is stored in register 4003 and can be displayed on the CRT at any time. Up to six references, in any order, can be displayed. Addresses for holding registers are from 4001 to 4126 (see Figure 8.8). When external button 1033 is released, N.C. contact 1033 is closed, and 10 s later coil 0062 is energized while coil 0063 is deenergized. This circuit represents an off delay energizing and off delay deenergizing timer.

Counters operate on the same principle as timers except that the counter increments its current count by 1 every time that the set input goes high. Output Q is high whenever the count equals the preset value; otherwise, $Q = 0$ (see Figure 8.9). Output Q goes high and coil 0062 will be energized after N.O. button 1033 has closed 10 times. The actual count is stored in register 4004 and can be displayed on the CRT.

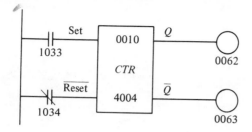

Figure 8.9

8.5 SEQUENCERS

Eight distinct sequencer circuits with up to 32 steps are available. These sequencers can be compared to the old stepper relay where every pulse advances the relay one step. Register addresses for these sequencers are from 4051 to 4058 and they are coded in the following manner: sequencer 1 has register 4051 with contacts 2101 to 2132; sequencer 8 has register 4058 with contacts 2801 to 2832. See Figure 8.10.

Sequencer 5 is shown with only six steps. Every time the input 1040 is energized the count in register 4055 is incremented by 1, and the corresponding output coil is energized. When the count equals 6, the sequencer

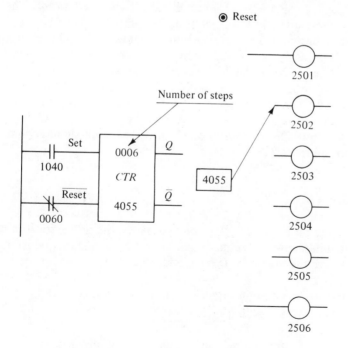

Figure 8.10 Equivalent Sequencer Circuit

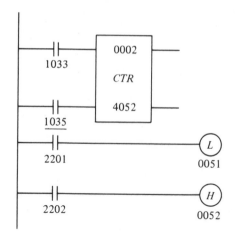

Figure 8.11

stops and the output coil 2506 remains energized until coil 0060 is energized; this will reset the counter and all outputs are deenergized. These output references can be used anywhere in the logic circuit in the same way as relay contacts.

Example 2. Design a control circuit for a two-speed motor. The motor can only be started in low speed; use sequencer 2 (see Figure 8.11). If N.O. start button 1033 is pressed once, contact 2201 closes, energizing coil 0051; pressing button 1033 again opens contact 2201 and closes contact 2202, energizing the high-speed contacter. The motor cannot be switched from high to low speed. N.C. stop button 1035 will reset the counter to 0 when actuated. This will stop the motor. What is the effect of depressing button 1033 three or more times in succession? This has no effect on coil 0052; it remains energized since the maximum count cannot exceed the preset value of 2.

8.6 ARITHMETIC FUNCTION

In addition to the various relay configurations, the processor is capable of data manipulations. The data can be stored in the various registers or addresses (see Figure 8.12). Typical addresses are

3001 to 3XXX, input registers (not used in this text)
4001 to 4126, holding registers

The maximum number that can be stored in any of these registers is 999. Refer to Figures 8.13 to 8.16 for diagrams of data manipulations.

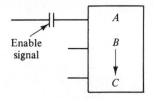

A can be data or a register (content of a register)

B can be data or a register

C is a holding register that contains the result of the operation

If the enable signal is high, the operation occurs at every scan of the processor.

Figure 8.12

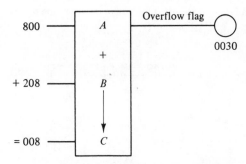

If an overflow occurs, output 0030 is energized and only the overflow is stored in register C (800 + 208 − 1000).

Figure 8.13 Addition

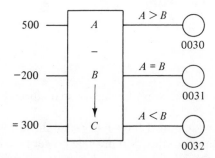

The result is stored in register C and output 0030 is energized (A > B). Only absolute numbers are used; e.g. 5 − 7 = 2 and 7 − 5 = 2.

Figure 8.14 Subtraction

Note: No output node can be energized unless the input node receives power.

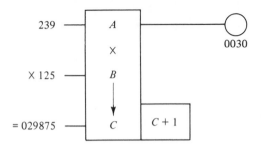

The answer is stored in two consecutive registers, C and $C + 1$.
Figure 8.15 Multiplication   The output 0030 is energized everytime the operation occurs.

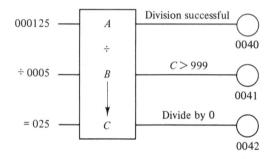

The contents of registers A and $A + 1$ are divided by 5. The result
is placed in register C. Coil 0040 is energized indicating that the
Figure 8.16 Division   division is successful.

8.7 MOVE

Data can be transferred between registers as shown in Figures 8.17 to
8.19.

The upper output will provide power flow every scan the input is enabled (supplying power). The lower output will provide power flow only if
a move is attempted (input enabled) with an illegal pointer value. Use of
either output is optional and depends upon the application requirements.

8.8 CONVERSION

This function is used to read BCD code in the case of digital meters or
straight binary numbers. It can also function as an output signal in the
case of a D/A converter, or an LED display (see Figure 8.20). When pro-

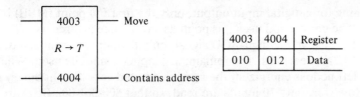

Content of register 4003 is moved to register 4012
(content of register 4004 has been used as a pointer).

Figure 8.17 Register to Table (table refers to the content of a register)

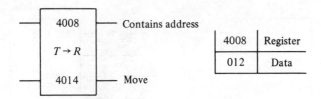

Content of register 4012 is moved to register 4014
(content of register 4008 has been used as a pointer).   **Figure 8.18** Table to Register

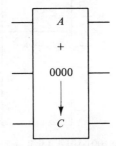

The content of register A is added to zero
and stored in register C.   **Figure 8.19** Register to Register Direct

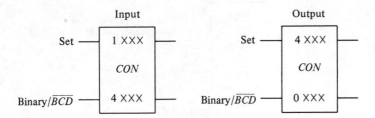

Figure 8.20

gramming for a digital input/output, only the first I/O point (MSB) is indicated. The conversion uses this point as a reference and reserves the next 11. In the case of a 2-digit BCD signal, only 8 inputs are connected, but a total of 12 are reserved. The remaining 4 inputs cannot be used. When the lower left node is energized, the signal is treated as a straight binary number. However, only 10 inputs are read, so that 800-801-802 BCD are read as 512 (see Figure 8.21).

It is possible to change the content of any register by means of the previously discussed arithmetic operations, whether these registers are used for sequencers, timers, or counters (see Figure 8.22). Signal 1033 increments the counter by 1, while signal 1036 increments the same counter

Input modules

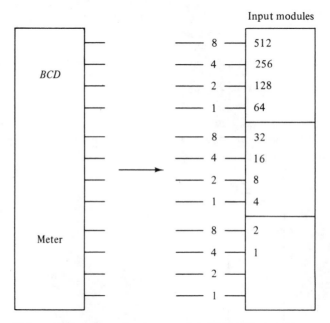

If the meter is connected to input module 1001-1012, it will be programmed as follows:

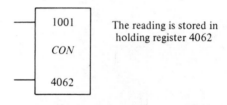

The reading is stored in holding register 4062

Figure 8.21

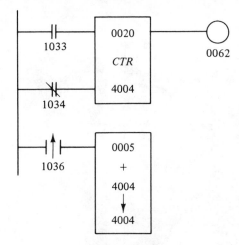

Figure 8.22

by 5. A transitional contact is used; otherwise, while signal 1036 is present, 5 will be added to the contents of register 4004 for every scan of the processor. A transitional contact will only allow one output pulse each time the push button is actuated.

8.9 POWER FAILURE

Any logic coil can be latched (——(L)—); that is, it will assume the same state after a power failure as before. It must be appreciated that any power failure, whether at the controller itself or in the field or at both locations, will cause all output signals to turn off. If no N.C. external buttons are used in the circuit, then the latching relay will operate as described. A regular stop/start motor circuit, however, will not be energized after a power failure, because the stop button is normally closed. The same holds true when using an N.C. button to reset a counter; it will cause the register to be reset to zero at a power failure (see Figure 8.23).

8.10 SKIP AND BYPASS I/O

The controller can be programmed to skip one or more consecutive networks. It is the only nonrelay function that uses only one element for programming. Into this element is placed the number of networks to be skipped, where a 1 indicates the present network. Reference can also be made to the content of a register. The status of the coils controlled by

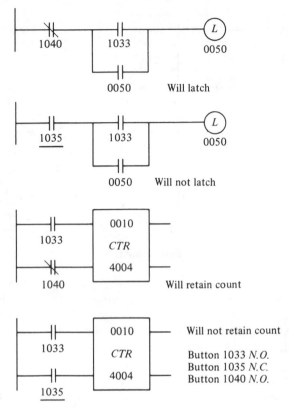

Figure 8.23

these networks will not change while they are being skipped; but the content of holding registers will be reset to zero or set to the preset value depending upon the application.

Register 4059 can be used to bypass input as well as output signals. This function is selected by disabling the last input available (1128) to ON and placing a number between 1 and 32 in this register.

8.11 TUTORIAL EXAMPLES

Example 1. Design a pulse train with 2-s intervals (see Figure 8.24). When the timer reaches 2 s, coil 0033 is energized, opening N.C. contact 0033. This resets the timer and contact 0033 is closed again.

Example 2. Design a circuit that flashes two lights alternately 3 s on and 3 s off (see Figure 8.25). If coil 0068 is deenergized, then timer 1 will ener-

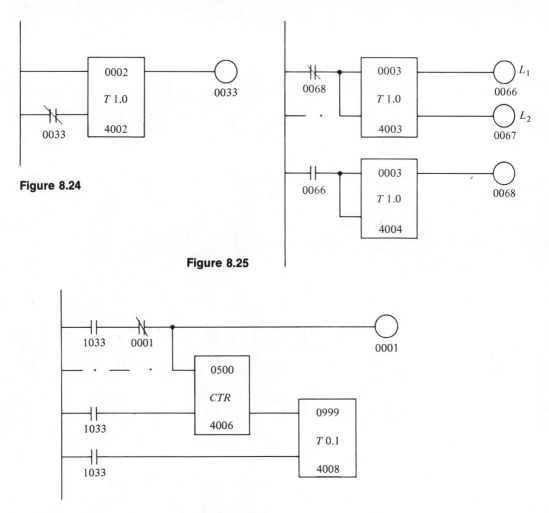

Figure 8.24

Figure 8.25

Figure 8.26

gize coil 0066 after 3 s, closing contact 0066. This will start timer 2, which will reset timer 1 after 3 s.

Example 3. Design a circuit that counts the scan rate of the controller (see Figure 8.26). When N.O. button 1033 is pressed, coil 0001 and its N.C. contact 0001 will alternate for every scan of the controller; this will cause the counter to increment its count by 1 for every two scans. When the counter reaches 500 (1000 scans), it will stop the timer. The time it takes for 1000 scans is stored in register 4008. A typical value is 548, that is, 5.48 ms per scan. Release of button 1033 resets the timer and counter.

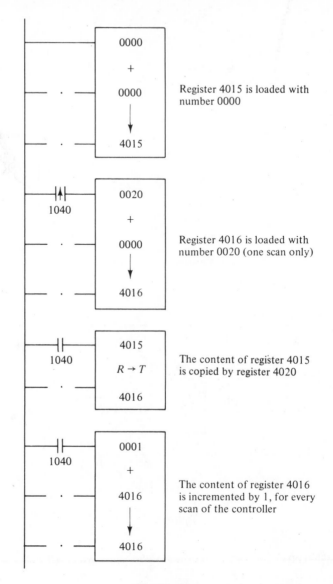

Figure 8.27

Example 4. Reset all holding registers above 4019 to zero (see Figure 8.27). When N.O. button 1040 is pressed, register 4016 is loaded with 20, and the content of register 4015 is moved (copied) to register 4020, after which register 4016 is incremented by 1. The cycle repeats itself.

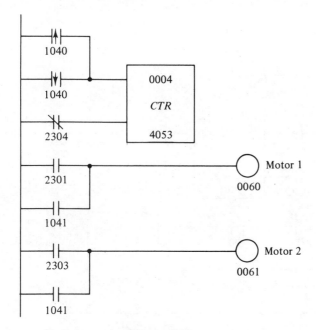

Figure 8.28

Example 5. Refer to Figure 2.39 and implement this circuit (see Figure 8.28). The float switch 1040 activates the sequencer when it makes and when it breaks its contacts. When the count reaches 4, the circuit resets itself. Float switch 1041 will start both motors.

Example 6. Design a control circuit for a motor-generator set that conforms to the following specifications:

1. Start the motor and with 10-s intervals connect the loads.
2. Read an ammeter at 10-s intervals, and if the motor current exceeds 750 A, shed load 3; if after 10 s the load still exceeds 750 A, shed load 2; repeat the above also for load 1.
3. No load is to be picked up again unless the current drops below 700 A, in which case all the load will be picked up at 10-s intervals. The circuit repeats itself. Refer to Figure 8.29.

Example 7. Design a circuit that will take a number of meter readings and store these readings in consecutive holding registers. The data can be displayed on an LED display, as first in first out (FIFO), or queuing (see Figure 8.30). Registers 4066 and 4068 are used as input and output point-

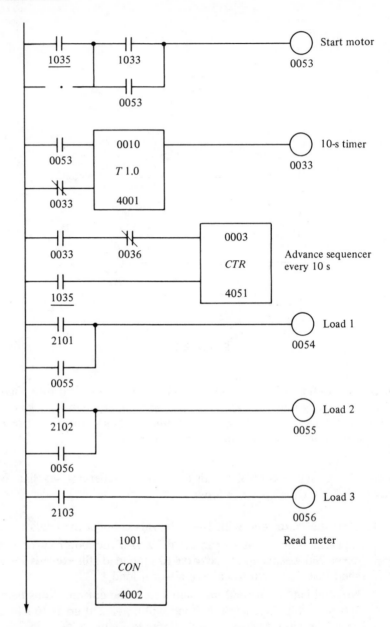

Figure 8.29

ers, respectively, and are initialized by signal 1038 so that both registers are pointing to the same address.

The meter is connected to input module 1001, and the readings are stored beginning with register 4070 and up every time signal 1040 is actuated.

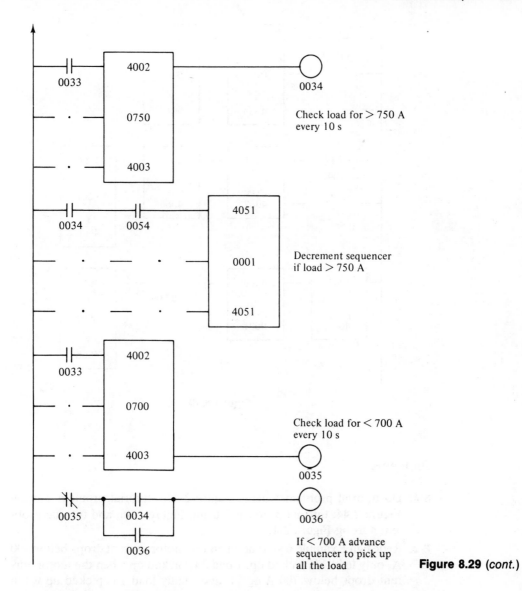

Figure 8.29 (cont.)

The LED display is connected to output module 0021, and the first reading is taken from register 4070; if button 1033 is pressed, then the next reading is displayed.

Example 8. The contents of registers 4007, 4008, and 4009 are 002, 017, and 500, respectively. What would be the result of the operations shown in Figure 8.31?

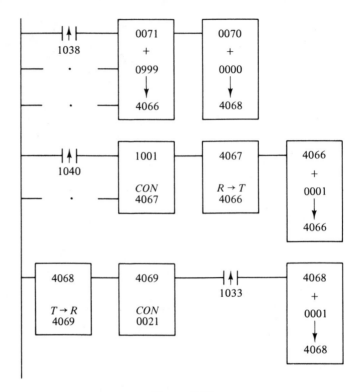

Figure 8.30

Problems

8-1. Do tutorial problem 1 using Figure 2.42; tutorial problem 2 using Figure 2.44; tutorial problem 3 using Figure 2.46; and tutorial problem 4 using Figure 2.47.

8-2. Redesign Example 6 so that when the motor current drops below 700 A, only load 3 is picked up. Load 2 is picked up when the motor current drops below 700 A again, and finally load 1 is picked up when the motor current once more drops below 700 A.

8-3. Design a circuit that will control a stepper motor to move 50 steps forward, wait for 40 s, and then cause the motor to reverse itself 50 steps.

8-4. Design an alarm circuit according to the following specifications:
    1. When a fault occurs, a bell rings and a light flashes at a steady rate. When the operator acknowledges the alarm, the bell is silenced and the light stays on steadily without flashing.

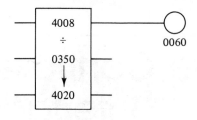

The answer 50 is stored in register 4020 and coil 0060 is energized, provided that the input node is also energized

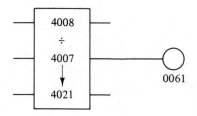

Coil 0061 is energized, because the answer is greater than 999

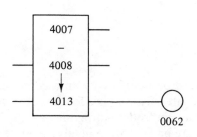

The answer 15 is stored in register 4013, and coil 0062 is energized $(A < B)$

Figure 8.31

2. When the operator presses a reset button, the light goes out if the fault is no longer present.
3. If, however, the fault is still present, the bell rings, the light flashes, and the operations are repeated.

8-5. Modify Example 7 so that when the last reading is displayed the pointers are reset, and if more than 100 readings are stored, an alarm will sound.

Chapter 9

Programmable Controller Interface

The 484 programmable controller can be interfaced with a computer or microprocessor via the J470 EIA adapter. This allows for data storage by a host computer and for changing the controller's program.

This section covers the hardware and software involved in interfacing the 484 controller with an MC 6800 microprocessor (see Figure 9.1). The output of the ACIA 6850 is 5-V TTL logic, while the input to the J470 is RS 232C $\pm$ 15 V. This requires the use of 1488/1489 line driver/receiver chips for proper interfacing. To get a read-out of the messages received, a teletype is connected to the ACIA via a 20-mA loop (Figure 9.1). The characters transmitted to and received from the 484 controller are 8 bit, even parity, 1 stop bit, while the characters sent to the teletype conform to 7-bit ASCII code. This requires a program to change 8-bit data received from the controller to 7-bit data transmitted to the teletype. The sequence of operations is as follows:

The processor sends a message to the controller; the starting address of this message is 0000. The controller responds and the processor stores this response at starting address 0020 before transmitting it to the teletype. The starting address of the main program is C000 (resides in EPROM). A transfer switch, built from 4-7400 NAND gates, is used to direct the data signals either to the Modicon controller or the CRT; a signal from the processor will enable or disable the logic gates (see Table 9.1).

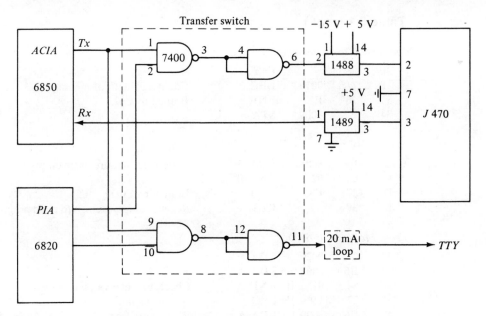

Figure 9.1 MC 6800 and Modicon 484 Interface

Table 9.1

From 6800 to Modicon

C000	86	03	LDA-A#	
02	B7	8008	STA-A	Master reset ACIA
05	CE	FF04	LDX#	
08	FF	8006	STX	Set up PIA for output
0B	86	01	LDA-A#	
0D	B7	8006	STA-A	Connect ACIA to J470 via electronic switch
10	86	59	LDA-A#	Set up ACIA for 8 bit, 1 stop, even parity
12	B7	8008	STA-A	
15	CE	0000	LDX#	Starting address of message
18	96	02	LDA-A	Length of message
1A	80	01	SUB	
1C	97	10	STA-A	Counter, length of message
1E	86	FF	LDA-A#	Make up check sum and store it at 0011
20	97	11	STA-A	
22	A6	00	LDA-A(O,X)	
24	B7	8009	STA-A	Transmit character via ACIA
27	BD	C0D0	JSR	Check for TDRE
2A	9B	11	ADD-A	Add characters and check sum and store the result
2C	97	11	STA-A	

Table 9.1 (*cont.*)

From 6800 to Modicon

2E	08		INX	
2F	7A	0010	DEC	Counter, length of message
32	26	EE	BNE	Branch to C022
34	B7	8009	STA-A	

Subroutine

C0D0	F6	8008	LDA-B	Is transmit data register empty?
D3	C4	02	AND-B	
D5	27	F9	BEQ	Branch to C0D0
D7	39		RTS	Register empty; return to main program

From Modicon to 6800

C037	CE	0020	LDX	
3A	B6	8008	LDA-A	
3D	84	01	AND-A	Check for receive data
3F	27	F9	BEQ	register full
41	B6	8009	LDA-A	Read data from ACIA and check for
44	81	02	CMP-A	02, start of message code
46	26	F2	BNE	
48	A7	00	STA-A(O,X)	Store first character at 0020
4A	C6	02	LDA-B#	Counter, length of message
4C	86	FF	LDA-A#	Temporary length of message
4E	97	22	STA-A	Length of message stored
50	B6	8008	LDA-A	
53	84	01	AND-A	Check for receive data
55	27	F9	BEQ	register full
57	B6	8009	LDA-A	Receive next character
5A	08		INX	
5B	A7	00	STA-A(O,X)	Store next character
5D	5C		INC-B	
5E	D1	22	CMP-B	Compare with length of message
60	23	EE	BLS	Branch to C050
62	01		NOP	

From 6800 to Teletype

C063	86	03	LDA-A#	Master reset ACIA
65	B7	8008	STA-A	
68	86	49	LDA-A#	Set up ACIA for 7-bit ACII code
6A	B7	8008	STA-A	
6D	86	02	LDA-A#	Connect 6800 to teletype via transfer
6F	B7	8006	STA-A	switch
72	0C		CLC	
73	C6	01	LDA-B	
75	D7	12	STA-B	Counter at 0012

Table 9.1 (*cont.*)

From 6800 to Teletype

77	CE	0020	LDX#	Load from storage area
7A	A6	00	LDA-A(O,X)	
7C	84	F0	AND-A	Get one character and rotate 4 times
7D	46-46-46-46		ROR	
82	81	09	CMP-A	Check for a letter or digit
84	22	04	BHI	Branch to C08A if letter
86	8B	30	ADD	Add 30 for correct ASCII code
88	20	02	BRA	Branch to C08C
8A	8B	37	ADD	Add 37 for correct ASCII code
8C	BD	C0D0	JSR	Check for TDRE
8F	B7	8009	STA-A	Transmit the data in Acc-A
92	A6	00	LDA-A(O,X)	
94	84	0F	AND-A	Get one character and check for letter
96	81	09	CMP	or digit
98	22	04	BHI	
9A	8B	30	ADD	
9C	20	02	BRA	
9E	8B	37	ADD	Add 37 to contents of Acc-A
A0	BD	C0D0	JSR	Check for TDRE
A3	B7	8009	STA-A	Transmit the data in Acc-A
A6	86	20	LDA-A#	ASCII code for blank
A8	BD	C0D0	JSR	Check for TDRE
AB	B7	8009	STA-A	Transmit data
AE	08		INX	
AF	7C	0012	INC	Increment counter
B2	D6	22	LDA-B	Length of message
B4	D1	12	CMP	Compare with counter
B6	2C	C2	BGE	Branch to C07A
B8	86	0A	LDA-A#	ASCII code for line feed
BA	BD	C0D0	JSR	Check for TDRE
BD	B7	8009	STA-A	
C0	86	0D	LDA-A#	ASCII code for carriage return
C2	BD	C0D0	JSR	Check for TDRE
C5	B7	8009	STA-A	Transmit the data in Acc-A
C8	3F		SWI	STOP

The program can be shortened by better use of the index mode; this, however, makes the explanation of the exact sequence of operation more complex.

9.1 PROTOCOL

The requests shown in Table 9.2 can be made from the processor; hex code is used in all the examples. Each message must conform to the following format: start of message, request, length of message, data, check sum.

The start of message is always 02, while the length of message indicates the total length, including the check sum. The check sum is determined by adding FF to the arithmetic sum of all the characters; overflows are neglected.

Table 9.2

Request	Code	
Read	1 N	
Write	2 N	
Insert start	5 N	
Insert continue	B N	
Delete general	6 N	where N is a number from 1 to 8 that cor-
Delete last	C N	responds to the number of bytes being
Search	30	requested or stored
Power flow	40	
Stop	80	
Go	90	
Initialize (clears memory)	A0	

Example 1. Illustrate the message format for a stop command. FF is not part of the transmitted message; it is included here only to illustrate how the check sum is determined.

	FF	1111	1111
Start of message	02	0000	0010
Request	80	1000	0000
Length of message	04	0000	0100
Check sum	85	1000	0101

This message halts the processor and turns all output signals off, causing all operating equipment, such as motors, to stop. The processor echoes the message if no errors are found; otherwise, it will send an error message indicating the problem in the following way:

Start of message	02
Not acknowledged	D0

Message length   05
Error code   —
Check sum   —

The error codes are as follows:

Parity/framing	01	Illegal command	06
Overrun	02	Invalid data	0A
Wrong check sum	03	Memory protect on	0B
Wrong address		System not stopped	0C
range	04	Wrong length	0D
Wrong boundary	05	Memory full	11

Under certain circumstances (e.g., wrong start of message), the controller may send the error message before the command or request message has been completely received. This causes a problem if the interfacing between processor and the modicon controller is half-duplex, since the processor is transmitting data, and simultaneously the controller is transmitting data on the *same pair of lines*.

Example 2.   Send a message to start the controller:

Start of message   02
Request   90
Length   04
Check sum   95

This will restart the controller.

Example 3.   Illustrate the Read request. Refer to Figure 9.2 and Table 9.3.

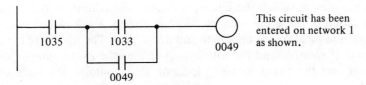

Figure 9.2

The following data are received:

$$00\text{-}00 \mid 0C\text{-}22 \mid 08\text{-}40 \mid 0C\text{-}20 \mid 0D\text{-}30 \mid 08\text{-}40 \mid 9D\text{-}30 \mid 04\text{-}00 \quad (9.1)$$

Table 9.3

Transmit		*Receive*	
Start of message	02	Start of message	02
Request	18 (8 pair)	Request	18
Length of message	06	Length of message	16
High address	00	High address	00
Low address	02	Low address	02
Check sum	21	High data	— } See (9.1)
		Low data	—
		Check sum	29

The high byte of the data is decoded in the following way:

Bit No.	7	6	5	4	3	2	1	0
	x	y	y	y	y	y	z	z

End of column	Element types see Table 9.4	Reference code

Bits 1, 0 reference code:

$\quad\quad\quad$ 00 = input signal $\quad\quad$ 10 = internal coil

$\quad\quad\quad$ 01 = output coil $\quad\quad$ 11 = sequencer steps

$\quad\quad$ Bits 6 ↔ 2 indicate the type of element (e.g., N.O. or N.C. contact).
$\quad\quad$ Bit 7 indicates whether an end of column is used (1) or not used (0). The controller scans the circuit column by column from top to bottom of one page and must encounter an end of column before it can move on to the next column. See Table 9.5. Address 0004 contains an N.O. contact (code 0C), while address 0006 contains a vertical connection that indicates the end of a column. The content of these address locations depends upon the way in which the circuit is drawn; the addresses only indicate the location of these elements within this circuit. Each element requires two address locations, a high byte and a low byte. The high byte indicates the *type of element* and the low byte indicates the *exact number* of this element and the exact memory location that contains the state of the signal.

Example 4

$\quad\quad$ 0C 22 $\quad$ N.O. contact $\quad$ 1035
$\quad\quad$ 0C 20 $\quad$ N.O. contact $\quad$ 1033

Table 9.4 Definition of Elements

Code (Binary)	Code (Decimal)	Significance	Discrete	Register	Constant	None	Other	Active	Inactive
				Z Bit Coding				Approx. Effect on Scan Time (μSec per element)	
00000	0	Start of Network				×		23	23
00001	1	End-of-Logic				×		10	10
00010	2	End-of-Column					×	16	16
00011	3	Normally Open Contact	×					9	9
00100	4	Normally Closed Contact	×					9	9
00101	5	Positive Transitional Contact	×					13	13
00110	6	Negative Transitional Contact	×					13	13
00111	7	Nonretentive Coil	×					10	10
01000	8	Retentive (Latch) Coil	×					10	10
01001	9	Disabled Nonretentive Coil	×					10	10
01010	10	Disabled Retentive (Latch) Coil	×					10	10
01011	11	Horizontal Open				×		9	9
01100	12	Horizontal Shunt				×		9	9
01101	13	Constant—T/C Preset or Arith. Upper			×			25	25
01110	14	Register—T/C or Move Preset or Arith. Upper		×				25	25
01111	15	Counter Storage		×				55	55
10000	16	Timer (Seconds) Storage		×				52	52
10001	17	Timer ($\frac{1}{10}$ Sec) Storage		×				52	52
10010	18	Timer ($\frac{1}{100}$ Sec) Storage		×				52	52
10011	19	Convert					×	35/25 (In/Out)	60/100 (In/Out)
10100	20	Constant Arithmetic Middle			×			20	20
10101	21	Register Arithmetic Middle		×				20	20
10110	22	Arithmetic Lower Storage		×				20	60/270/200 ($\pm/\times/\div$)
10111	23	Null Element				×		5	5
11000	24	Table to Register Move				×		12	12
11001	25	Register to Table Move				×		55	55
11010	26	Not Assigned							
11011	27	Not Assigned							
11100	28	Not Assigned							
11101	29	Not Assigned							
11110	30	Not Assigned							
11111	31	Skip					×	16	16

Table 9.5

Address	Data Received (9.1)		End of Column	Element Type	Z Bit	
	High Byte	Low Byte	7	654 32	10	
			X	YYY YY	ZZ	
00 02	00	00	0	000 00	00	Start of network
00 04	0C	22	0	000 11	00	N.O. contact, input signal
00 06	08	40	0	000 10	00	End of column
00 08	0C	20	0	000 11	00	N.O. contact, input signal
00 0A	0D	30	0	000 11	01	N.O. contact, output coil
00 0C	08	40	0	000 10	00	End of column
00 0E	9D	30	1	001 11	01	End of column, nonretentive coil output
00 10	04	00	0	000 01	00	End of logic

These low-byte numbers are obtained as follows:

$$\text{Contact } 1035: 1035 - 1 = 1034$$
$$34_{10} = 22_{16}, \qquad \text{address} = 2022$$
$$\text{Contact } 1033: 1033 - 1 = 1032$$
$$32_{10} = 20_{16}, \qquad \text{address } 2020$$

The complete decoding of the data is shown in Table 9.6 and Figure 9.3.

Table 9.6

0002	00 00	Start of network
0004	0C 22	N.O. contact 1035, exact address 2022
0006	08 40	End of column, vertical connection rung 6-7
0008	0C 20	N.O. contact 1033, exact address 2020
000A	0D 30	N.O. contact 0049, exact address 1030
000C	08 40	End of column, vertical connection rung 6-7
000E	9D 30	End of column, output coil 0049, exact address 2030
0010	04 00	End of logic

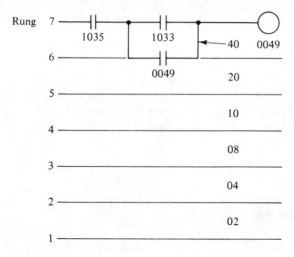

Figure 9.3

If contact 1035 is not followed by a vertical connection and if it is the only contact in that column, then its coding would be $\underline{8}$ C 22 to indicate the end of column.

To obtain the state or condition of these input or output signals, it is necessary to read the exact memory locations.

Example 5. Read the condition of contact 1035; that is, is it on or off? See Table 9.7.

These data are decoded in Table 9.8. Contacts 1035 and 1036 are both high and enabled, as indicated by bits 2 and 3.

Table 9.7

Request		Response	
Start of message	02	02	
Request	11	11	(read 1 pair)
Message length	06	08	
High address	20	20	Address
Low address	22	22	Address
Check sum	5A	44	Data for contact 2022
		44	Data for contact 2023
		E4	

Example 6. Power Flow

This is to determine which part of a circuit is energized. Refer to Table 9.9 and Figure 9.4. Column 1, rung 80 is energized.

Table 9.8

Bit No.	*Previous Scan* 7	6	5	4	3	*Present Scan* 2	1	0
Address	Not used	Contact	Coil	Internal coil	Disable contact	Contact	Coil	Internal coil
2022		1035	0035	0291	1035	1035	0035	0291
2023		1036	0036	0292	1036	1036	0036	0292
Data 44 →	0	1	0	0	0	1	0	0

Add 256

Table 9.9

	Request		*Response*
Start of message	02	02	
Request	40	40	
Length of message	06	06	
High network	00	00	
Low network	01	01	
Check sum	48	—	} See Figure 9.4
		—	
		D3	

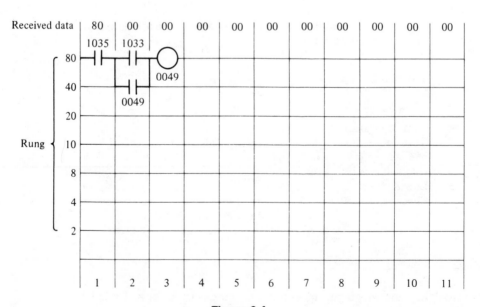

Figure 9.4

When button 1033 is pressed, the following data are obtained:

80   C0   80   00   00   etc.

This indicates that

Rung 80, column 1 is energized
Rungs 80 and 40, column 2 are energized
Rung 80, column 3 is energized

Release of button 1033 results in the following data:

80   40   80

This indicates that

Rung 80, column 1 is energized
Rung 40, column 2 is energized
Rung 80, column 3 is energized

It must be kept in mind that the coil 0049 is not located in column 11 as indicated on the CRT, but in column 3. Its address is 000E and the address of the vertical connector is 000C (see Example 4).

Example 7.   Search (Code 30)

This will indicate where in the circuit a particular contact or coil is located. Refer to Table 9.10 and Figure 9.5.

Table 9.10

Request		*Response*	
Start of message	02	02	
Request	30	30	
Length of message	0A	0A	
Start address	⌈00	00 ⌉ Address of	
for the search	⌊02	0A ⌋ contact	
Data high	0D	0D ⌉ Type of	
Data low	32	32 ⌋ contact	
Mask bit high	80 (bit 7)	00 ⌉ Same page	
Mask bit low	00	01 ⌋	
Check sum	FC	85	

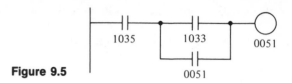

Figure 9.5

Holding contact 0051 (0D32) is located at address 00A0 on the same page as the starting address of the search. The exact description of the element to be found is necessary; otherwise, the response will indicate FFFF (not found). If it is not certain that the element is also at the end of a column, then bit 7 can be masked; this bit will now be ignored during the search.

If the contact is used more than once, the search can be continued with a starting address beyond the one indicated in the response (e.g., 00C0).

Example 8. Write (Code 2N)

This command can be used to change the content of a register or to change a N.O. contact to a N.C. Refer to Table 9.11 and Figure 9.6.

Table 9.11

	Request		*Response*	
Start of message	02	02		
Request	21	21		
Length of message	0A	0A		
Address	00-04	00-04	See Example 3	
Data	10-22	10-22	N.C. contact 1035	
Mask	00-00	00-00		
Check sum	62	62		

Before

After

Figure 9.6

Example 9. Write (Code 2N)

It is possible to change more than one contact, as this example will illustrate. Refer to Table 9.12 and Figure 9.7.

Table 9.12

Request		Response	
Start of message	02	02	
Request	24	24	
Length of message	10	10	
Starting address	00-04	00-04	
Data	10-22	10-22	N.C. contact 1035
Data	08-40	08-40	Vertical connection
Data	10-20	10-20	N.C. contact 1033
Data	0D-30	0D-30	N.O. contact 0049
Mask	00-00	00-00	
Check sum	20	20	

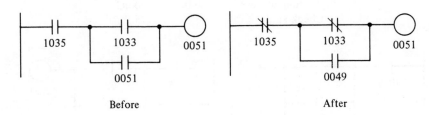

Before   After

Figure 9.7

Example 10. Delete General (Code 6N)

This command removes up to eight elements from a column, except the last element. Code C_ is used for that purpose. See Table 9.13 and Figure 9.8.

Table 9.13

Request		Response	
Start of message	02	02	
Request	61	61	
Length of message	06	06	
Address	00-0A	00-0A	
Check sum	72	72	

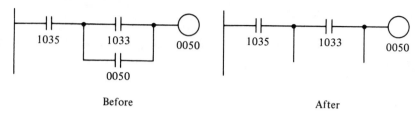

Before

After

Figure 9.8

Example 11. Delete contacts 1033 and 1034. Refer to Table 9.14 and Figure 9.9.

Table 9.14

	Request		*Response*
Start of message	02		02
Request	62		62
Length of message	06		06
Starting address	00-0C		00-0C
Check sum	75		75

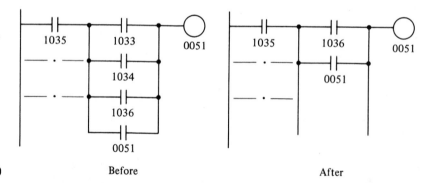

Figure 9.9   Before   After

Delete General will not delete end of column (vertical connection).

Example 12. Delete Last (Code CN)

This message will also delete the last element of a column. Refer to Table 9.15 and Figure 9.10.

The delete last command (code C_) will set bit 7 of the N.O. contact 1033 to indicate that this contact is now at the end of a column. For example, 0C-20 has become 8C-20.

Table 9.15

	Request	Response
Start of message	02	02
Request	C4	C4 (4 pairs)
Length of message	06	06
Starting address	00-0E	00-0E
Check sum	D9	D9

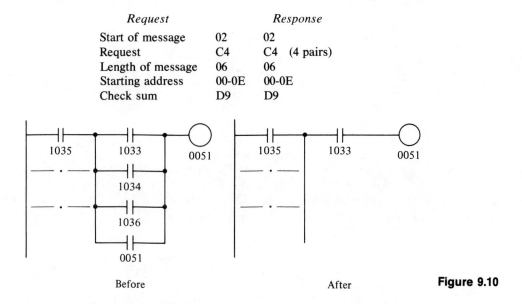

Before   After   **Figure 9.10**

Example 13.   Insert Start (Code 5N)

This example will illustrate how a contact can be inserted in an existing circuit. Refer to Table 9.16 and Figure 9.11.

Table 9.16

	Request	Response
Start of message	02	02
Request	51	51
Length of message	08	08
Address	00-0A	00-0A
Element to be inserted	0D-32	0D-32
Check sum	A3	A3

Before   After   **Figure 9.11**

Example 14. Insert a contact between the vertical connection and the output coil 0051 of Example 13. Refer to Table 9.17 and Figure 9.12.

Code 5N does not change bit 7 of a preceding element.

Table 9.17

	Request		*Response*
Start of message	02	02	
Request	51	51	
Length of message	08	08	
Address	00-0E	00-0E	
Element	8C-21	8C-21	(end of column)
Check sum	15	15	

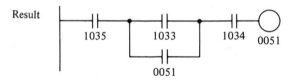

Figure 9.12

Example 15. Insert Continue (Code BN)

This command changes the end of column bit of the preceding element while inserting a contact in an existing circuit. Refer to Table 9.18 and Figure 9.13.

Bit 7 of contact 1035 has been reset by this instruction, and the end of column bit of contact 1033 can be reset by another insert or write command.

Table 9.18

	Request	*Response*
Start of message	02	02
Request	B2	B2
Length of message	0A	0A
Address	00-06	00-06
Elements to	08-40	08-40
be inserted	0D-32	0D-32
Check sum	4A	4A

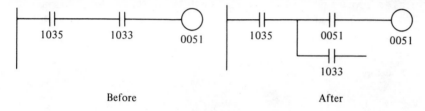

Before   After

Figure 9.13

Example 16. Insert a vertical connection to the right of contact 0051.

Request = Response

Start of message	02
Request	B1
Length of message	08
Address	00-0C
Element	08-40
Check sum	0E

The result is shown in Figure 9.14. Bit 7 of contact 1033 has been reset.

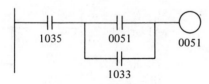

Figure 9.14

Example 17. Insert a new column (see Figure 9.15).

Request = Response

Start	02
Request	B4
Length of Message	0E
Address	00-06
Element 1	08-40
Element 2	0C-20
Element 3	0D-32
Element 4	08-40
Check sum	C4

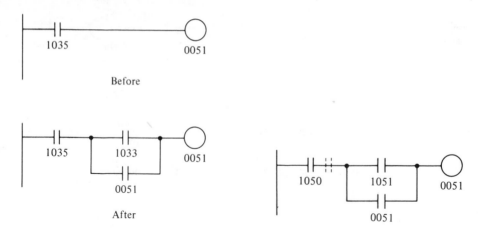

Figure 9.15

Figure 9.16

It is impossible to insert a contact prior to a vertical connection with only one instruction (e.g., open a new column), because an insert command does not set bit 7 of a preceding element (see Figure 9.16).

Example 18.   Insert a contact in the position shown. This request will result in the circuit of Figure 9.17.

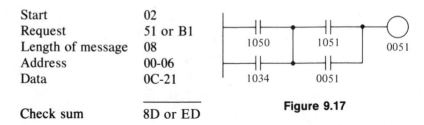

Start	02
Request	51 or B1
Length of message	08
Address	00-06
Data	0C-21
Check sum	8D or ED

Figure 9.17

Holding registers, which are used for counters and arithmetic operations, have as address 4002 to 40FF and are coded according to Table 9.4, in the same way as contacts.

Example 19.   The following data were obtained from a read request: 37-E6; 54-01; DB-11.

	Data	*Code*	*Element*
998	37-E6	13	Constant
÷			
4002	54-01	21	Register, arithmetic middle
4018	DB-11	22	Arithmetic lower storage

Code 13 indicates a constant, and in order to be able to store 998 it is necessary to use bits 1 and 0 (Z bit) as well ($998_{10} = \underline{3}\ E6_{16}$).

$$
\begin{array}{r|c|cc|c|cc}
 & x & \multicolumn{2}{c|}{y} & z & \multicolumn{2}{c}{} \\
37\text{-}E6 = & 0 & 011 & 01 & 11 & 1110 & 0110 \\
 & 0 & \multicolumn{2}{c|}{13_{10}} & \underline{3} & E & 6 \\
54\text{-}01 = & 0 & 101 & 01 & 00 & 0000 & 0001 \\
DB\text{-}11 = & 1 & 101 & 10 & 11 & 0001 & 0001 \\
\end{array}
$$

The Z bit for code 21 is 00 to indicate a holding register, while the Z bit for code 22 has the following significance:

$$00 = \text{add} \qquad 10 = \text{multiply}$$

$$01 = \text{subtract} \qquad 11 = \text{divide}$$

The contents of registers 4002 and 4018 are stored at addresses 4003 and 4013.

Example 20. Change the content of holding register 4062 to 02CB (715_{10}).

Request = Response

Start of message	02
Request	21
Length of message	0A
Address of register	40-3F
Data	02-CB
Mask bit	FA-00
Check sum	72

The mask bit prevents any changes of the high bits. If an attempt is made to place a number greater than 999 in a register, an "invalid" message will flash on the screen.

The special sequencer registers (4051 to 4058) have their contacts coded as shown in Table 9.19.

Example 21. Decode the circuit of Figure 9.18. The result is

N.O. contact 0066 = 0D 41

N.C. contact 2805 = 13 E4

Table 9.19

Sequencers	Contacts	Code
4051	2101-2132	00-1F
4052	2201-2232	20-3F
4053	2301-2332	40-5F
4054	2401-2432	60-7F
4055	2501-2532	80-9F
4056	2601-2632	A0-BF
4057	2701-2732	C0-DF
4058	2801-2832	E0-FF

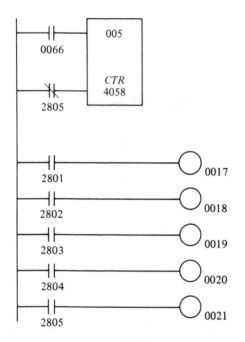

N.O. contact 2801 = 0F E0

N.O. contact 2802 = 0F E1

N.O. contact 2803 = 0F E2

N.O. contact 2804 = 0F E3

N.O. contact 2805 = 8F E4   (end of column)

Constant 0005 = 3405

Counter = 3C 39

Coil 0017 = ID 10

Coil 0018 = ID 11

Coil 0019 = ID 12

Coil 0020 = ID 13

Coil 0021 = 9D 14 end of column

Figure 9.18

Code 19 is used with convert operations. The Z bit is decoded as follows:

00 = source, discrete inputs   10 = destination, discrete coils

01 = source, holding register   11 = destination, holding register

Example 22

1001	*Data*	*Code*	*Element*
Conv	4C-00	19	Convert
4002	4F-01	19	Convert

4003			
Conv	4D-02	19	Convert
1001	CE-00	19	Convert

	X	*Y*	*Z*	
4C-00 =	0	10011	00	0000 0000
4F-01 =	0	10011	11	0000 0001
4D-02 =	0	10011	01	0000 0010
CE-00 =	1	10011	10	0000 0000

9.2 TUTORIAL EXAMPLES

Example 1

Write the necessary code to energize the circuit of Figure 9.19.

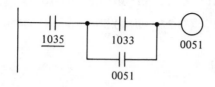

Figure 9.19

Solution

Start of message	02	
Request	21	
Length of message	0A	
Address	20-32	Address of coil 0051
Data	22-00	See Table 9.8
Mask	00-FF	Mask coil 0052
Check sum	9F	

Example 2. Disable the coil in Figure 9.19.

Solution

Start of message	02	
Request	21	
Length of message	0A	
Address	00-0E	
Element	A5-32	
Mask	00-00	
Check sum	11	

Example 3. Insert an SKP 001 command in parallel with the 1035 contact.

Solution

Start of message	02
Request	51
Length of message	08
Address	00-06
Data	7E-01
Check sum	DF

Example 4. Decode the following data, which were obtained from a read request, and draw the circuit.

00 00   0C 20   0C 21   0C 22   0D 30   08 70   10 23   90 24

34 0A   C0 31   10 25   0C 27   08 40   8C 26   8C 27   9D 30

Solution. See Figure 9.20.

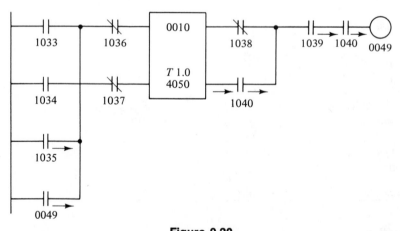

Figure 9.20

Example 5. Refer to Example 4 and indicate which contacts are closed if a power flow request resulted in the following response:

30 00 40 40 80 80 80

Solution. See arrows in Figure 9.20.

Example 6. Read the contents of register 4010 to 4014 inclusive.

Solution

Start of message	02
Request	15
Length of message	06
Address	40-0B
Check sum	67

Example 7. Load holding registers 4120 to 4123 with numbers 120 to 123.

Solution

Start of message	02
Request	24
Length of message	10
Address	40-79
Data	00-78 00-79 00-7A 00-7B
Mask	FA-00
Check sum	CE

Example 8. Insert a contact prior to the vertical connection in Figure 9.16.

Solution

This request must be done in two stages: (1) delete the vertical connection, and (2) insert the contact and the vertical connection.

1.
Start of message	02
Request	C1
Length of message	06
Address	00-06
Check sum	CE

2.
Start of message	02
Request	52
Length of message	0A
Address	00-06
Data	0C-21 08-40
Check sum	D8

Example 9. Interpret the following data, which were received from the 484 controller, regarding contacts 1036 and 1037.

a. 02 11 08 20 23 44 08 A9
b. 02 11 08 20 23 00 4C A9
c. 02 11 08 20 23 44 4C ED
d. 02 11 08 20 23 44 00 A1

Solution. See Table 9.8.

44	08	Contact 1036 on	Contact 1037 disabled off
00	4C	Contact 1036 off	Contact 1037 disabled on
44	4C	Contact 1036 on	Contact 1037 disabled on
44	00	Contact 1036 on	Contact 1037 enabled off

Problems

9-1. Write the necessary code for the following problems:
   1. Read the contents of registers 4062 to 4100.
   2. Add 5 to the content of register 4020.
   3. Add the content of register 4025 to the content of register 4026 and store the result in register 4030.

9-2. Write the code for Figure 9.17.

Chapter 10

Interfacing Power Controllers, Microprocessors, Noise, and Grounding

The final chapter of this text will discuss data-conversion devices and techniques that link the real world of analog signals with the digital world of microprocessors and computers. These devices include not only analog-to-digital (A/D) converters and digital-to-analog (D/A) converters, but also a number of auxiliary components, such as analog multiplexers, digital multiplexers, amplifiers, and filters. Also, the solid-state devices that are replacing the relay and the magnetic motor starter will be discussed, devices such as the SCR (silicon-controlled rectifier), the triac, and the power transistor. These devices are rapidly replacing the relay, contactors, and magnetic starters because they are fast operating, have small size, low power consumption, and are maintenance-free. However, they are affected by the following:

1. High temperature, since silicon deteriorates at junction temperatures above 150°C
2. Electromagnetic interference
3. Overvoltage

Therefore, special precautions must be taken when using electronic-type devices in power circuits.

The SCR, triac, and power transistor are virtually chunks of the crystalline material silicon, and as such they are susceptible to damage due to electrical overloading, temperature, atmospheric pollution, electrical transients, electrical interference, and so on. These devices are currently being used as (power) controllers for motors, and when the power is off to the motor, there is no actual mechanical separation of supply line and motor line, merely a very strong electrical field that inhibits the flow of electrical charge carriers. Thus the failure mechanism is different from that of the relay or motor starter contacts, and must therefore be taken into account in fail-safe environments.

Finally, noise problems and the transmission of electrical signals via an electrically noisy environment will be considered. Noise generation from the standpoint of how it is produced and how problems created by it can be reduced will also be examined.

10.1 DATA ACQUISITION AND CONVERSION

The real world of physical devices and analog electrical signals must be interfaced with the digital world of microprocessors and computers. Computerized feedback control systems and the advent of microprocessors have caused rapid development in this field, and greater productivity per unit of investment has been the result. Thus the microprocessor is appearing in many applications, from the control of the intensity of light sources in work shops and offices to the control of the frost level in a domestic refrigerator, from the automatic control of motors and generators to the programmable logic controller.

Analog data are acquired in digital form for any or all of the following reasons:

1. Storage
2. Transmission
3. Processing
4. Display

Data acquisition can be divided into at least two categories: (1) electrically quiet environments (laboratories, classrooms, etc.) and (2) electrically noisy environments (factories, vehicles, remote installations, etc.). Category 2 includes industrial process systems such as temperature control from tanks, boilers, pipelines, bearings, and the like, and often these devices are spread over a large geographical area, such as steel mills, power stations, or automobile manufacturing plants. Any of the applications mentioned are characterized by the vulnerability of the data signals to electrical noise, off ground voltages, megawatt load changes, and so on. Thus data-acquisition systems may exist in hostile surroundings and may require devices capable of operation under wide temperature range, with large common-mode voltages and large electromagnetic interference signals.

The simplest digital system is a single analog-to-digital converter (ADC) performing repetitive conversions at a free running, internally determined rate. Its outputs are (1) digital code word, (2) status output to indicate when the conversion is complete, and (3) "over range" indication.

One of the best known converters is the digital meter; however, the method used in digital meters has two major shortcomings: (1) it is slow, and (2) its BCD digital coding must be changed into hexadecimal binary form if it is to be processed.

However, industrial data-acquisition systems may be designed to handle a small number of data channels in the case of a small plant, to several thousand channels in a steel works, power station, or petrochemical complex. Basically, all data-acquisition systems share a common form, shown in Figure 10.1.

The electrical signals generated by the physical system parameters are usually analog and are generated by devices called *transducers,* which convert some physical quantity or parameter into electrical signals. These signals may be in the microvolt or millivolt range, and must be processed electronically, as is shown in Figure 10.1. The output from the transducer is fed into an amplifier, which amplifies the signal to a level that is useful for further processing. Also, the amplifier may perform the function of impedance matching, since the transducer output impedance can be vastly different from the input impedance of the filter.

Generally, following the amplifier is a low-pass *active* filter, which reduces high-frequency noise, unwanted electrical interference, and electronic component noise from the actual analog signal.

The analog signal then passes to the *multiplexer,* which is nothing more than a multiposition electronic switch that connects a specified input, or the next sequential input, to its output terminal. The switch will remain in this position until it is instructed to move by the external control line or address lines. During the period that the input signal is connected to the multiplexer output line, the sample-hold circuit will take the signal and hold it at the level that exists at some specified point in time during the sampling period of the multiplexer; the point at which the sample is taken is determined by the sequence controller.

The output from the sample-hold circuit is therefore held steady during the conversion cycle of the A/D converter. The A/D converter converts the analog signal into its equivalent digital form, and then informs the microprocessor that the conversion is completed. The processor then reads in the data and processses them.

The analog multiplexer and sample-hold circuit time share the A/D

TYPICAL DATA ACQUISITION SYSTEM BLOCK DIAGRAM

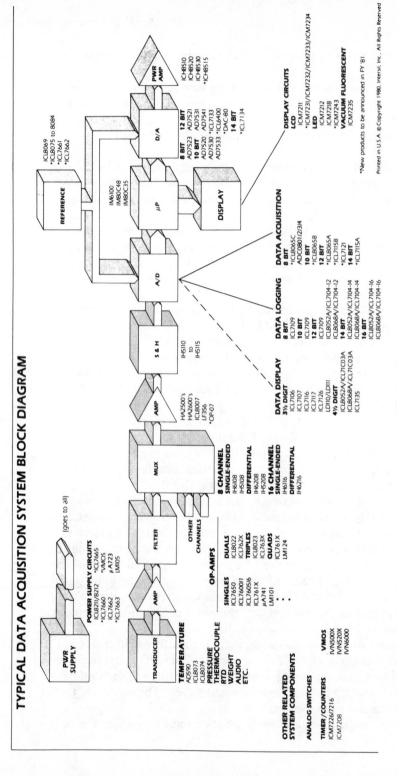

Figure 10.1

converter with the other analog channels coming in; thus timing is very important and this task also falls to the program sequence controller, which could be, in the case of a large system, a dedicated microprocessor. In the case of a smaller system it may well be a group of logic gates and/or flip-flops.

Other methods exist for achieving the task shown in Figure 10.1:

1. Convert the analog signal to digital form adjacent to the transducer, and send the data serially via an ACIA to the processor.
2. Multiplex the signals from the transducer and use only one amplifier in the output of the multiplexer. The gain of the amplifier, in this case, should be variable under program control as the signal level of different transducers may vary widely.

Only method 1 will be considered in addition to the method of Figure 10.1.

The reverse of the data-acquisition system is also required, that is, controlled outputs. These outputs from the microprocessor are in digital form and must be converted into analog form in order to operate the controlled process. The conversion from digital to analog is performed by D/A converters; each converter is connected to the microprocessor via a latched register (see Figure 10.2). Once again timing is very important to connect the correct actuator to the data bus either sequentially or on command. The process actuator will modify the process parameter (i.e., temperature, flow, acceleration, etc.), in accordance with the data read into

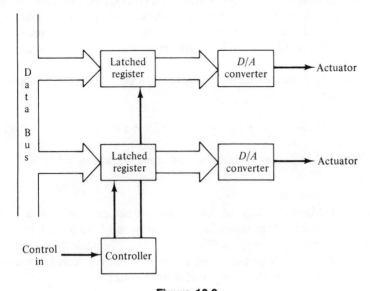

Figure 10.2

the processor via the data-acquisition system. The control loop is therefore closed on the process and the result is a complete automatic process control system under computer or microprocessor control.

10.1.1 Quantizing Theory

Analog-to-digital conversion basically is a two-step process: (1) quantizing, and (2) coding. Quantizing is the process of transforming a continuous analog signal into a set of discrete output states. Coding is the process of assigning a digital code word to each of the output states.

Consider initially an ideal eight-state quantizer; it will have eight output states, one of which is 000, and it will have 3 data bits to represent each output state. In other words, it is a 3-bit A/D converter. The voltage range of this quantizer is $+10$ V, and since there are eight output states the output voltage levels will be

Voltage level	0.00	1.25	2.50	3.75	5.00	6.25	7.50	8.75	(10.1)
Output code	000	001	010	011	100	101	110	111	

The voltage levels are established in the following way: the number of output states $= 2^n$, where $n =$ number of bits; in this case $n = 3$.

$$\therefore \quad 2^n = 2^3 = 8 \tag{10.2}$$

Actually, there are $2^n - 1$ output states, as (10.1) illustrates, since 0.00 is also an output state. The output curve is shown in Figure 10.3. The output voltages (10.1) are the center points of each output and the curve of the output states is a staircase, where the center voltages form the center of the stair tread.

The output curve changes its level at a point halfway between the output voltages; that is,

$$0.625, \ 1.875, \ 3.125, \ 4.375, \ 5.625, \ 6.875, \ \text{and} \ 8.125 \tag{10.3}$$

Q: What will the output code be for an input voltage of 0.622 V?

A: 000

Similarly, for input voltage of 3.14 V the output code will be 011. The staircase is the best approximation that can be made to a straight line as can be seen from Figure 10.3. The sections of the staircase above and below the center voltage curve are in error; however, this error can be reduced by increasing the number of output states.

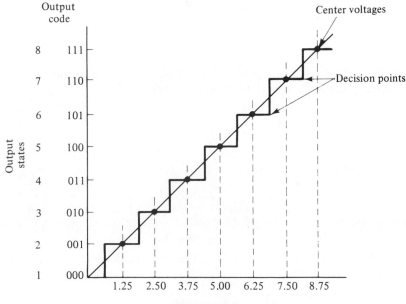

Figure 10.3

Notice that the top of the range, 10 V, does not have a code, since 8.75 V is coded 111. The decision points are precisely set in the quantizer by the manufacturer to divide the input analog voltage into its correct quantized values.

Furthermore, it has been noticed that the output code word is the same for a range of voltages; that is, all the voltages between 0.625 and 1.875 have the same output code: 001.

The voltage range between the decision points is called the *quantization size* or *quantum Q*, where Q is determined by the simple formula

$$Q = \frac{\text{full-scale deflection}}{2^n} \qquad (10.4)$$

In the case of 10 V full-scale deflection and a 3-bit output code,

$$Q = \frac{10}{8} = 1.25 \qquad (10.5)$$

Thus the steps are 1.25 V wide, and the center points are 1.25 V apart. The reader is now aware of the error that exists in the output for this particular converter; to reduce the error a larger code word could be used, say 8 bits:

$$\therefore \quad Q = \frac{10}{2^8} = \frac{10}{256} = 0.039 \tag{10.6}$$

$$= 39 \text{ mV}$$

Thus the output error is greatly reduced by increasing the number of output states. Similarly, if a 12-bit quantizer were used,

$$Q = \frac{10}{2^{12}} = \frac{10}{4096} = 2.44 \text{ mV} \tag{10.7}$$

reducing the error even further.

10.1.2 Coding

The digital codes used by A/D and D/A converters are binary codes in their fractional form; that is,

$$N = a_1 2^{-1} + a_2 2^{-2} + a_3 2^{-3} + \cdots + a_n 2^{-n} \tag{10.8}$$

where a is either 1 or 0 and N is a number between 0 and 1, and

$$2^{-1} = \frac{1}{2} = 0.5, \quad 2^{-2} = \frac{1}{2^2} = 0.25, \quad 2^{-n} = \frac{1}{2^n} \tag{10.9}$$

Consider the binary code word 10010001; this is equivalent to

$$1 \times \frac{1}{2} + 0 \times \frac{1}{2^2} + 0 \times \frac{1}{2^3} + 1 \times \frac{1}{2^4} + 0 \times \frac{1}{2^5} + 0$$

$$\times \frac{1}{2^6} + 0 \times \frac{1}{2^7} + 1 \times \frac{1}{2^8} \tag{10.10}$$

$$= 1 \times 0.5 + 1 \times 0.125 + 1.0.0039625$$
$$= 0.62890625 \quad \text{or} \quad 62.89\% \text{ of the full-range value}$$

If the full-range value is 10 V the code word represents 6.289 V and the

$$\text{LSB} = \frac{10}{2^8} = \frac{10}{256} = 39.625 \approx 40 \text{ mV}$$

The first bit on the left is the most significant bit $\left(2^{-1} = \frac{1}{2}\right)$ and is equal to 0.5 of the full range; the last bit on the right is the least significant bit $\left(2^{-n} = \dfrac{1}{2^n}\right)$ of the full range.

Some A/D converters use other codes, such as binary-coded decimal, 2's complement codes, and offset binary code. The BCD code is used where digital displays are required and 2's complement codes where computer arithmetic operations are required. The offset binary code is used with bipolar analog measurements; that is, the analog range is offset by a half-scale, as Figure 10.4 illustrates.

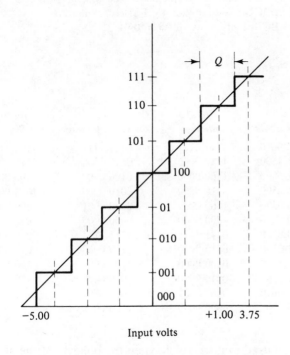

Input volts   **Figure 10.4**

$$\text{Code 000 corresponds to } -5 \text{ V} \qquad (10.11)$$

$$100 \text{ corresponds to } 0 \text{ V}$$

$$111 \text{ corresponds to } +3.75 \text{ V}$$

Tables 10.1 and 10.2 compare the various codes for 8-bit unipolar and 8-bit bipolar converters.

The sign magnitude binary has two code words for 0, as shown:

$$0+ \quad 1000\ 0000 \qquad (10.12)$$

$$0- \quad 0000\ 0000$$

Table 10.1 8-Bit Unipolar

Fraction of Full Scale	+10 Volts Full Scale	Straight Binary	BCD
+FS − LSB	9.99	1111 1111	1001 1001 1001
$\frac{3}{4}$ FS	7.50	1100 0000	0111 0101 0000
$\frac{1}{2}$ FS	5.00	1000 0000	0101 0000 0000
$\frac{1}{4}$ FS	2.50	0100 0000	0010 0101 0000
$\frac{1}{8}$ FS	1.25	0010 0000	0001 0010 0101
LSB	0.01	0000 0001	0000 0000 0001

LSB in BCD is determined by FSD/10^d, where d = number of BCD digits; thus $10/10^3 = 0.01$

Table 10.2 8-Bit Polar

Fraction of Full Scale	±5 Full Scale	Offset Binary	2's Complement	Sign Magnitude Binary
+FS − LSB	+4.9610	1111 1111	0111 1111	1111 1111
+$\frac{3}{4}$ FS	+3.7500	1110 0000	0110 0000	1110 0000
+$\frac{1}{2}$ FS	+2.5000	1100 0000	0100 0000	1100 0000
+$\frac{1}{4}$ FS	+1.2500	1010 0000	0010 0000	1010 0000
+LSB	+0.0390	1000 0001	0000 0001	1000 0001
0	0.0000	1000 0000	0000 0000	1000 0000*
−LSB	−0.0390	0111 1111	1111 1111	0000 0001
−$\frac{1}{4}$ FS	−1.2500	0110 0000	1110 0000	0010 0000
−$\frac{1}{2}$ FS	−2.5000	0100 0000	1100 0000	0100 0000
−$\frac{3}{4}$ FS	−3.7500	0010 0000	1010 0000	0110 0000
−FS + LSB	−4.9610	0000 0001	1000 0001	0111 1111
−FS	−5.000	0000 0000	1000 0000	—

The two codes can be used to distinguish when the polarity of the signal changes sign. Furthermore, because of this the sign magnitude binary code cannot reach either + full scale or − full scale.

The big disadvantage with offset binary and 2's complement code when used in bipolar converters is that *all* the bits change when passing through the zero point. This can be the cause of "glitches" (large spikes) in the output wave form.

10.1.3 Amplifiers

The front end of a data-acquisition system takes the output signal from the transducer; it then amplifies and filters it. The amplifier must perform one or more of the following functions:

1. Amplify the signal to higher levels.
2. Act as a buffer or impedance matching device.
3. Convert a current signal into a voltage signal.
4. Extract any differential signal from the common-mode noise.

The most popular amplifier used to perform these functions is the operational amplifier or op-amp. It can be connected in many different ways to achieve the desired results. Figure 10.5 illustrates a few. The op-amp is used for single-ended signals that are to be converted from current to voltage, voltage to current, amplified, or buffered.

If the signal is differential (no common ground connection), the instrumentation amplifier is a better choice because of its high impedance at the differential inputs, its high common-mode rejection ratio, and its low input bias currents, to state a few reasons. Figure 10.6 illustrates the instrumentation amplifier connected to a transducer with a common-mode signal included.

Common-mode rejection ratio is one of the most important parameters of the differential amplifier, since it is important that the differential amplifier only amplify the voltage difference between its two input terminals. The input signal common to both terminals may be small, but nevertheless it is still there and causes an error in the output.

The common-mode rejection ratio (CMRR), that is, the ability of the amplifier to reject this common signal, is expressed in decibels and is determined by the formula

$$CMRR = 20 \log_{10} \frac{\text{differential voltage gain}}{\text{common-mode voltage gain}} \qquad (10.13)$$

Common-mode rejection is also affected by source resistance and shunt capacitance.

A special amplifier used extensively in data-acquisition systems is the chopper-stabilized amplifier; it is used to accurately amplify microvolt-level signals to the amplitude required.

10.1.4 Filters

In data-acquisition systems, the purpose of filtering is to remove or at least reduce transmitted, inherent, or induced noise. The low-pass filter frequently follows the amplifier to reduce the effects of (1) electrical noise due to electrical components, such as resistors and transistors, and (2) electrical interference (man-made noise) caused by machines starting/stopping and industrial operations.

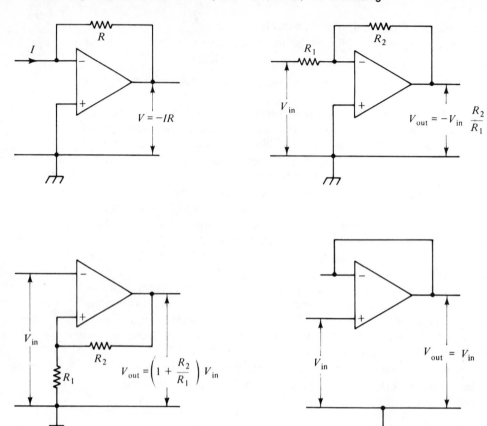

$$V = -IR$$

$$V_{out} = -V_{in}\ \frac{R_2}{R_1}$$

$$V_{out} = \left(1 + \frac{R_2}{R_1}\right) V_{in}$$

$$V_{out} = V_{in}$$

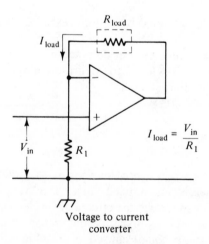

$$I_{load} = \frac{V_{in}}{R_1}$$

Voltage to current
converter

Figure 10.5

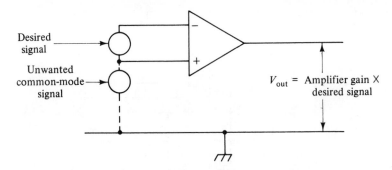

Figure 10.6

Passive *RLC* filters are seldom used because of the undesirable effects of the inductor. Active filters using capacitors, resistors, and op-amps are generally used since they permit very accurate filter characteristics to be established; furthermore, they have very low signal loss due to the insertion of a filter in the system.

10.1.5 Analog-to-Digital Converters

Analog/digital converters basically fall into two groups: (1) the successive-approximation converter, and (2) the integrating dual-slope converter. The successive-approximation type is the most popular since it is capable of high resolution and high speed and is reasonable in cost. Its weakness is that it cannot tolerate high rates of change of input during the conversion process. To overcome this difficulty a sample-hold circuit can be fitted to its front end. Thus the signal is held constant during the conversion process.

The successive-approximation converter is a feedback type of device and is based upon the principle of a laboratory scale using standard weights, that is, 0.5, 0.25, 0.125, and so on. Initially, the 0.5 weight is put in the scale pan. If it is too heavy, it is removed and the next smaller weight (0.25) is used; if it is too light, the next smaller weight is added. The weights are continually added and removed until balance is achieved or until the smallest weight is used. After a decision is made as to whether the smallest weight is to be added or deleted, the total of the standard weights is determined; this gives the closest approximation to the unknown. Figure 10.7 illustrates the technique.

The successive-approximation register (SAR) controls the A/D converter in the following manner. The most significant bit (0.5) is initially turned on, and a decision is made by the comparator as to whether the bit is to be left on or turned off. That is, if the current flowing into the D/A

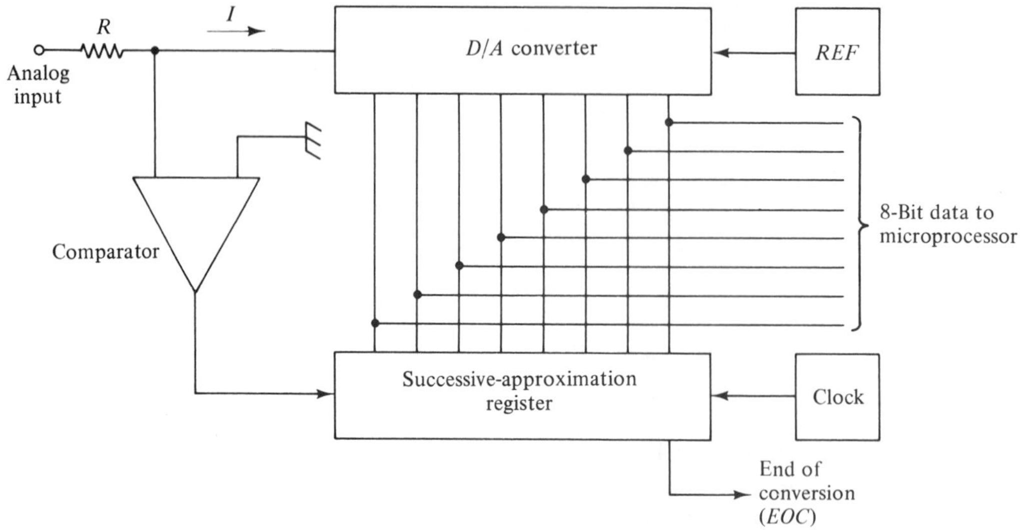

Figure 10.7

converter is still in the same direction as is shown, the comparator terminal is positive; therefore, the unknown analog signal is greater than 0.5 of the reference. Should the comparator terminal go negative, then the unknown analog signal is smaller than 0.5 of the reference. Thus a decision is made by the comparator to leave the bit on or turn it off, after which bit 2 is turned on and a second comparison is made, and so on. After the nth comparison, the successive-approximation register indicates to the microprocessor that the conversion is complete by changing the state of the end-of-conversion (EOC) line; the processor then loads the data in. A clock controls the timing of the SAR. The output of an A/D converter during a typical conversion is illustrated in Figure 10.8.

The integrating dual-slope type is an excellent choice for converting analog signals that vary slowly, such as battery voltages and temperature, especially in the presence of noise. The integrating type of A/D converter operates by the indirect conversion method. The unknown input voltage is converted into a time period that is measured by a clock and a counter. The integrator is the basic technique and leads to a variation of methods: (1) dual-slope A/D conversion, and (2) triple-slope A/D conversion (Figure 10.9).

Only the dual-slope integration method will be considered as it appears to be the most popular (Figure 10.10). The unknown voltage is switched into the integrator; a counter simultaneously starts to count the incoming clock pulses. When the counter overflows, the internal circuitry switches the integrator to a negative reference voltage supply, and the in-

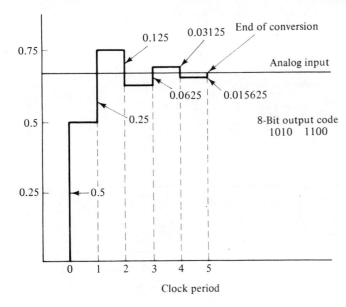

Figure 10.8

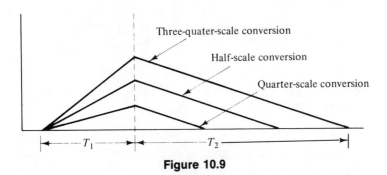

Figure 10.9

tegrator output starts to go back down toward zero; simultaneously the counter is continuously decremented by the clock pulses. When the integrator output passes through zero, the countdown is stopped and the count remaining in the counter is converted into a digital signal. Thus V_{in} (analog signal) is now known and can be converted by the following formula;

$$V_{in} = \frac{T_2 \times V_{ref}}{T_1} \qquad (10.14)$$

T_1 is a constant; since the counter fills to overflow, V_{ref} is a constant.

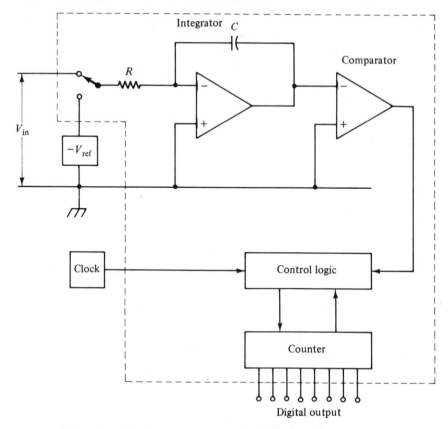

Figure 10.10 The section inside the dashed line is contained within the A/D converter chip.

$$\therefore \quad V_{in} = KT_2, \quad \left(K = \frac{V_{ref}}{T_1} \right) \tag{10.15}$$

T_2 is the digital word that represents the input analog voltage.

This type of converter is very accurate and depends primarily upon the stability of the clock, the integrating capacitor, and the reference supply. The noise rejection is also extremely good.

There are other types of A/D converters than the types described, but they will not be discussed here.

10.1.6 Digital-to-Analog Converters

These are the devices by which microprocessors communicate with the outside world. Principally, they fall into two categories, parallel conver-

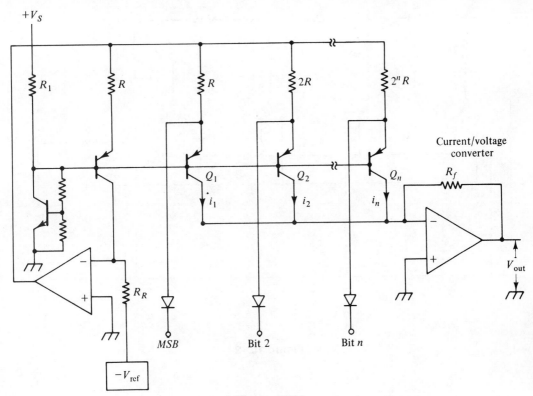

Figure 10.11

sion and serial conversion. Parallel converters cause the output to change almost instantaneously when the data are written in. The serial converter, on the other hand, waits until all the data are received before beginning its conversion.

Digital-to-analog conversion techniques fall into two basic groups: (1) weighted current source converter, and (2) R-$2R$ converter.

The weighted current source converter (Figure 10.11) is basically an array of switched transistor current sources. The binary weighting is achieved by using emitter resistors, in the individual transistors, that are binary related (i.e., R, $2R$, $4R$, $8R$, etc.). The composite current caused by the individual currents flowing through the selected resistors is the addition of the aggregate.

When a data input line is high, current is flowing through the selected resistor; when the line is low, current flow via the transistor is zero. The R-$2R$ D/A converter uses a ladder network method. The resistors forming the ladder are of values R and $2R$, as shown in Figure 10.12. At the bottom of each resistor is a single-pole double-throw electronic

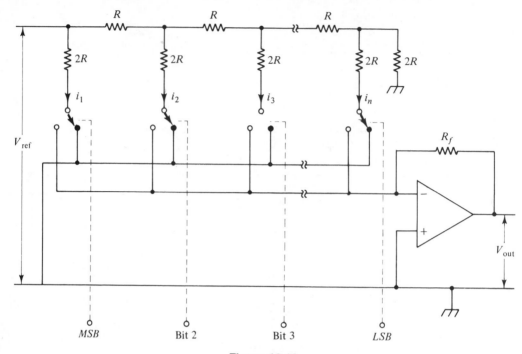

Figure 10.12

switch, which is controlled by a digital data line and which connects the resistor either to the summing point of the op-amp or to ground. The ladder network is designed in this manner to cause binary division of current as it flows down the ladder. The result is a binary weighted current at the input to the current-to-voltage converting op-amp.

The output of the D/A converters is connected to transistors, power transistors, or firing circuits for SCRs and triacs, which ultimately modify the process being controlled.

10.1.7 Analog Multiplexers

Analog multiplexers are the devices that time share an A/D converter with a number of different analog channels. Basically, they are multiposition switches, as shown in Figure 10.13. A truth table for this figure is given in Table 10.3.

If the multiplexer is switching low-level signals, that is, microvolt signals direct from the transducer, the problems encountered can become very involved, with the introduction of cross-talk, common-mode voltages

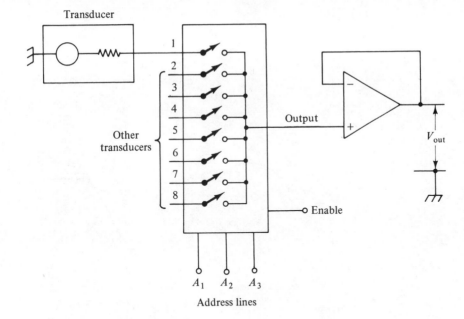

Figure 10.13

Table 10.3 Truth Table

A_1	A_2	A_3	Enable	Output Channel No.
$\times$	$\times$	$\times$	0	None
0	0	0	1	1
0	0	1	1	2
0	1	0	1	3
0	1	1	1	4
1	0	0	1	5
1	0	1	1	6
1	1	0	1	7
1	1	1	1	8

in the cable shields, thermocouple voltages, and so on. Therefore, this topic will not be discussed further in this text.

10.1.8 Sample-Hold Circuits

The sample-hold module has two steady-state operating modes:

1. Sample, in which it acquires the input signal as rapidly as possible and tracks it faithfully until instructed to:

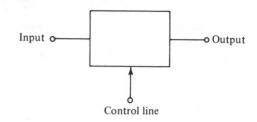

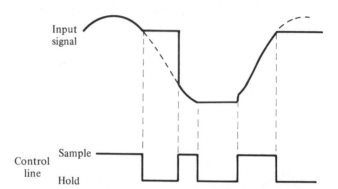

Figure 10.14

2. Hold; it then retains the last value of input signal that it had at the time the control signal called for a mode change.

Figure 10.14 illustrates the method.

Sample-hold modules are used to "freeze" fast-moving signals or to store analog multiplexer outputs while signal conversion is taking place. The basic operation of a sample-hold circuit is shown in Figure 10.15. When the switch is closed, the capacitor changes exponentially to the input voltage via op-amp 1; the output of op-amp 2 follows the capacitor voltage. The electronic switch is an FET, and the input stage to op-amp 2 is an FET input.

10.2 INTERFACING DATA CONVERTERS AND MICROPROCESSORS

Basically there are two methods of interfacing a data system with a microprocessor: (1) parallel transfer, for close-in operation, and (2) serial transfer, for remote operations. If high sampling rates are required, the data converter should be located near the microprocessor and a parallel interface used. If data are gathered some hundreds of meters away from the microprocessor, a twisted-pair shielded-cable serial data transmission should be used.

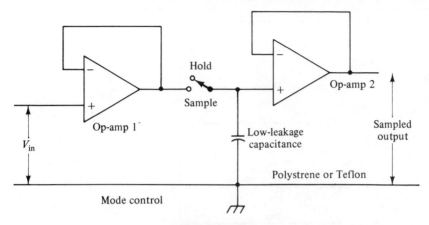

Figure 10.15

Whether successive-approximation or integrating A/D converters are used is no problem as both can be interfaced with a suitable logic circuit to move the data out serially. The movement of data into the microprocessor creates no problem in the case of the 6800 or the 6805 microprocessor since they are both memory-mapped structures, and serial and/or parallel I/O ports are readily available and interfaced. For example, to interface the National ADC0816/ADC0817 data-acquisition system module is simplicity in itself, as Figures 10.16 and 10.17 show. The block diagram and pin out are shown in Figure 10.18.

The 6800 microprocessor will be used to illustrate the technique of loading data in from the converter. Conversion starts when the processor causes the start line to make a negative transition; approximately 8 clock cycles later the end of conversion line makes a low to high transition, which interrupts the processor via line CA_1 port A. The processor enters the interrupt routine and signals the A/D converter via the interface logic and pin 21 of the converter to place the digital data on the data lines.

Data are now on the eight data lines. The microprocessor reads the data in the usual way and then resets the address lines of the multiplexer to the address of the next transducer to be read. The processor once again causes the start line to make a negative transition, and the whole cycle repeats itself. The tristate logic (see Figure 10.17) is essential as it informs the tristate buffer on board the A/D converter when it is time to place the data on the data bus. The data are placed on data lines when a positive transition occurs on the interface logic output.

The data are read in via the LDA-A instruction B68004 when the processor is in its interrupt cycle. The timing to enable the data to be placed on the data lines and read into Acc-A via the same instruction is shown in the timing diagram for ϕ_2 in Figure 10.19.

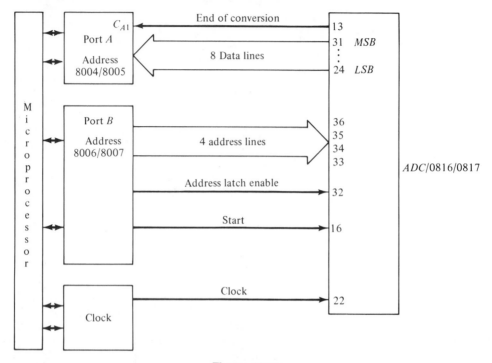

Figure 10.16

In multichannel conversion systems, certain elements may be shared by two or more inputs; this sharing or multiplexing may occur in a number of different ways, depending upon the property to be multiplexed. Two methods are given:

Method 1: share a single A/D converter (i.e., multiplex analog signals).

Method 2: multichannel conversion, use a digital multiplexer.

Method 2 is usually used in industrial data-acquisition systems where strain gauges, flow rates, thermocouples, thermisters, and the like are spread out over a large area. Digitizing the signals at source and transmitting the digital data back to the data center, either serially or in parallel, decreases the pickup due to line frequency as well as ground loop interference. Also, digital signals can be transformer or optically coupled to gain complete electrical isolation, and low-impedance digital drive and receiving circuits reduce the vulnerability to noise even further. In addition to this, on-site data processing via local microprocessors may take place

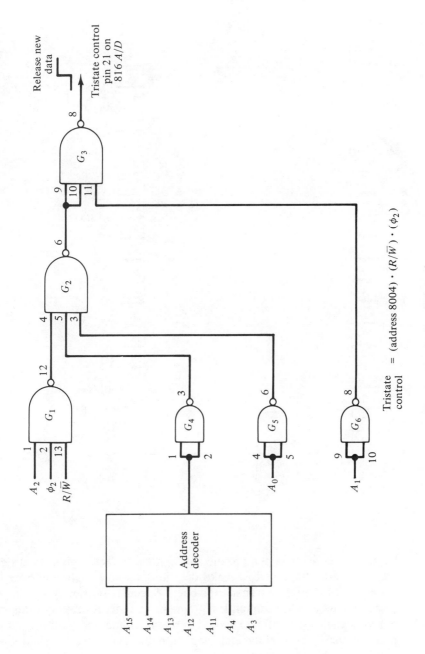

Figure 10.17 Logic Interface

$$\text{Tristate control} = (\text{address } 8004) \cdot (R/\overline{W}) \cdot (\phi_2)$$

Dual-in line package

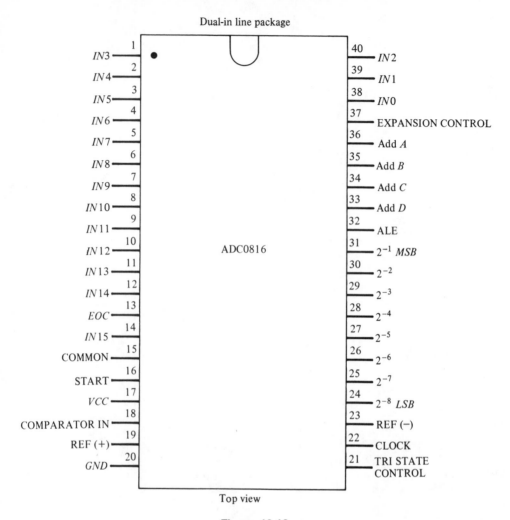

Top view

Figure 10.18

before the data are shipped to the data center. In fact, digital subsystems are frequently employed to make decisions on what data and how frequently they should be transmitted back. Specifically, slowly varying devices may be accessed much less frequently, and yet other devices may not vary their output for extended periods of time and then for a short period experience rapidly changing outputs. Therefore, flexibility and versatility are gained by transferring the interfacing process from analog multiplexing to digital multiplexing.

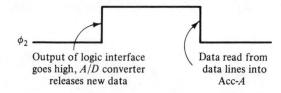

Output of logic interface
goes high, *A/D* converter
releases new data

Data read from
data lines into
Acc-*A*

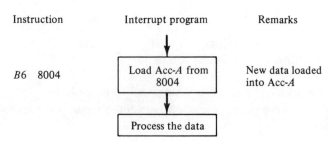

Instruction   Interrupt program   Remarks

B6 8004   Load Acc-*A* from
                    8004   New data loaded
                                                into Acc-*A*

                    Process the data

Figure 10.19

10.3 SOLID-STATE CIRCUITS

The application of solid-state devices for power control started in the 1950s with primitive diodes and work continued with the signal transistor. But the real start of applications of solid-state devices came in December 1957 when the first SCR was introduced. Since that time, many thyrister devices (a generic term given to a wide variety of four-layer semiconductor devices) have been developed. The SCR, diac, and triac are all capable of being stimulated into conduction by the right combination of signals. The use of solid-state components in industrial control applications is rapidly expanding and replacing many of the functions previously carried out by relay logic. While solid-state controls and electromechanical relay controls are basically capable of conducting identical machine operations, there are a number of significant differences among these devices requiring some special application considerations. Solid-state devices provide many advantages, such as high speed, small size, ability to handle extremely complex functions, and have low power consumption; they differ primarily in electrical ruggedness and are sensitive to certain environmental influences, such as elevated temperature and electromagnetic interference. They are less "forgiving" than relays when overstressed. In addition, they exhibit unique failure mechanisms that must be taken into account in critical and/or potentially hazardous applications.

It will be noted that numerous potential problems are presented. However, relatively few, if any, will be encountered in a well-designed and properly installed control system in the usual industrial environment. The discussion presented here is intended to point out areas in which solid-state controls are unique, to point out various operating limits, and to provide some application guides where solid state is employed. The input and output stages are the areas of the control system most vulnerable to damage since they are directly exposed to external influences; thus they are discussed separately.

10.3.1 The SCR

The SCR (Figure 10.20) is a four-layer, two-state (on/off) PNPN device that can block voltages in either direction, and can be stimulated into conduction when the following conditions are met:

AND OR OR

1. Its anode is positive with respect to its cathode
2. A positive pulse is applied between its gate and cathode
3. Increased junction temperature
4. Applied voltage exceeds the breakover point, V_{BO}

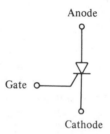

Figure 10.20

Finally, if the rate of change of voltage from anode to cathode exceeds specified limits, the device can also be stimulated into conduction. Once the SCR is conducting, it is latched into the ON state, and the only way that conduction can be stopped is to reduce the current flowing through it below a critical value called $I_{holding}$ (I_h). When the current falls below I_h, the latch is opened and conduction ceases. Normally, the SCR is operated on ac circuits and is triggered into conduction via a gate pulse, which must be applied each positive half-cycle since conduction ceases when the ac cycle passes through zero. Furthermore, the device cannot conduct when the negative half-cycle is present, because the anode is negative with respect to the cathode. Therefore, if the microprocessor-controlled process is a variable direct-current supply to some equipment, then the controlling SCRs must be triggered each positive half-cycle at the appropriate point in the cycle. Normally, the microprocessor will control the gate firing network and advance or retard the firing (Figure 10.21).

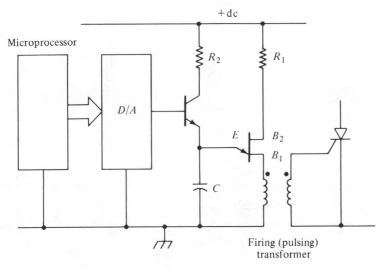

+dc

Microprocessor

R_2 R_1

D/A

E B_2

B_1

C

Firing (pulsing)
transformer

Figure 10.21

Overtemperature turn on can be reduced and/or eliminated by good heat sinking methods. The heat sink must be of adequate size with the heat fins parallel to the air flow; also the SCR must be seated in heat conducting grease before it is torqued down.

To prevent turn on due to overvoltage or line transients, a varistor (clip cell) is connected in parallel with the SCR. This also prevents the SCR from being damaged by overvoltage during its OFF state. If the voltage applied to the anode changes very abruptly, the SCR can turn on and start conducting; this is because of the effect of the rapidly changing voltage upon the SCR's junction capacitance. To reduce this effect, a combination series resistor/capacitor network is connected in parallel with the SCR (Figure 10.22). The parallel branch is called a *snubber*. Its use is twofold:

1. The parallel capacitor branch appears as a temporary conduction circuit to a rapidly changing anode voltage. This circuit is in parallel with the internal junction capacitance, and if the external capacitor is of the correct size, it virtually shorts the SCR during the period of rapid change. Thus the false turn on is eliminated.

2. When an SCR is controlling an inductive circuit, the rise of current in the circuit is delayed due to the effects of the induction. Thus the growth of the conducting surface area, within the SCR, to the application of the trigger pulse and the flow of load current is limited, and the energy contained within the pulse is not sufficient to establish conduction. However, when a snubber network

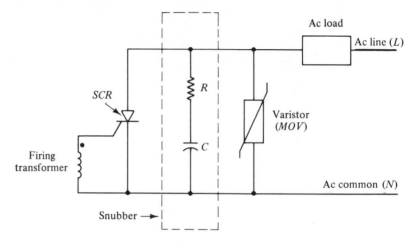

Figure 10.22

is present, the external capacitor will probably be already changed, and at the time the firing pulse is applied the capacitor discharges via the SCR and aids in establishing current flow.

The SCR can be damaged by continuous overcurrent or large surge currents, which can occur in capacitive loads or dc motor loads during start-up. Microprocessor/SCR control is being used to a very large degree in traction motor control and electric car control.

10.3.2 The Triac

The triac is virtually a bidirectional SCR or two SCRs back to back; its principal application is ac motor loads since it can conduct on both half-cycles of an ac wave (Figure 10.23).

Unlike relay contacts, the triac is more sensitive to applied voltages, currents, and internal power dissipation. The triac is limited to a maximum peak "off-state" voltage. Exceeding this ac peak voltage results in a "dielectric type breakdown," which may result in a permanent short circuit failure. In the case of an electromechanical contact, an arc-over might occur; but usually the contacts would not be seriously damaged and the device could continue to be used. These peak voltages may be the result of transients and noise on an output line. Often a varistor is placed across the triac to limit the peak voltage to some value below the maximum rating, and in some instances the *RC* snubber alone may adequately protect the triac. The power control circuit for ac motors usually consists of a triac (in some instances SCRs are used). Proper application of the

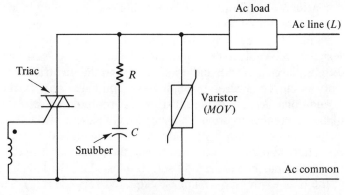

Figure 10.23

control is associated with the observance of a number of absolute maximum ratings and awareness of various electrical characteristics of the output triac. Carefully following the recommendations on the manufacturer's product data sheets allows operation below the absolute maximum ratings.

Special precautions must be taken as, for example, when driving lamp loads where the cold resistance of a lamp is often one-tenth of the hot resistance. The turn-on current surge must be within the maximum peak nonrepetitive on-state current. There is also a maximum rms on-state current that can safely be carried continuously. This is the rating by which the various triacs are categorized by their manufacturer. Exceeding the on-state current rating again may result in a permanent short circuit failure due to over dissipation.

Under certain conditions where load resistances are low, stray circuit inductances are low, and power line impedance is low, it may be necessary to limit the rate of turn on of the triac. This parameter, as given in triac data sheets, is the rate of change of the turn-on current (di/dt) expressed in amperes per microsecond. If conditions are available to exceed this rating, localized heating occurs within the triac, resulting in damage to the triac. Subsequent operation generally results in overdissipation under normal loading conditions and early short-circuit failures. When found necessary, an inductance should be connected in series with the triac, which will limit the rate of change of current to within the maximum rating. This parameter is sometimes exceeded due to improper selection of the RC snubber, where one is provided. Selection of the proper snubber is beyond the scope of this discussion, and sources of information are readily available [RCA Thyristers/Rectifiers, SSD-206C, 1975, pp. 421–425, p. 442].

Of considerable importance where loads may be highly inductive is the critical rate of rise of commutation voltage (dV/dt), expressed in triac data sheets in volts per microsecond. Solid-state output circuits are generally designed to be able to drive inductive loads such as electromechanical loads (e.g., large motors), and it may be found that the triac is not able to turn off as current passes through zero (commutation). In these instances, additional RC snubbing must be incorporated as selected from the available references to prevent against loss of gate-controlled triac operation. Such loss of control generally does not degrade the triac.

Solid-state outputs are often fused, and the particular fuse employed or recommended has been carefully selected based on the fusing current rating of the triac expressed as amperes squared-seconds (I^2T). This rating is seen to incorporate a fuse opening time along with a current rating. For a particular overload current, it is necessary to open the circuit within a certain time to avoid damage to the triac; thus, fuses must be selected that have the proper current-time characteristic. The recommendations in the product data sheet must be followed when replacing fuses, or in instances where outputs are not provided with a fuse, the triac rating must be determined and taken into consideration. Fusing of solid-state devices is effective only in protecting the semiconductor in the event of a load failure or accidental short circuit. It must be noted that if, for example, the load is a motor and a triac failure results a fuse will not be effective in removing power from the load. Certain types of indicating fuses can have system diagnostic value in warning of a load malfunction.

Triacs have several characteristics that are not observed in electromechanical contacts. Since the solid-state "contact" is a solid block of material, its on and off states are actually low and high resistance levels, respectively. A small leakage current flows through the triac and RC snubber network in its off state. This is usually on the order of a few microamperes to several milliamperes and should present little problem, except in the case where sensitive instruments, such as a volt-ohmmeter, are used to check for "contact continuity" where false indications are likely or instrument damage could occur. When the triac is on, there is an on-state voltage developed across it of up to 1.5 V. The product of this voltage and the rms current being carried is approximately the power dissipation within the device, which is the prime factor used in the determining current-carrying capability of the "contact." The on-state voltage divided by the rms load current can be considered a *contact resistance*. Triacs also exhibit a minimum holding current presented as the *dc holding current*. When the instantaneous load current (each half-cycle) falls below this value (e.g., 25 to 100 mA), the triac ceases conduction and passes only its leakage current until again triggered. Thus, it may not be possible to turn on or commutate full power for very light (high ohmic) loads. In these instances, a bleeder resistor may be used to provide a minimum load. In some systems, special circuitry is provided to overcome this problem.

Triacs are capable of carrying surge currents an order of magnitude larger than their continuous rating, but only for short duration and under nonrepetitive conditions. This rating, referred to as the peak nonrepetitive on-state current, must not be exceeded in magnitude (usually given for one cycle at 60 Hz) since the device may overdissipate, resulting in a permanent short circuit.

10.3.3 Power Transistor

Transistors, like triacs, are sensitive to excessive applied voltages and heavy surge currents. One of the most critical parameters requiring attention in a transistor output stage is the absolute maximum collector to emitter voltage. This is generally referred to as a *breakdown* voltage and appears across the solid-state "contact" when open. When this voltage is exceeded, the transistor can be destroyed, often resulting in a permanent short-circuit type of failure. The most common cause of excessive collector to emitter voltage is transients on the output load line and on the power supply line and switching of inductive loads. In the latter case, as load current is interrupted during switching, extremely high voltages appear across the inductive load if the load is not protected and provided with a means to allow the discharge of the electromagnetic field. In most cases, protection is provided in the form of a free-wheeling diode, zener diode, or *RC* network across the load terminals. The diode must be capable of withstanding full-rated load current and have a reverse breakdown voltage in excess of the dc supply voltage. The dc power circuit shown in Figure 10.24 comprises a power transistor protected by a free-wheeling diode against inductive loads. Proper observation of absolute maximum ratings is essential.

Transistors are also sensitive to heavy surges of collector or load current, more so than the triac used in the ac output. Excessive collector currents result in overdissipation and failure to a shorted condition.

Transistor switches also exhibit certain characteristics not encountered in electromechanical or "hard contact" switches. When turned on, a small forward voltage drop $V_{ce\,sat}$ (typically 0.35 V) appears across the transistor. This must be taken into account where supply voltages are low, and it is also this voltage which is responsible for transistor heat dissipation under load. The maximum load current given in product data sheets for a particular control takes into account the product of load current and the forward voltage drop of the transistor used, which deter-

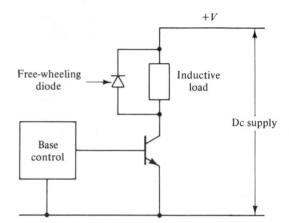

Figure 10.24

mines the internal heat generated. This recommendation must be followed in all instances.

As in the case of the triac, transistors exhibit a small leakage current when in the off state. This current is generally much lower than that of the triac, usually in the microampere range, and must be taken into account when switching low currents or when using sensitive instruments during setup and checkout.

Transistor outputs are sometimes fused. This fuse, as in the case of the triac, is generally used to protect the transistor during moderate overloads. This fuse must be capable of opening very quickly before excessive heat can build up within the transistor. Here, again, a fuse does not disconnect the load in the event of a transistor failure. When used, the manufacturer's recommendations for fuse type must be observed.

Unlike triacs, transistors are not as easily turned on inadvertently by electrical transients since base current must be applied continuously to maintain the output in an on state. Transistors may, however, be turned off or turned on momentarily by such transients. Thus, these effects should be evaluated for possible problems in a particular application.

Transistors exhibit finite turn-on and turn-off times, which must be considered where "contact" overlap problems could occur.

Power transistors find great application in pulse-width-modulated (PWM) variable-frequency drives, as well as in switched-mode power supplies.

10.4 NOISE

Electrical noise is capable of causing various types of malfunctions of solid-state controlled equipment. This noise is defined as any unwanted electrical signal that enters the equipment through various means. It may cover the entire spectrum of frequencies and exhibit any wave shape. Solid-state devices are especially sensitive to noise since they operate at low signal levels. Also, the speed of solid-state components allows them to respond to relatively high frequencies.

Electrical noise entering a solid-state control is often incapable of damaging components directly unless extremely high-energy and/or high-voltage levels are encountered. Most of the malfunctions due to noise are temporary nuisance-type occurrences or operating errors, but could result in hazardous machine operations in certain applications. Noise entering triac output circuits generally results in non-gate-controlled operation, which can be of short duration (half-cycle or less) or long term (more than one cycle), depending on the nature of the noise signal. DC outputs may be turned on due to rectification in the output transistor. Data circuits may be susceptible to all forms of noise, and their response may produce a turn-on signal to the triac, SCR, or output transistor control circuit. Larger systems containing complex logic circuitry may be expected to respond by turning on various outputs in a random fashion and occasionally tripping self-check circuits, which result in nuisance shutdown of machinery.

Noise reduction in analog or digital circuits is an ongoing problem, since noise is never eliminated; it basically takes three forms:

1. Transmitted noise, inherent in the original signal.
2. Inherent noise, generated within the elements of the data-acquisition system.
3. Induced noise, picked up from power supplies, magnetic coupling, electrostatic coupling, and galvanic coupling.

Noise may be categorized into two forms:

1. Random, generated within electrical components (e.g., resistors, semiconductors junctions, transformer cores).
2. Coherent, locally generated by chopper-stabilized converters and amplifiers, switching of inductive loads such as motors and transformers.

One major source of noise is *induced* noise; in this case, early preamplification and conversion is essential, as well as isolation via optical or transformer coupling.

In noisy environments, digital conversion near the signal source is usually the most advantageous, since it can easily be seen that, with an 8-bit converter ($\frac{1}{256}$ resolution) and a 10-V full-scale signal, if the peak-to-peak noise is greater than 40 mV, then the LSB will be affected, whereas standard TTL noise immunity is approximately 1.8 V (2.0 = binary 1 and 0.8 V = binary 0).

Active filters are essential elements in a data-acquisition system to minimize the unwanted effects of noise in the analog signal.

Noise enters solid-state controls by various means, usually through input lines, output lines, and power supply lines; it may be coupled into the lines electrostatically through capacitance between these lines and lines carrying the noise signals. Here a high potential is usually required or long, closely spaced conductors are necessary. Magnetic coupling is also quite common when control lines are closely spaced to lines carrying large currents. The signals in this case are coupled through the mutual inductances as in a transformer. Electrostatic and magnetic noise may also be directly coupled into the control logic circuitry. This generally appears in unenclosed, unshielded control electronics and requires a strong noise field. Finally, noise can occur in the form of electromagnetic radiation from remote sources. Close coupling is not required, and here the various lines entering the control area act as receiving antennas. Occasionally, the control circuitry itself is sufficiently sensitive to detect a radiated signal and respond to it. This type of noise is troublesome when it is encountered since it is usually of high frequency and occasionally difficult to filter and shield against on a generalized basis. A particular installation may require special treatment, since the various coupling elements exhibit unpredictable characteristics, some of which could render built-in filters ineffective. Metal enclosures are effective shields where adequate electrical bonding around doors and bolted surfaces is provided.

Coupled electrical noise can be viewed as an unwanted signal originating in other wiring. To reduce the coupling, solid-state control lines, especially TTL input lines, should be separated from noise sources by routing them through their own separate conduit. Thus, due to separation, there is little opportunity to couple from lines carrying 60-Hz power and other lines carrying rapidly changing currents. Also, twisted, shielded-pair wiring, as shown in Figure 10.26, provides a very effective shield against electrostatic and magnetic coupling. The twisted wire should have approximately a 1-inch lay or 12 twists per foot. The magnetic field induced between 1-inch sections is alternately 180° out of phase, providing effective cancellation of the coupling effect. Eddy currents induced in the shield produce a magnetic field opposing the original magnetic field. The shield is also effective in electrostatically decoupling the noise source since it places a conductive barrier between the signal lines and the noise source. The shield must be connected to control ground at *only one point,* as shown in Figure 10.25, and shield continuity must be maintained over the entire length of the cable. Care must be taken to adequately insulate the shield over its entire length to maintain the ''one-point'' connection. Cables should also be routed around rather than through high noise areas.

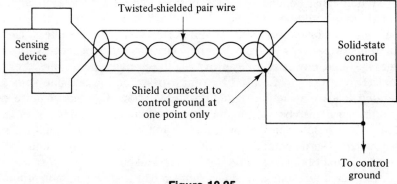

Figure 10.25

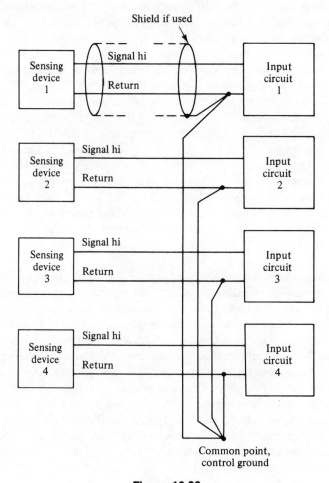

Figure 10.26

Where noise conducted into a control circuit is encountered, appropriate filters should be employed. In the case of power lines where transients are anticipated or noted, commercially available line filters, isolation transformers, or voltage limiting varistors may be necessary. In larger systems, power supplies providing logic power are usually already protected but the raw ac used by triacs to switch various line-voltage-operated loads may require additional protection. Conducted noise on input lines may also be dealt with by use of isolation transformers, or isolated power supplies in dc applications, and occasionally signal filters. Also, where "hard contact" input devices are employed, such sources of noise as contact bounce should not be overlooked.

Impedance coupling, where systems might share a common ground, is often a source of conducted noise and must be avoided. Noise coupling in this case is due to voltage drops appearing across any impedances that are common to several circuits. These impedances are usually the result of the inductances of common wires. Whenever possible, or when found to be necessary, individual signal return wires should be used as in Figure 10.26. Voltage drop across common wires may also be reduced by the use of large conductor diameters where signal paths are shared. Often, this line takes the form of a heavy bus bar. In instances where several controls share a common course of power (60-Hz line for ac controls; dc power supplied for dc controls), it must be determined whether the line impedance, or current capabilities of dc supplies, is adequate to maintain rated voltage under all loading conditions. Voltage drops across the internal power line impedance or deregulation of a dc power supply results in interference between the systems similar to that experienced with an impedance coupling. The internal impedances of the power supplies here act as the common coupling element. Whenever possible, separate dc power supplies or separate ac branch circuits should be used for each control system.

Various loads, especially inductive devices, are capable of generating excessive noise, which may affect other systems through coupling or radiation and may feed back unwanted signals to their own source of control. In these instances, some form of load suppression should be employed. The methods commonly used act to limit the transient voltage (and thus the noise current) across the load. Voltage limiters generally consist of *RC* snubbers, varistors, and back-to-back zener diodes for ac applications, and in cases of less severe noise, resistors. Direct-current loads are generally suppressed by reverse-biased or reverse-connected diodes. It should be noted, however, that use of these devices may reduce the speed of operation of electromechanical loads significantly. In addition, use of lightning arrestors from each power line to ground should not be overlooked.

Special Application Considerations. Solid-state components exhibit a high degree of reliability when operated within their ratings. For example, a triac might have an average life of 450,000 hours or 50 years under typical operating conditions, but it fails at random even when operated well within its ratings. The lifetime given is an average. The time of failure of an individual device cannot be predicted by observation, as in the case of a relay where patterns of wear might be watched. It is thus advisable to provide some sort of independent check on the operation of individual devices when they are controlling a critical or potentially hazardous operation. In addition, the predominant mode of failure of a solid-state output is in the "on" or short-circuit mode. This must also be considered in certain critical applications.

In relay systems, a relay having a mechanically linked normally open and normally closed contact can be wired so as to check itself for proper operation. This feature is sometimes used in certain safety circuits, such as the antirepeat circuit for a press control. Monitoring of a solid-state component is not quite as simple. A dependable mutually exclusive N.O., N.C. arrangement is not available. Thus, in such applications external circuitry must be employed that can sample the input signal and "contact" state and compare them to determine whether the control is functioning properly. By proper design, however, the critical machine operation can be shut down in the event of either a solid-state control failure or a failure of the monitor circuit.

In large solid-state control systems containing many independent output stages, a need for built-in diagnostic circuitry becomes apparent. Fault monitoring can be performed rather easily in system areas such as microprocessors, which can be serial in operation. All information shares a common data bus, and a diagnostic such as a parity check is quite effective, along with a few circuits to exercise certain critical program functions. However, in other areas, especially input and output channels, effective monitoring can only be performed on an individual channel basis. This becomes extremely costly and, if implemented, would severely lower the overall system reliability, resulting in many "nuisance" shutdowns. In practice, only a minimum amount of monitoring is provided for these circuits, and in the event that a particular signal channel is responsible for control of a potentially hazardous machine operation, external monitoring or redundancy must be provided for that channel. Here, for example, an input channel might be reserved to receive a signal from the auxiliary contact of a starter to check on its operation. This input channel and the starter output channel can then be examined for occurrence of mutually exclusive events and shut down system power when a failed condition is noted. This method is often quite practical when only a few outputs are classified as critical.

For any size of solid-state control capable of performing potentially hazardous machine operations, emergency circuits for stopping operation must be routed outside the controller. For example, devices such as end of travel limit switches or emergency stop pull cord switches should operate motor starters directly without being processed through control logic. This forms a reliable, redundant means of control and should be implemented using a minimum number of simple, highly dependable components of an electromechanical nature if possible. Thus, in the event of a complete controller failure an independent rapid shutdown means is available.

Additional safety precautions should also be observed. Systems should incorporate a feature that guards against hazardous machine operation upon application of power or restoration of power following a shutdown. A convenient means of disconnecting the critical or potentially hazardous portions of a machine from the controller should be provided for use during troubleshooting or setup following maintenance.

10.5 GROUNDING

Good engineering practice, as well as electrical code requirements, requires that the chassis of all electrical equipment be grounded (i.e., be at the same potential). However, in spite of the grounding of electrical equipment, various items of the systems may be separated by several hundred millivolts, and ground currents may be flowing between these items.

The ideal grounding system for an electrical system, that is dc supplies, signal sources, and the like, should be an isolated and insulated ground bus that is connected to a ground plate or ground rod at only one point. Furthermore, the signal ground rod should be outside the sphere of influence of the ac power ground (Figure 10.27).

In solid-state control systems, the grounding practices employed have a significant effect on noise immunity. Each ground should be connected to its respective reference point by no more than *one wire* (single-point grounding). Under no circumstances should two or more systems share a common single ground wire, either equipment ground or control common. Since all wires are somewhat resistive and inductive, this would allow the system to share a common impedance. Variations in current flowing from one system would tend to modulate the zero reference level of the other and possibly result in erratic operation. Minimum wire sizes, color coding, safety practices, and so on, should comply with the National Electrical Code and any local codes or laws. The recommendations found in product data sheets should always be followed since the manufacturer has determined, by testing, the adequacy of the particular

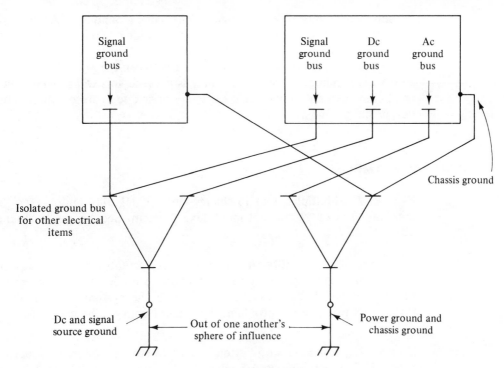

Figure 10.27

grounding arrangement in a typical industrial environment. (For further information on grounding practices, see IEEE Standard 142-1972, Recommended Practices for Grounding of Industrial and Commercial Power Systems.) The resistance between the buses, when entirely isolated should be at least 10 MΩ.

All power return wires should be run with the power supply wires to reduce magnetic interference. One of the biggest problems when designing and building a data acquisition/microprocessor system is the analog and digital grounding and the emission of noise from the digitally actuated microprocessor system. The analog system depends upon the accuracy of voltage levels and must operate in the vicinity of a noisy digital system, since analog and digital chips are adjacent elements in the data-acquisition system.

Care must be taken when laying out the printed circuit board or other wiring method that the analog section be kept isolated from the digital section by as much physical separation as can be spared.

The analog and digital grounds *must* be separate networks, that is, an isolated analog ground bus and a digital ground bus. Since both systems share a common power supply, the two grounds must be common at the supply as Figure 10.28 illustrates. The analog ground bus must be kept short and of physically large size (i.e., much wider than it is thick) to reduce inductance effects.

Clamping diodes should be installed between the grounds on every card where analog and digital circuitry is present; this will prevent damage to the circuitry should excess voltages arise due to accidental separation of the grounds.

10.6 TUTORIAL EXAMPLES

Example 1. Multiplex two 16-channel Intersil 1H 6116 multiplexers to make a 1 out of 32 channel multiplexer system. See Figure 10.29 and Table 10.4.

The submultiplexer is fitted to reduce leakage currents and output capacitance, which can degrade system performance if allowed to exist, since the rate at which data can be put through the system is reduced. If throughput rate is not a problem, the 1H5041 chip can be deleted.

Example 2. A master stop button is used to stop an industrial process. When the button is activated it will stop all motors and the like by deenergizing a master relay, which supplies power to all the operating circuits. At the same time it will reset the microprocessor.

Show the wiring and the interface to the processor (see Figure 10.30). The microprocessor is in its reset or initialized state after the master stop has been actuated.

This example illustrates the application of an ac/dc interface. The circuit could be modified to include automatic start-up when the start button is depressed. This is left to the reader in the problems.

Table 10.4

A_4	A_3	A_2	A_1	A_0	Channel Switched to Output
0	0	0	0	0	S1
0	0	0	0	1	S2
0	0	0	1	0	S3
⋮	⋮	⋮	⋮	⋮	⋮
1	1	1	1	0	S31
1	1	1	1	1	S32

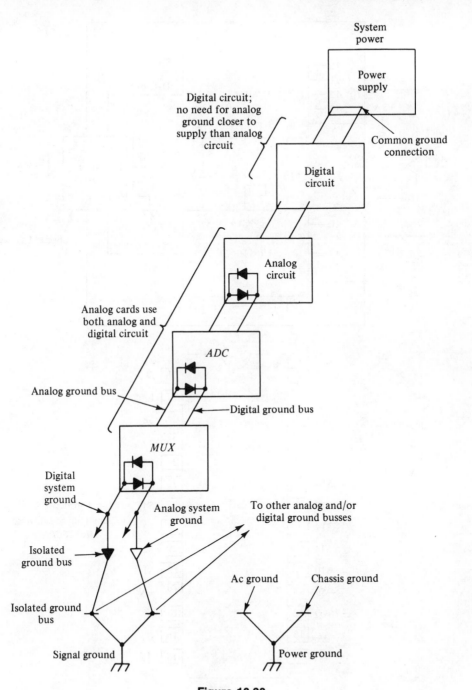

System
power

Power
supply

Common ground
connection

Digital circuit;
no need for analog
ground closer to
supply than analog
circuit

Digital
circuit

Analog
circuit

Analog cards use
both analog and
digital circuit

ADC

Analog ground bus

Digital ground bus

MUX

Digital
system
ground

Analog system
ground

To other analog and/or
digital ground busses

Isolated
ground bus

Ac ground

Chassis ground

Isolated ground
bus

Signal ground

Power ground

Figure 10.28

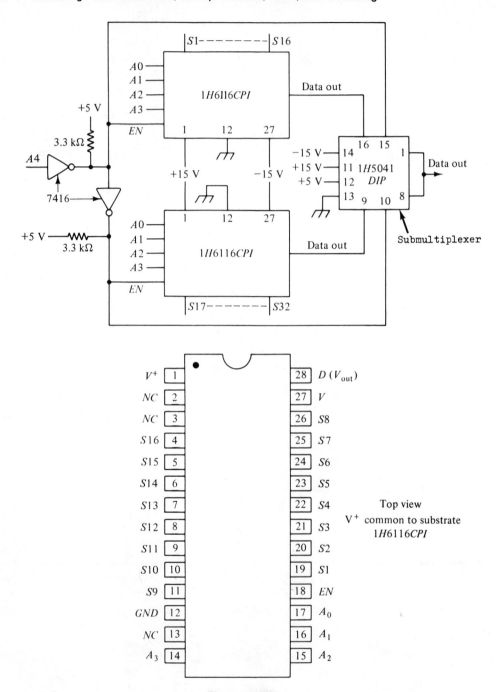

Figure 10.29

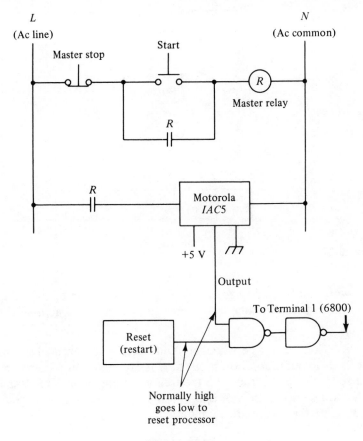

Figure 10.30

Example 3. A microprocessor-controlled unit is being used in conjunction with several electronic instruments. Several instruments operate entirely isolated from ground, whereas the microprocessor has its negative line grounded; discuss the connection between these two pieces of equipment. See Figure 10.31. The interface should be either an (1) isolating transformer or (2) optical isolator (see Figure 7.9).

Example 4. Demonstrate the correct method of connecting and grounding a transducer used in a data-acquisition system (see Figure 10.32).

Example 5. A microprocessor-controlled unit is installed in a plant that contains many relay-controlled applications. The inductive effect of the

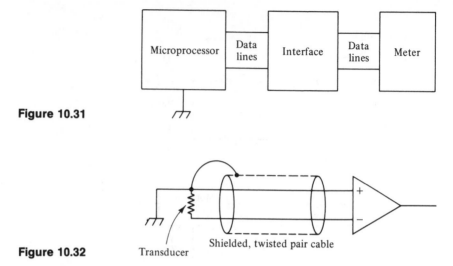

Figure 10.31

Figure 10.32

Transducer

Shielded, twisted pair cable

relays when they deenergize is affecting the operation of the new unit. What action should be taken? Assume the relays are dc actuated.

Solution. Connect a free-wheeling diode across its terminals. This will affect its drop-out time; if drop-out time is an important parameter, connect a zener diode in series with the free-wheeling diode. The zener voltage should be greater than or equal to the coils operating voltage.
Assume the relays are ac actuated.

Solution. Connect an *RC* snubber across the coil. To calculate the size of the snubber, see the RCA Thyristor/Rectifier Data Book SSD-206C.

Example 6. Discuss the effects of electrical interference due to electrostatic coupling and electromagnetic coupling, the mechanisms by which they are produced and the techniques in existence to reduce their effects.
Electrostatic coupling occurs when a changing voltage in one circuit causes a current to flow in an adjacent circuit. The coupling is due to the mutual capacitance that exists between circuits, since a capacitor can be described as any two conductors separated by a dielectric. Therefore, among the varous physical factors that determine the magnitude of the interference are the following:

1. Distance between the wires.
2. Length of the parallel run (i.e., wires running parallel to one another).

3. Distance between the wires and a grounded surface.
4. Frequency of the signal carried by the cables.
5. Ratio of the impedance in the two circuits.

The induced current can be calculated by $C(dV/dt)$, where C is the capacitance between the circuits, and V is the voltage across the capacitor. Thus, the induced current will appear as "spikes" during times of high dV/dt conditions. To reduce the interference, the affected wires must be shielded and the shield must be grounded at *only 1 point* (e.g., coaxial cable). If coaxial cable is not available, twisted pair cable with the unused lead grounded at one end is a good substitute.

Electromagnetic coupling occurs when energy is transferred from one circuit to another via mutual inductance. The magnitude of the induced voltage depends upon the following:

1. Magnitude of the current producing the magnetic flux coupling.
2. Length of the parallel run.
3. Frequency of the signal.
4. Distance between conductors.

The amplitude of the noise can be significant if the rate of change of current in the conductor is high (i.e., if di/dt is high).

Electromagnetic coupling is reduced by twisting conductors together, since the voltage induced into a twisted pair line is proportional to the area of the loop between the conductors and the rate of change of the magnetic field (see Figure 10.33).

The loop area cannot be zero as the wires are insulated; coaxial cable, however, does provide an effective zero loop area since one conductor completely surrounds the other. The screen is grounded at one end and is also used as the return conductor. The twisting not only reduces electromagnetic pickup; it also reduces electromagnetic radiation due to

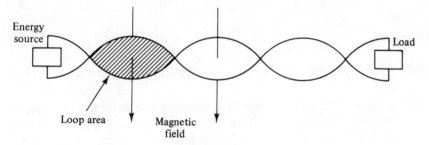

Figure 10.33

the current in the wires. Furthermore, twisted pair cable is very effective for frequencies below 5 kHz; above this frequency copper braided cable should be used. See Appendix E for the FCC rules concerning the use of equipment at frequencies above 10 kHz.

Example 7. Draw the interface and write the program to read the BCD data signals from a 3-digit digital meter; display the digits on a three-section, seven-segment LED display (see Figure 10.34 and Table 10.5).

Example 8. Two electronic/electrical control devices require ac, dc, and signal grounds; illustrate how these devices should be grounded (see Figure 10.35). The ground terminals ac, dc, and signal are insulated from the metal chassis, as ground currents in metal chassis are not predictable because of paint, corrosion, and the like. Also, the chassis is usually used as an electrostatic or electromagnetic shield, and it cannot act as an effective shield if it is carrying ground currents.

Example 9. Draw the wiring diagram and write the program to operate and read converted analog data for a 16-channel data-acquisition system, similar to Figure 10.1. The microprocessor sends out all timing signals as well as address data, and operates at 1.843 MHz (see Figure 10.36). The wait times quoted are the minimum times required to allow the electronic equipment time to settle and reach steady operating levels.

Example 10. Wire the A/D converter ADC0816 used in Section 10.2 to the microprocessor via port A in the handshake mode to receive the converted data, and via port B in the master/slave mode to change the address lines and so on.

Example 11. Ground the negative line of each load and the chassis of each electrical device in Figure 10.37.

Problems

10.1. Write a program to convert a 3-digit BCD code into a hexadecimal code.

10.2. Design the circuit and interface diagram and write the program to monitor the output voltage of a three-phase solid-state motor starter. If line voltage to the motor is present and the signal to energize the motor has not been given, sound an alarm.

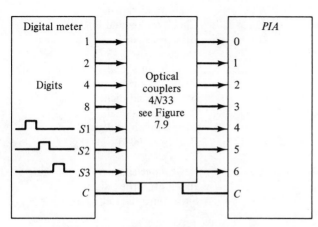

During strobe periods $S1$, $S2$, and $S3$, the first (LSB), second, and third digits are present at the output of the meter

(a)

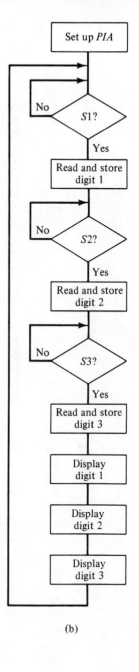

(b)

Figure 10.34

10.3. The alarm in Problem 10.2 is motor actuated e.g. a siren, and draws 2 A at 110 Vac; select a suitable interface.

Table 10.5

Read meter and store the readings

0020	CE	0004	LDX⎫	Initialize PIA
0023	FF	8004	STX⎭	
0026	B6	8004	LDA-A	Read input signal
0029	84	10	AND-A#	Check for first digit
002B	27	F9	BEQ	Branch to 0026
002D	BD	0005	JSR	Delay
0030	B6	8004	LDA-A	Read input signal
0033	84	0F	AND-A#	Blank off strobe signal
0035	97	00	STA-A	Store first digit at address 0000
0037	B6	8004	LDA-A	Read input signal
003A	84	20	AND-A#	Check for second digit
003C	27	F9	BEQ	Branch to 0037
003D	BD	0005	JSR	Delay
0041	B6	8004	LDA-A	Read input signal
0044	84	0F	AND-A#	Blank off strobe signal
0046	97	01	STA-A	Store second digit
0048	B6	8004	LDA-A	Read input signal
004B	84	40	AND-A#	Check for third digit
004D	27	F9	BEQ	Branch to 0048
004F	BD	0005	JSR	Delay
0052	B6	8004	LDA-A	Read input signal
0055	84	0F	AND-A#	Blank off strobe signal
0057	97	02	STA-A	Store third digit at address 0002

Delay subroutine to ensure that the signal is read in the middle of the strobe pulse

0005	86	01	LDA-A#	
0007	4A		DEC-A	
0008	26	FD	BNE	
000A	39		RTS	

Display of stored signals

0059	96	00	LDA-A	Load A with first digit
005B	C6	08	LDA-B	Location of the display
005D	BD	0080	JSR	Display subroutine
0060	96	01	LDA-A	Load A with second digit
0062	C6	10	LDA-B	Location of the display
0064	BD	0080	JSR	Display subroutine
0067	96	02	LDA-A	Load A with third digit
0069	C6	20	LDA-B	Location of the display
006B	BD	0080	JSR	Display subroutine
006E	7E	0020	JMP	Jump to the beginning of the program

Table 10.5 (cont.)

Display subroutine

0080	CE	E3C9	LDX	Display address in J bug
0083	08		INX	Match digit with look-up table
0084	4A		DEC-A	
0085	2C	FC	BGT	Branch to 0080
0087	A6	00	LDA-A(O,X)	Load from table
0089	F7	8022	STA-B	Address of the display (digits)
008C	B7	8020	STA-A	Address of the display (segments)
008F	C6	FF	LDA-B#	Delay for proper timing of the display
0091	5A		DEC-B	
0092	26	FD	BNE	
0094	39		RTS	

The PIA has been initialized by the J bug.

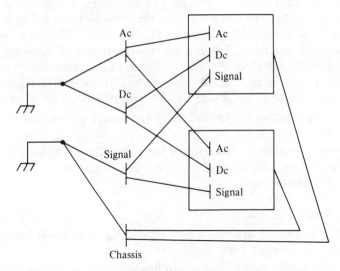

Figure 10.35

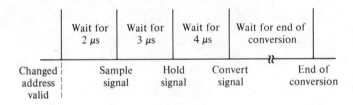

Figure 10.36

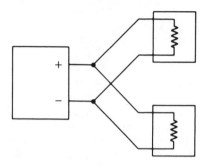

Figure 10.37

10.4. Wire in the limit switches on the planer (Example 8, Chapter 7) to the motor starter so that the limit switches actually *deenergize* the forward or reverse operating coils, rather than send a signal to the microprocessor, which then sends a deenergizing signal to the coils. Switching the operating coils via the limit switches offers a higher degree of security. The signals to energize the coils still come from the microprocessor. Select a suitable interface and show the wiring diagram to interface the limit switches to the microprocessor. The wiring method must be fail-safe; that is, if the interfacing lines become defective, the planer will not restart at the end of its travel.

10.5. Write a program to convert an 8-digit binary number into a 3-digit BCD code.

10.6. Write a program to determine if the output from a bipolar A/D converter is positive or negative. Assume the output code is:
a. 2's complement.
b. Offset binary.
c. Sign magnitude.

10.7. Write a program to convert 8-bit bipolar sign magnitude binary code into offset binary code.

10.8. High-security data signals are to be run across an electrically noisy environment; illustrate the wiring techniques used. Assume the peak-to-peak signal noise is *not* greater than $\frac{1}{2}$ LSB.

Appendix A

6800 Instruction Set and Summary

The following nomenclature is used in the subsequent definitions.

(a) *Operators*

( )	= contents of
←	= is transferred to
↑	= "is pulled from stack"
↓	= "is pushed into stack"
·	= Boolean AND
⊙	= Boolean (Inclusive) OR
⊕	= Exclusive OR
≈	= Boolean NOT

(b) *Registers in the MPU*

ACCA	= Accumulator A
ACCB	= Accumulator B
ACCX	= Accumulator ACCA or ACCB
CC	= Condition codes register
IX	= Index register, 16 bits
IXH	= Index register, higher order 8 bits
IXL	= Index register, lower order 8 bits
PC	= Program counter, 16 bits
PCH	= Program counter, higher order 8 bits

PCL = Program counter, lower order 8 bits
SP = Stack pointer
SPH = Stack pointer high
SPL = Stack pointer low

(c) *Memory and Addressing*

M = A memory location (one byte)
M + 1 = The byte of memory at 0001 plus the address of the memory location indicated by "M"
Rel = Relative address (i.e., the 2's complement number stored in the second byte of machine code corresponding to a branch instruction)

(d) *Bits 0 through 5 of the Condition Codes Register*

C = Carry—borrow   bit 0
V = Two's complement overflow
      indicator   bit 1
Z = Zero indicator   bit 2
N = Negative indicator   bit 3
I = Interrupt mask   bit 4
H = Half-carry   bit 5

MC 6800 Microprocessor Instruct Set Summary

The instruction set of the 6800 can be divided into four specific groups.

1. Data transfer instructions
2. Arithmetic/logic instructions
3. Control transfer instructions
4. Processor control instructions

Each of these groups will now be considered, and the instructions pertaining to each group will be itemized in alphabetical order in the mnemonic form of the instruction set.

The data sources of the 6800 are as follows:

a. I/O devices
b. Immediate
c. Memory locations
d. Eight-bit registers A, B and condition codes
e. Sixteen-bit registers X, PC and stack pointer

1. Data transfer instructions can be divided into four basic groups: (a) register to register, (b) register to memory, (c) immediate, and (d) I/O.

a. Register-to-register transfers: TAX, TXA, TAP, TPA, TXS, TSX.

b. Register to memory: LDA, LDX, LDS, STA, STX, STS, PSH, PUL.

c. Immediate: LDA, LDX, LDS.

d. I/O: uses the PIA port and/or the ACIA as a data source/data destination.

2. Arithmetic/logic instructions: any instruction that can be used to modify the contents of an internal register or flag.

a. Two-byte instructions: ADC, ADD, AND, BIT, CMP, CPX, EOR, ORA, SBC, SUB.

b. One-byte instructions: ABA, ASL, ASR, CBA, CLC, CLR, CLV, COM, DAA, DEC, DES, DEX, INC, INS, INX, LSR, NEG, ROL, ROR, SEC, SEV, TST.

3. Transfer of control instructions is any instruction that is used to transfer the execution of the program from its current location to some other location. These instructions can be further subdivided into *returning* and *nonreturning* instructions.

a. Returning instructions: BSR, JSR.

b. Nonreturning instructions: BCC, BCS, BEQ, BGE, BGT, BHI, BLE, BLS, BLT, BMI, BNE, BPL, BRA, BVC, BVS, JMP, RTI, RTS.

4. Processor control instructions, used to control the processor's operation: CLI, NOP, SEI, SWI, WAI.

Add Accumulator B to Accumulator A   **ABA**

Operation:	ACCA ← (ACCA) + (ACCB)
Description:	Adds the contents of ACCB to the contents of ACCA and places the result in ACCA.
Condition Codes:	H: Set if there was a carry from bit 3; cleared otherwise.
	I: Not affected.
	N: Set if most significant bit of the result is set; cleared otherwise.
	Z: Set if all bits of the result are cleared; cleared otherwise.
	V: Set if there was two's complement overflow as a result of the operation; cleared otherwise.
	C: Set if there was a carry from the most significant bit of the result; cleared otherwise.

Boolean Formulae for Condition Codes:

$$H = A_3 \cdot B_3 + B_3 \cdot \bar{R}_3 + \bar{R}_3 \cdot A_3$$
$$N = R_7$$
$$Z = \bar{R}_7 \cdot \bar{R}_6 \cdot \bar{R}_5 \cdot \bar{R}_4 \cdot \bar{R}_3 \cdot \bar{R}_2 \cdot \bar{R}_1 \cdot \bar{R}_0$$
$$V = A_7 \cdot B_7 \cdot \bar{R}_7 + \bar{A}_7 \cdot \bar{B}_7 \cdot R_7$$
$$C = A_7 \cdot B_7 + B_7 \cdot \bar{R}_7 + \bar{R}_7 \cdot A_7$$

Addressing Modes, Execution Time, and Machine Code (hexadecimal / octal / decimal):

Addressing Modes	Execution Time (No. of cycles)	Number of bytes of machine code	Coding of First (or only) byte of machine code		
			HEX.	OCT.	DEC.
Inherent	2	1	1B	033	027

Add with Carry   ADC

Operation:	ACCX ← (ACCX) + (M) + (C)
Description:	Adds the contents of the C bit to the sum of the contents of ACCX and M, and places the result in ACCX.

Condition Codes: H: Set if there was a carry from bit 3; cleared otherwise.
                      I: Not affected.
                      N: Set if most significant bit of the result is set; cleared otherwise.
                      Z: Set if all bits of the result are cleared; cleared otherwise.
                      V: Set if there was two's complement overflow as a result of the operation; cleared otherwise.
                      C: Set if there was a carry from the most significant bit of the result; cleared otherwise.

Boolean Formulae for Condition Codes:

$$H = X_3 \cdot M_3 + M_3 \cdot \bar{R}_3 + \bar{R}_3 \cdot X_3$$
$$N = R_7$$
$$Z = \bar{R}_7 \cdot \bar{R}_6 \cdot \bar{R}_5 \cdot \bar{R}_4 \cdot \bar{R}_3 \cdot \bar{R}_2 \cdot \bar{R}_1 \cdot \bar{R}_0$$
$$V = X_7 \cdot M_7 \cdot \bar{R}_7 + \bar{X}_7 \cdot \bar{M}_7 \cdot R_7$$
$$C = X_7 \cdot M_7 + M_7 \cdot \bar{R}_7 + \bar{R}_7 \cdot X_7$$

Addressing Modes, Execution Time, and Machine Code (hexadecimal / octal / decimal):

(DUAL OPERAND)

Addressing Modes	Execution Time (No. of cycles)	Number of bytes of machine code	Coding of First (or only) byte of machine code		
			HEX.	OCT.	DEC.
A IMM	2	2	89	211	137
A DIR	3	2	99	231	153
A EXT	4	3	B9	271	185
A IND	5	2	A9	251	169
B IMM	2	2	C9	311	201
B DIR	3	2	D9	331	217
B EXT	4	3	F9	371	249
B IND	5	2	E9	351	233

Add Without Carry

ADD

Operation: $\quad$ ACCX $\leftarrow$ (ACCX) + (M)

Description: $\quad$ Adds the contents of ACCX and the contents of M and places the result in ACCX.

Condition Codes:
- H: Set if there was a carry from bit 3; cleared otherwise.
- I: Not affected.
- N: Set if most significant bit of the result is set; cleared otherwise.
- Z: Set if all bits of the result are cleared; cleared otherwise.
- V: Set if there was two's complement overflow as a result of the operation; cleared otherwise.
- C: Set if there was a carry from the most significant bit of the result; cleared otherwise.

Boolean Formulae for Condition Codes:

$H = X_3 \cdot M_3 + M_3 \cdot \bar{R}_3 + \bar{R}_3 \cdot X_3$

$N = R_7$

$Z = \bar{R}_7 \cdot \bar{R}_6 \cdot \bar{R}_5 \cdot \bar{R}_4 \cdot \bar{R}_3 \cdot \bar{R}_2 \cdot \bar{R}_1 \cdot \bar{R}_0$

$V = X_7 \cdot M_7 \cdot \bar{R}_7 + \bar{X}_7 \cdot \bar{M}_7 \cdot R_7$

$C = X_7 \cdot M_7 + M_7 \cdot \bar{R}_7 + \bar{R}_7 \cdot X_7$

Addressing Modes, Execution Time, and Machine Code (hexadecimal/ octal/ decimal):

(DUAL OPERAND)

Addressing Modes	Execution Time (No. of cycles)	Number of bytes of machine code	Coding of First (or only) byte of machine code		
			HEX.	OCT.	DEC.
A IMM	2	2	8B	213	139
A DIR	3	2	9B	233	155
A EXT	4	3	BB	273	187
A IND	5	2	AB	253	171
B IMM	2	2	CB	313	203
B DIR	3	2	DB	333	219
B EXT	4	3	FB	373	251
B IND	5	2	EB	353	235

Logical AND

AND

Operation: $\quad$ ACCX $\leftarrow$ (ACCX) $\cdot$ (M)

Description: $\quad$ Performs logical "AND" between the contents of ACCX and the contents of M and places the result in ACCX. (Each bit of ACCX after the operation will be the logical "AND" of the corresponding bits of M and of ACCX before the operation.)

Condition Codes:
- H: Not affected.
- I: Not affected.
- N: Set if most significant bit of the result is set; cleared otherwise.
- Z: Set if all bits of the result are cleared; cleared otherwise.
- V: Cleared.
- C: Not affected.

Boolean Formulae for Condition Codes:

$N = R_7$

$Z = \bar{R}_7 \cdot \bar{R}_6 \cdot \bar{R}_5 \cdot \bar{R}_4 \cdot \bar{R}_3 \cdot \bar{R}_2 \cdot \bar{R}_1 \cdot \bar{R}_0$

$V = 0$

Addressing Modes, Execution Time, and Machine Code (hexadecimal / octal / decimal):

Addressing Modes	Execution Time (No. of cycles)	Number of bytes of machine code	Coding of First (or only) byte of machine code		
			HEX.	OCT.	DEC.
A IMM	2	2	84	204	132
A DIR	3	2	94	224	148
A EXT	4	3	B4	264	180
A IND	5	2	A4	244	164
B IMM	2	2	C4	304	196
B DIR	3	2	D4	324	212
B EXT	4	3	F4	364	244
B IND	5	2	E4	344	228

Arithmetic Shift Left   **ASL**

Operation:

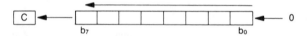

Description: Shifts all bits of the ACCX or M one place to the left. Bit 0 is loaded with a zero. The C bit is loaded from the most significant bit of ACCX or M.

Condition Codes: H: Not affected.
                I: Not affected.
                N: Set if most significant bit of the result is set; cleared otherwise.
                Z: Set if all bits of the result are cleared; cleared otherwise.
                V: Set if, after the completion of the shift operation, EITHER (N is set and C is cleared) OR (N is cleared and C is set); cleared otherwise.
                C: Set if, before the operation, the most significant bit of the ACCX or M was set; cleared otherwise.

Boolean Formulae for Condition Codes:

$$N = R_7$$
$$Z = \overline{R}_7 \cdot \overline{R}_6 \cdot \overline{R}_5 \cdot \overline{R}_4 \cdot \overline{R}_3 \cdot \overline{R}_2 \cdot \overline{R}_1 \cdot \overline{R}_0$$
$$V = N \oplus C = [N \cdot \overline{C}] \odot [\overline{N} \cdot C]$$

(the foregoing formula assumes values of N and C after the shift operation)

$$C = M_7$$

Addressing Modes, Execution Time, and Machine Code (hexadecimal / octal / decimal):

Addressing Modes	Execution Time (No. of cycles)	Number of bytes of machine code	Coding of First (or only) byte of machine code		
			HEX.	OCT.	DEC.
A	2	1	48	110	072
B	2	1	58	130	088
EXT	6	3	78	170	120
IND	7	2	68	150	104

Arithmetic Shift Right   **ASR**

Operation:

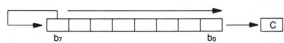

Description: Shifts all bits of ACCX or M one place to the right. Bit 7 is held constant. Bit 0 is loaded into the C bit.

Condition Codes:
H: Not affected.
I: Not affected.
N: Set if the most significant bit of the result is set; cleared otherwise.
Z: Set if all bits of the result are cleared; cleared otherwise.
V: Set if, after the completion of the shift operation, EITHER (N is set and C is cleared) OR (N is cleared and C is set); cleared otherwise.
C: Set if, before the operation, the least significant bit of the ACCX or M was set; cleared otherwise.

Boolean Formulae for Condition Codes:

$N = R_7$

$Z = \overline{R}_7 \cdot \overline{R}_6 \cdot \overline{R}_5 \cdot \overline{R}_4 \cdot \overline{R}_3 \cdot \overline{R}_2 \cdot \overline{R}_1 \cdot \overline{R}_0$

$V = N \oplus C = [N \cdot \overline{C}] \odot [\overline{N} \cdot C]$

(the foregoing formula assumes values of N and C after the shift operation)

$C = M_0$

Addressing Modes, Execution Time, and Machine Code (hexadecimal/octal/decimal):

Addressing Modes	Execution Time (No. of cycles)	Number of bytes of machine code	Coding of First (or only) byte of machine code		
			HEX.	OCT.	DEC.
A	2	1	47	107	071
B	2	1	57	127	087
EXT	6	3	77	167	119
IND	7	2	67	147	103

Branch if Carry Clear   **BCC**

Operation: PC ← (PC) + 0002 + Rel if (C)=0

Description: Tests the state of the C bit and causes a branch if C is clear.

See BRA instruction for further details of the execution of the branch.

Condition Codes: Not affected.

Addressing Modes, Execution Time, and Machine Code (hexadecimal/octal/decimal):

Addressing Modes	Execution Time (No. of cycles)	Number of bytes of machine code	Coding of First (or only) byte of machine code		
			HEX.	OCT.	DEC.
REL	4	2	24	044	036

Branch if Carry Set   **BCS**

Operation: PC ← (PC) + 0002 + Rel if (C)=1

Description: Tests the state of the C bit and causes a branch if C is set.

                   See BRA instruction for further details of the execution of the branch.

Condition Codes: Not affected.

Addressing Modes, Execution Time, and Machine Code (hexadecimal/octal/decimal):

Addressing Modes	Execution Time (No. of cycles)	Number of bytes of machine code	Coding of First (or only) byte of machine code		
			HEX.	OCT.	DEC.
REL	4	2	25	045	037

Branch if Equal   **BEQ**

Operation: PC ← (PC) + 0002 + Rel if (Z)=1

Description: Tests the state of the Z bit and causes a branch if the Z bit is set.

                   See BRA instruction for further details of the execution of the branch.

Condition Codes: Not affected.

Addressing Modes, Execution Time, and Machine Code (hexadecimal/octal/decimal):

Addressing Modes	Execution Time (No. of cycles)	Number of bytes of machine code	Coding of First (or only) byte of machine code		
			HEX.	OCT.	DEC.
REL	4	2	27	047	039

Branch if Greater than or Equal to Zero   **BGE**

Operation: PC ← (PC) + 0002 + Rel if (N) $\oplus$ (V) = 0

                          i.e. if (ACCX) ≥ (M)
                          (Two's complement numbers)

Description: Causes a branch if (N is set and V is set) OR (N is clear and V is clear).

                   If the BGE instruction is executed immediately after execution of any of the instructions CBA, CMP, SBA, or SUB, the branch will occur if and only if the two's complement number represented by the minuend (i.e. ACCX) was greater than or equal to the two's complement number represented by the subtrahend (i.e. M).

                   See BRA instruction for details of the branch.

Condition Codes: Not affected.

Addressing Modes, Execution Time, and Machine Code (hexadecimal/octal/decimal):

Addressing Modes	Execution Time (No. of cycles)	Number of bytes of machine code	Coding of First (or only) byte of machine code		
			HEX.	OCT.	DEC.
REL	4	2	2C	054	044

Branch if Greater than Zero   BGT

Operation: PC ← (PC) + 0002 + Rel if (Z) ⊙ [(N) ⊕ (V)] = 0

i.e. if (ACCX) > (M)

(two's complement numbers)

Description: Causes a branch if [Z is clear] AND [(N is set and V is set) OR (N is clear and V is clear)].

If the BGT instruction is executed immediately after execution of any of the instructions CBA, CMP, SBA, or SUB, the branch will occur if and only if the two's complement number represented by the minuend (i.e. ACCX) was greater than the two's complement number represented by the subtrahend (i.e. M).

See BRA instruction for details of the branch.

Condition Codes: Not affected.

Addressing Modes, Execution Time, and Machine Code (hexadecimal/octal/decimal):

Addressing Modes	Execution Time (No. of cycles)	Number of bytes of machine code	Coding of First (or only) byte of machine code		
			HEX.	OCT.	DEC.
REL	4	2	2E	056	046

Branch if Higher   BHI

Operation: PC ← (PC) + 0002 + Rel if (C) · (Z)=0

i.e. if (ACCX) > (M)

(unsigned binary numbers)

Description: Causes a branch if (C is clear) AND (Z is clear).

If the BHI instruction is executed immediately after execution of any of the instructions CBA, CMP, SBA, or SUB, the branch will occur if and only if the unsigned binary number represented by the minuend (i.e. ACCX) was greater than the unsigned binary number represented by the subtrahend (i.e. M).

See BRA instruction for details of the execution of the branch.

Condition Codes: Not affected.

Addressing Modes, Execution Time, and Machine Code (hexadecimal/octal/decimal):

Addressing Modes	Execution Time (No. of cycles)	Number of bytes of machine code	Coding of First (or only) byte of machine code		
			HEX.	OCT.	DEC.
REL	4	2	22	042	034

Bit Test

BIT

Operation: (ACCX) · (M)

Description: Performs the logical "AND" comparison of the contents of ACCX and the contents of M and modifies condition codes accordingly. Neither the contents of ACCX or M operands are affected. (Each bit of the result of the "AND" would be the logical "AND" of the corresponding bits of M and ACCX.)

Condition Codes: H: Not affected.
                  I: Not affected.
                  N: Set if the most significant bit of the result of the "AND" would be set; cleared otherwise.
                  Z: Set if all bits of the result of the "AND" would be cleared; cleared otherwise.
                  V: Cleared.
                  C: Not affected.

Boolean Formulae for Condition Codes:

$$N = R_7$$
$$Z = \overline{R_7} \cdot \overline{R_6} \cdot \overline{R_5} \cdot \overline{R_4} \cdot \overline{R_3} \cdot \overline{R_2} \cdot \overline{R_1} \cdot \overline{R_0}$$
$$V = 0$$

Addressing Modes, Execution Time, and Machine Code (hexadecimal/octal/decimal):

Addressing Modes	Execution Time (No. of cycles)	Number of bytes of machine code	Coding of First (or only) byte of machine code		
			HEX.	OCT.	DEC.
A IMM	2	2	85	205	133
A DIR	3	2	95	225	149
A EXT	4	3	B5	265	181
A IND	5	2	A5	245	165
B IMM	2	2	C5	305	197
B DIR	3	2	D5	325	213
B EXT	4	3	F5	365	245
B IND	5	2	E5	345	229

Branch if Less than or Equal to Zero

BLE

Operation: PC ← (PC) + 0002 + Rel if (Z)⊙[(N) ⊕ (V)]=1

                        i.e. if (ACCX) ≤ (M)
                    (two's complement numbers)

Description: Causes a branch if [Z is set] OR [(N is set and V is clear) OR (N is clear and V is set)].

If the BLE instruction is executed immediately after execution of any of the instructions CBA, CMP, SBA, or SUB, the branch will occur if and only if the two's complement number represented by the minuend (i.e. ACCX) was less then or equal to the two's complement number represented by the subtrahend (i.e. M).

See BRA instruction for details of the branch.

Condition Codes: Not affected.

Addressing Modes, Execution Time, and Machine Code (hexadecimal / octal / decimal):

Addressing Modes	Execution Time (No. of cycles)	Number of bytes of machine code	Coding of First (or only) byte of machine code		
			HEX.	OCT.	DEC.
REL	4	2	2F	057	047

Branch if Lower or Same BLS

Operation: $PC \leftarrow (PC) + 0002 + Rel$ if $(C) \odot (Z) = 1$

i.e. if $(ACCX) \leq (M)$

(unsigned binary numbers)

Description: Causes a branch if (C is set) OR (Z is set).

If the BLS instruction is executed immediately after execution of any of the instructions CBA, CMP, SBA, or SUB, the branch will occur if and only if the unsigned binary number represented by the minuend (i.e. ACCX) was less than or equal to the unsigned binary number represented by the subtrahend (i.e. M).

See BRA instruction for details of the execution of the branch.

Condition Codes: Not affected.

Addressing Modes, Execution Time, and Machine Code (hexadecimal / octal / decimal):

Addressing Modes	Execution Time (No. of cycles)	Number of bytes of machine code	Coding of First (or only) byte of machine code		
			HEX.	OCT.	DEC.
REL	4	2	23	043	035

Branch if Less than Zero BLT

Operation: $PC \leftarrow (PC) + 0002 + Rel$ if $(N) \oplus (V) = 1$

i.e. if $(ACCX) < (M)$

(two's complement numbers)

Description: Causes a branch if (N is set and V is clear) OR (N is clear and V is set).

If the BLT instruction is executed immediately after execution of any of the instructions CBA, CMP, SBA, or SUB, the branch will occur if and only if the two's complement number represented by the minuend (i.e. ACCX) was less than the two's complement number represented by the subtrahend (i.e. M).

See BRA instruction for details of the branch.

Condition Codes: Not affected.

Addressing Modes, Execution Time, and Machine Code (hexadecimal/octal/decimal):

Addressing Modes	Execution Time (No. of cycles)	Number of bytes of machine code	Coding of First (or only) byte of machine code		
			HEX.	OCT.	DEC.
REL	4	2	2D	055	045

Branch if Minus   BMI

Operation: $PC \leftarrow (PC) + 0002 + Rel$ if $(N) = 1$

Description: Tests the state of the N bit and causes a branch if N is set.

See BRA instruction for details of the execution of the branch.

Condition Codes: Not affected.

Addressing Modes, Execution Time, and Machine Code (hexadecimal/octal/decimal):

Addressing Modes	Execution Time (No. of cycles)	Number of bytes of machine code	Coding of First (or only) byte of machine code		
			HEX.	OCT.	DEC.
REL	4	2	2B	053	043

Branch if Not Equal   BNE

Operation: $PC \leftarrow (PC) + 0002 + Rel$ if $(Z) = 0$

Description: Tests the state of the Z bit and causes a branch if the Z bit is clear.

See BRA instruction for details of the execution of the branch.

Condition Codes: Not affected.

Addressing Modes, Execution Time, and Machine Code (hexadecimal/octal/decimal):

Addressing Modes	Execution Time (No. of cycles)	Number of bytes of machine code	Coding of First (or only) byte of machine code		
			HEX.	OCT.	DEC.
REL	4	2	26	046	038

Branch if Plus

BPL

Operation: PC ← (PC) + 0002 + Rel if (N) = 0

Description: Tests the state of the N bit and causes a branch if N is clear.

                   See BRA instruction for details of the execution of the branch.

Condition Codes: Not affected.

Addressing Modes, Execution Time, and Machine Code (hexadecimal/octal/decimal):

Addressing Modes	Execution Time (No. of cycles)	Number of bytes of machine code	Coding of First (or only) byte of machine code		
			HEX.	OCT.	DEC.
REL	4	2	2A	052	042

Branch Always

BRA

Operation: PC ← (PC) + 0002 + Rel

Description: Unconditional branch to the address given by the foregoing formula, in which R is the relative address stored as a two's complement number in the second byte of machine code corresponding to the branch instruction.

                   Note: The source program specifies the destination of any branch instruction by its absolute address, either as a numerical value or as a symbol or expression which can be numerically evaluated by the assembler. The assembler obtains the relative address R from the absolute address and the current value of the program counter PC.

Condition Codes: Not affected.

Addressing Modes, Execution Time, and Machine Code (hexadecimal/octal/decimal):

Addressing Modes	Execution Time (No. of cycles)	Number of bytes of machine code	Coding of First (or only) byte of machine code		
			HEX.	OCT.	DEC.
REL	4	2	20	040	032

Branch to Subroutine

BSR

Operation: PC ← (PC) + 0002
                   ↓ (PCL)
                   SP ← (SP) − 0001
                   ↓ (PCH)
                   SP ← (SP) − 0001
                   PC ← (PC) + Rel

Description: The program counter is incremented by 2. The less significant byte of the contents of the program counter is pushed into the stack. The stack pointer is then decremented (by 1). The more significant byte of the contents of the program counter is then pushed into the stack. The stack pointer is again decremented (by 1). A branch then occurs to the location specified by the program.

See BRA instruction for details of the execution of the branch.

Condition Codes: Not affected.

Addressing Modes, Execution Time, and Machine Code (hexadecimal/ octal/ decimal):

Addressing Modes	Execution Time (No. of cycles)	Number of bytes of machine code	Coding of First (or only) byte of machine code		
			HEX.	OCT.	DEC.
REL	8	2	8D	215	141

Branch if Overflow Clear   **BVC**

Operation: $PC \leftarrow (PC) + 0002 + Rel$ if $(V) = 0$

Description: Tests the state of the V bit and causes a branch if the V bit is clear.

See BRA instruction for details of the execution of the branch.

Condition Codes: Not affected.

Addressing Modes, Execution Time, and Machine Code (hexadecimal/ octal/ decimal):

Addressing Modes	Execution Time (No. of cycles)	Number of bytes of machine code	Coding of First (or only) byte of machine code		
			HEX.	OCT.	DEC.
REL	4	2	28	050	040

Branch if Overflow Set   **BVS**

Operation: $PC \leftarrow (PC) + 0002 + Rel$ if $(V) = 1$

Description: Tests the state of the V bit and causes a branch if the V bit is set.

See BRA instruction for details of the execution of the branch.

Condition Codes: Not affected.

Addressing Modes, Execution Time, and Machine Code (hexadecimal/octal/decimal):

Addressing Modes	Execution Time (No. of cycles)	Number of bytes of machine code	Coding of First (or only) byte of machine code		
			HEX.	OCT.	DEC.
REL	4	2	29	051	041

Compare Accumulators   **CBA**

Operation: (ACCA) − (ACCB)

Description: Compares the contents of ACCA and the contents of ACCB and sets the condition codes, which may be used for arithmetic and logical conditional branches. Both operands are unaffected.

Condition Codes: H: Not affected.
                   I: Not affected.
                   N: Set if the most significant bit of the result of the subtraction would be set; cleared otherwise.
                   Z: Set if all bits of the result of the subtraction would be cleared; cleared otherwise.
                   V: Set if the subtraction would cause two's complement overflow; cleared otherwise.
                   C: Set if the subtraction would require a borrow into the most significant bit of the result; clear otherwise.

Boolean Formulae for Condition Codes:

$$N = R_7$$
$$Z = \overline{R_7} \cdot \overline{R_6} \cdot \overline{R_5} \cdot \overline{R_4} \cdot \overline{R_3} \cdot \overline{R_2} \cdot \overline{R_1} \cdot \overline{R_0}$$
$$V = A_7 \cdot \overline{B_7} \cdot \overline{R_7} + \overline{A_7} \cdot B_7 \cdot R_7$$
$$C = \overline{A_7} \cdot B_7 + B_7 \cdot R_7 + R_7 \cdot \overline{A_7}$$

Addressing Modes, Execution Time, and Machine Code (hexadecimal/octal/decimal):

Addressing Modes	Execution Time (No. of cycles)	Number of bytes of machine code	Coding of First (or only) byte of machine code		
			HEX.	OCT.	DEC.
INHERENT	2	1	11	021	017

Clear Carry   **CLC**

Operation: C bit ← 0

Description: Clears the carry bit in the processor condition codes register.

Condition Codes: H: Not affected.
                   I: Not affected.
                   N: Not affected.
                   Z: Not affected.
                   V: Not affected.
                   C: Cleared

Boolean Formulae for Condition Codes:
                   $C = 0$

Addressing Modes, Execution Time, and Machine Code (hexadecimal/octal/decimal):

Addressing Modes	Execution Time (No. of cycles)	Number of bytes of machine code	Coding of First (or only) byte of machine code		
			HEX.	OCT.	DEC.
INHERENT	2	1	0C	014	012

CLI

Clear Interrupt Mask

Operation: I bit ← 0

Description: Clears the interrupt mask bit in the processor condition codes register. This enables the microprocessor to service an interrupt from a peripheral device if signalled by a high state of the "Interrupt Request" control input.

Condition Codes: H: Not affected.   Z: Not affected.
                 I: Cleared.   V: Not affected.
                 N: Not affected.   C: Not affected.

Boolean Formulae for Condition Codes:
    I = 0

Addressing Modes, Execution Time, and Machine Code (hexadecimal/octal/decimal):

Addressing Modes	Execution Time (No. of cycles)	Number of bytes of machine code	Coding of First (or only) byte of machine code		
			HEX.	OCT.	DEC.
INHERENT	2	1	0E	016	014

CLR

Clear

Operation: ACCX ← 00
or: M ← 00

Description: The contents of ACCX or M are replaced with zeros.

Condition Codes: H: Not affected.   Z: Set
                 I: Not affected.   V: Cleared
                 N: Cleared   C: Cleared

Boolean Formulae for Condition Codes:
    N = 0   V = 0
    Z = 1   C = 0

Addressing Modes, Execution Time, and Machine Code (hexadecimal/octal/decimal):

Addressing Modes	Execution Time (No. of cycles)	Number of bytes of machine code	Coding of First (or only) byte of machine code		
			HEX.	OCT.	DEC.
A	2	1	4F	117	079
B	2	1	5F	137	095
EXT	6	3	7F	177	127
IND	7	2	6F	157	111

Clear Two's Complement Overflow Bit   **CLV**

Operation: V bit ← 0

Description: Clears the two's complement overflow bit in the processor condition codes register.

Condition Codes: H: Not affected.   Z: Not affected.
                     I: Not affected.   V: Cleared.
                     N: Not affected.   C: Not affected.

Boolean Formulae for Condition Codes:
     V = 0

Addressing Modes, Execution Time, and Machine Code (hexadecimal/octal/decimal):

Addressing Modes	Execution Time (No. of cycles)	Number of bytes of machine code	Coding of First (or only) byte of machine code		
			HEX.	OCT.	DEC.
INHERENT	2	1	0A	012	010

Compare   **CMP**

Operation: (ACCX) − (M)

Description: Compares the contents of ACCX and the contents of M and determines the condition codes, which may be used subsequently for controlling conditional branching. Both operands are unaffected.

Condition Codes: H: Not affected.
                     I: Not affected.
                     N: Set if the most significant bit of the result of the subtraction would be set; cleared otherwise.
                     Z: Set if all bits of the result of the subtraction would be cleared; cleared otherwise.
                     V: Set if the subtraction would cause two's complement overflow; cleared otherwise.
                     C: Carry is set if the absolute value of the contents of memory is larger than the absolute value of the accumulator; reset otherwise.

Boolean Formulae for Condition Codes:
$$N = R_7$$
$$Z = \overline{R}_7 \cdot \overline{R}_6 \cdot \overline{R}_5 \cdot \overline{R}_4 \cdot \overline{R}_3 \cdot \overline{R}_2 \cdot \overline{R}_1 \cdot \overline{R}_0$$
$$V = X_7 \cdot \overline{M}_7 \cdot \overline{R}_7 + \overline{X}_7 \cdot M_7 \cdot R_7$$
$$C = \overline{X}_7 \cdot M_7 + M_7 \cdot R_7 + R_7 \cdot \overline{X}_7$$

Addressing Modes, Execution Time, and Machine Code (hexadecimal/octal/decimal):

(DUAL OPERAND)

Addressing Modes	Execution Time (No. of cycles)	Number of bytes of machine code	Coding of First (or only) byte of machine code		
			HEX.	OCT.	DEC.
A IMM	2	2	81	201	129
A DIR	3	2	91	221	145
A EXT	4	3	B1	261	177
A IND	5	2	A1	241	161
B IMM	2	2	C1	301	193
B DIR	3	2	D1	321	209
B EXT	4	3	F1	361	241
B IND	5	2	E1	341	225

Complement

COM

Operation: $ACCX \leftarrow \approx (ACCX) = FF - (ACCX)$

or: $M \leftarrow \approx (M) = FF - (M)$

Description: Replaces the contents of ACCX or M with its one's complement. (Each bit of the contents of ACCX or M is replaced with the complement of that bit.)

Condition Codes:
- H: Not affected.
- I: Not affected.
- N: Set if most significant bit of the result is set; cleared otherwise.
- Z: Set if all bits of the result are cleared; cleared otherwise.
- V: Cleared.
- C: Set.

Boolean Formulae for Condition Codes:

$N = R_7$

$Z = \overline{R_7} \cdot \overline{R_6} \cdot \overline{R_5} \cdot \overline{R_4} \cdot \overline{R_3} \cdot \overline{R_2} \cdot \overline{R_1} \cdot \overline{R_0}$

$V = 0$

$C = 1$

Addressing Modes, Execution Time, and Machine Code (hexadecimal/octal/decimal):

Addressing Modes	Execution Time (No. of cycles)	Number of bytes of machine code	Coding of First (or only) byte of machine code		
			HEX.	OCT.	DEC.
A	2	1	43	103	067
B	2	1	53	123	083
EXT	6	3	73	163	115
IND	7	2	63	143	099

Compare Index Register

CPX

Operation: $(IXL) - (M+1)$

                  $(IXH) - (M)$

Description: The more significant byte of the contents of the index register is compared with the contents of the byte of memory at the address specified by the program. The less significant byte of the contents of the index register is compared with the contents of the next byte of memory, at one plus the address specified by the program. The Z bit is set or reset according to the results of these comparisons, and may be used subsequently for conditional branching.

The N and V bits, though determined by this operation, are not intended for conditional branching.

The C bit is not affected by this operation.

Condition Codes:
- H: Not affected.
- I: Not affected.
- N: Set if the most significant bit of the result of the subtraction from the more significant byte of the index register would be set; cleared otherwise.
- Z: Set if all bits of the results of both subtractions would be cleared; cleared otherwise.
- V: Set if the subtraction from the more significant byte of the index register would cause two's complement overflow; cleared otherwise.
- C: Not affected.

Boolean Formulae for Condition Codes:

$$N = RH_7$$
$$Z = (\overline{RH_7} \cdot \overline{RH_6} \cdot \overline{RH_5} \cdot \overline{RH_4} \cdot \overline{RH_3} \cdot \overline{RH_2} \cdot \overline{RH_1} \cdot \overline{RH_0}) \cdot$$
$$(\overline{RL_7} \cdot \overline{RL_6} \cdot \overline{RL_5} \cdot \overline{RL_4} \cdot \overline{RL_3} \cdot \overline{RL_2} \cdot \overline{RL_1} \cdot \overline{RL_0})$$
$$V = IXH_7 \cdot \overline{M_7} \cdot \overline{RH_7} + \overline{IXH_7} \cdot M_7 \cdot RH_7$$

Addressing Modes, Execution Time, and Machine Code (hexadecimal / octal / decimal):

Addressing Modes	Execution Time (No. of cycles)	Number of bytes of machine code	Coding of First (or only) byte of machine code		
			HEX.	OCT.	DEC.
IMM	3	3	8C	214	140
DIR	4	2	9C	234	156
EXT	5	3	BC	274	188
IND	6	2	AC	254	172

Decimal Adjust ACCA   # DAA

Operation: Adds hexadecimal numbers 00, 06, 60, or 66 to ACCA, and may also set the carry bit, as indicated in the following table:

State of C-bit before DAA (Col. 1)	Upper Half-byte (bits 4-7) (Col. 2)	Initial Half-carry H-bit (Col.3)	Lower to ACCA (bits 0-3) (Col. 4)	Number Added after by DAA (Col. 5)	State of C-bit after DAA (Col. 6)
0	0-9	0	0-9	00	0
0	0-8	0	A-F	06	0
0	0-9	1	0-3	06	0
0	A-F	0	0-9	60	1
0	9-F	0	A-F	66	1
0	A-F	1	0-3	66	1
1	0-2	0	0-9	60	1
1	0-2	0	A-F	66	1
1	0-3	1	0-3	66	1

Note: Columns (1) through (4) of the above table represent all possible cases which can result from any of the operations ABA, ADD, or ADC, with initial carry either set or clear, applied to two binary-coded-decimal operands. The table shows hexadecimal values.

Description: If the contents of ACCA and the state of the carry-borrow bit C and the half-carry bit H are all the result of applying any of the operations ABA, ADD, or ADC to binary-coded-decimal operands, with or without an initial carry, the DAA operation will function as follows.

Subject to the above condition, the DAA operation will adjust the contents of ACCA and the C bit to represent the correct binary-coded-decimal sum and the correct state of the carry.

Condition Codes: H: Not affected.
               I: Not affected.
               N: Set if most significant bit of the result is set; cleared otherwise.
               Z: Set if all bits of the result are cleared; cleared otherwise.
               V: Not defined.
               C: Set or reset according to the same rule as if the DAA and an immediately preceding ABA, ADD, or ADC were replaced by a hypothetical binary-coded-decimal addition.

Boolean Formulae for Condition Codes:
$$N = R_7 \quad Z = \bar{R}_7 \cdot \bar{R}_6 \cdot \bar{R}_5 \cdot \bar{R}_4 \cdot \bar{R}_3 \cdot \bar{R}_2 \cdot \bar{R}_1 \cdot \bar{R}_0 \quad C = \text{See table above.}$$

Addressing Modes, Execution Time, and Machine Code (hexadecimal/octal/decimal):

Addressing Modes	Execution Time (No. of cycles)	Number of bytes of machine code	Coding of First (or only) byte of machine code		
			HEX.	OCT.	DEC.
INHERENT	2	1	19	031	025

Decrement

DEC

Operation: ACCX ← (ACCX) − 01
or: M ← (M) − 01

Description: Subtract one from the contents of ACCX or M.

The N, Z, and V condition codes are set or reset according to the results of this operation.

The C bit is not affected by the operation.

Condition Codes: H: Not affected.
               I: Not affected.
               N: Set if most significant bit of the result is set; cleared otherwise.
               Z: Set if all bits of the result are cleared; cleared otherwise.
               V: Set if there was two's complement overflow as a result of the operation; cleared otherwise. Two's complement overflow occurs if and only if (ACCX) or (M) was 80 before the operation.
               C: Not affected.

Boolean Formulae for Condition Codes:
$$N = R_7$$
$$Z = \bar{R}_7 \cdot \bar{R}_6 \cdot \bar{R}_5 \cdot \bar{R}_5 \cdot \bar{R}_4 \cdot \bar{R}_3 \cdot \bar{R}_2 \cdot \bar{R}_1 \cdot \bar{R}_0$$
$$V = X_7 \cdot \bar{X}_6 \cdot \bar{X}_5 \cdot \bar{X}_4 \cdot \bar{X}_3 \cdot \bar{X}_2 \cdot \bar{X}_0 = \bar{R}_7 \cdot R_6 \cdot R_5 \cdot R_4 \cdot R_3 \cdot R_2 \cdot R_1 \cdot R_0$$

Addressing Modes, Execution Time, and Machine Code (hexadecimal/octal/decimal):

Addressing Modes	Execution Time (No. of cycles)	Number of bytes of machine code	Coding of First (or only) byte of machine code		
			HEX.	OCT.	DEC.
A	2	1	4A	112	074
B	2	1	5A	132	090
EXT	6	3	7A	172	122
IND	7	2	6A	152	106

Decrement Stack Pointer

DES

Operation: $SP \leftarrow (SP) - 0001$

Description: Subtract one from the stack pointer.

Condition Codes: Not affected.

Addressing Modes, Execution Time, and Machine Code (hexadecimal/octal/decimal):

Addressing Modes	Execution Time (No. of cycles)	Number of bytes of machine code	Coding of First (or only) byte of machine code		
			HEX.	OCT.	DEC.
INHERENT	4	1	34	064	052

Decrement Index Register

DEX

Operation: $IX \leftarrow (IX) - 0001$

Description: Subtract one from the index register.

     Only the Z bit is set or reset according to the result of this operation.

Condition Codes: H: Not affected.

                   I: Not affected.

                   N: Not affected.

                   Z: Set if all bits of the result are cleared; cleared otherwise.

                   V: Not affected.

                   C: Not affected.

Boolean Formulae for Condition Codes:

$$Z = (\overline{RH_7} \cdot \overline{RH_6} \cdot \overline{RH_5} \cdot \overline{RH_4} \cdot \overline{RH_3} \cdot \overline{RH_2} \cdot \overline{RH_1} \cdot \overline{RH_0}) \cdot$$
$$(\overline{RL_7} \cdot \overline{RL_6} \cdot \overline{RL_5} \cdot \overline{RL_4} \cdot \overline{RL_3} \cdot \overline{RL_2} \cdot \overline{RL_1} \cdot \overline{RL_0})$$

Addressing Modes, Execution Time, and Machine Code (hexadecimal/octal/decimal):

Addressing Modes	Execution Time (No. of cycles)	Number of bytes of machine code	Coding of First (or only) byte of machine code		
			HEX.	OCT.	DEC.
INHERENT	4	1	09	011	009

Exclusive OR

EOR

Operation: $ACCX \leftarrow (ACCX) \oplus (M)$

Description: Perform logical "EXCLUSIVE OR" between the contents of ACCX and the contents of M, and place the result in ACCX. (Each bit of ACCX after the operation will be the logical "EXCLUSIVE OR" of the corresponding bit of M and ACCX before the operation.)

Condition Codes: H: Not affected.

                   I: Not affected.

                   N: Set if most significant bit of the result is set; cleared otherwise.

                   Z: Set if all bits of the result are cleared; cleared otherwise.

                   V: Cleared

                   C: Not affected.

Boolean Formulae for Condition Codes:

$$N = R_7$$
$$Z = \bar{R}_7 \cdot \bar{R}_6 \cdot \bar{R}_5 \cdot \bar{R}_4 \cdot \bar{R}_3 \cdot \bar{R}_2 \cdot \bar{R}_1 \cdot \bar{R}_0$$
$$V = 0$$

Addressing Modes, Execution Time, and Machine Code (hexadecimal/octal/decimal):

Addressing Modes	Execution Time (No. of cycles)	Number of bytes of machine code	Coding of First (or only) byte of machine code		
			HEX.	OCT.	DEC.
A IMM	2	2	88	210	136
A DIR	3	2	98	230	152
A EXT	4	3	B8	270	184
A IND	5	2	A8	250	168
B IMM	2	2	C8	310	200
B DIR	3	2	D8	330	216
B EXT	4	3	F8	370	248
B IND	5	2	E8	350	232

Increment

INC

Operation:	ACCX ← (ACCX) + 01
or:	M ← (M) + 01
Description:	Add one to the contents of ACCX or M.

The N, Z, and V condition codes are set or reset according to the results of this operation.

The C bit is not affected by the operation.

Condition Codes: H: Not affected.

I: Not affected.

N: Set if most significant bit of the result is set; cleared otherwise.

Z: Set if all bits of the result are cleared; cleared otherwise.

V: Set if there was two's complement overflow as a result of the operation; cleared otherwise. Two's complement overflow will occur if and only if (ACCX) or (M) was 7F before the operation.

C: Not affected.

Boolean Formulae for Condition Codes:

$$N = R_7$$
$$Z = \bar{R}_7 \cdot \bar{R}_6 \cdot \bar{R}_5 \cdot \bar{R}_4 \cdot \bar{R}_3 \cdot \bar{R}_2 \cdot \bar{R}_1 \cdot \bar{R}_0$$
$$V = \bar{X}_7 \cdot X_6 \cdot X_5 \cdot X_4 \cdot X_3 \cdot X_2 \cdot X_1 \cdot X_0$$
$$C = \bar{R}_7 \cdot \bar{R}_6 \cdot \bar{R}_5 \cdot \bar{R}_4 \cdot \bar{R}_3 \cdot \bar{R}_2 \cdot \bar{R}_1 \cdot \bar{R}_0$$

Addressing Modes, Execution Time, and Machine Code (hexadecimal/octal/decimal):

Addressing Modes	Execution Time (No. of cycles)	Number of bytes of machine code	Coding of First (or only) byte of machine code		
			HEX.	OCT.	DEC.
A	2	1	4C	114	076
B	2	1	5C	134	092
EXT	6	3	7C	174	124
IND	7	2	6C	154	108

INS

Increment Stack Pointer

Operation: $SP \leftarrow (SP) + 0001$

Description: Add one to the stack pointer.

Condition Codes: Not affected.

Addressing Modes, Execution Time, and Machine Code (hexadecimal/octal/decimal):

Addressing Modes	Execution Time (No. of cycles)	Number of bytes of machine code	Coding of First (or only) byte of machine code		
			HEX.	OCT.	DEC.
INHERENT	4	1	31	061	049

INX

Increment Index Register

Operation: $IX \leftarrow (IX) + 0001$

Description: Add one to the index register.

                  Only the Z bit is set or reset according to the result of this operation.

Condition Codes: H: Not affected.

                     I: Not affected.

                     N: Not affected.

                     Z: Set if all 16 bits of the result are cleared; cleared otherwise.

                     V: Not affected.

                     C: Not affected.

Boolean Formulae for Condition Codes:

$$Z = (\overline{RH_7} \cdot \overline{RH_6} \cdot \overline{RH_5} \cdot \overline{RH_4} \cdot \overline{RH_3} \cdot \overline{RH_2} \cdot \overline{RH_1} \cdot \overline{RH_0}) \cdot$$
$$(\overline{RL_7} \cdot \overline{RL_6} \cdot \overline{RL_5} \cdot \overline{RL_4} \cdot \overline{RL_3} \cdot \overline{RL_2} \cdot \overline{RL_1} \cdot \overline{RL_0})$$

Addressing Modes, Execution Time, and Machine Code (hexadecimal/octal/decimal):

Addressing Modes	Execution Time (No. of cycles)	Number of bytes of machine code	Coding of First (or only) byte of machine code		
			HEX.	OCT.	DEC.
INHERENT	4	1	08	010	008

JMP

Jump

Operation: $PC \leftarrow$ numerical address

Description: A jump occurs to the instruction stored at the numerical address. The numerical address is obtained according to the rules for EXTended or INDexed addressing.

Condition Codes: Not affected.

Addressing Modes, Execution Time, and Machine Code (hexadecimal/octal/decimal):

Addressing Modes	Execution Time (No. of cycles)	Number of bytes of machine code	Coding of First (or only) byte of machine code		
			HEX.	OCT.	DEC.
EXT	3	3	7E	176	126
IND	4	2	6E	156	110

Jump to Subroutine

JSR

Operation:

Either: PC ← (PC) + 0003 (for EXTended addressing)

or: PC ← (PC) + 0002 (for INDexed addressing)

Then: ↓ (PCL)

                            SP ← (SP) − 0001

                            ↓ (PCH)

                            SP ← (SP) − 0001

                            PC ← numerical address

Description: The program counter is incremented by 3 or by 2, depending on the addressing mode, and is then pushed onto the stack, eight bits at a time. The stack pointer points to the next empty location in the stack. A jump occurs to the instruction stored at the numerical address. The numerical address is obtained according to the rules for EXTended or INDexed addressing.

Condition Codes: Not affected.

Addressing Modes, Execution Time, and Machine Code (hexadecimal/octal/decimal):

Addressing Modes	Execution Time (No. of cycles)	Number of bytes of machine code	Coding of First (or only) byte of machine code		
			HEX.	OCT.	DEC.
EXT	9	3	BD	275	189
IND	8	2	AD	255	173

Load Accumulator

LDA

Operation: ACCX ← (M)

Description: Loads the contents of memory into the accumulator. The condition codes are set according to the data.

Condition Codes: H: Not affected.

                             I: Not affected.

                             N: Set if most significant bit of the result is set; cleared otherwise.

                             Z: Set if all bits of the result are cleared; cleared otherwise.

                             V: Cleared.

                             C: Not affected.

Boolean Formulae for Condition Codes:

$$N = R_7$$
$$Z = \overline{R_7} \cdot \overline{R_6} \cdot \overline{R_5} \cdot \overline{R_4} \cdot \overline{R_3} \cdot \overline{R_2} \cdot \overline{R_1} \cdot \overline{R_0}$$
$$V = 0$$

Addressing Modes, Execution Time, and Machine Code (hexadecimal/octal/decimal):

(DUAL OPERAND)

Addressing Modes	Execution Time (No. of cycles)	Number of bytes of machine code	Coding of First (or only) byte of machine code		
			HEX.	OCT.	DEC.
A IMM	2	2	86	206	134
A DIR	3	2	96	226	150
A EXT	4	3	B6	266	182
A IND	5	2	A6	246	166
B IMM	2	2	C6	306	198
B DIR	3	2	D6	326	214
B EXT	4	3	F6	366	246
B IND	5	2	E6	346	230

LDS

Load Stack Pointer

Operation: SPH ← (M)
SPL ← (M+1)

Description: Loads the more significant byte of the stack pointer from the byte of memory at the address specified by the program, and loads the less significant byte of the stack pointer from the next byte of memory, at one plus the address specified by the program.

Condition Codes: H: Not affected.
I: Not affected.
N: Set if the most significant bit of the stack pointer is set by the operation; cleared otherwise.
Z: Set if all bits of the stack pointer are cleared by the operation; cleared otherwise.
V: Cleared.
C: Not affected.

Boolean Formulae for Condition Codes:

$N = RH_7$
$Z = (\overline{RH_7} \cdot \overline{RH_6} \cdot \overline{RH_5} \cdot \overline{RH_4} \cdot \overline{RH_3} \cdot \overline{RH_2} \cdot \overline{RH_1} \cdot \overline{RH_0}) \cdot$
$(\overline{RL_7} \cdot \overline{RL_6} \cdot \overline{RL_5} \cdot \overline{RL_4} \cdot \overline{RL_3} \cdot \overline{RL_2} \cdot \overline{RL_1} \cdot \overline{RL_0})$
$V = 0$

Addressing Modes, Execution Time, and Machine Code (hexadecimal/octal/decimal):

Addressing Modes	Execution Time (No. of cycles)	Number of bytes of machine code	Coding of First (or only) byte of machine code		
			HEX.	OCT.	DEC.
IMM	3	3	8E	216	142
DIR	4	2	9E	236	158
EXT	5	3	BE	276	190
IND	6	2	AE	256	174

LDX

Load Index Register

Operation: IXH ← (M)
IXL ← (M+1)

Description: Loads the more significant byte of the index register from the byte of memory at the address specified by the program, and loads the less significant byte of the index register from the next byte of memory, at one plus the address specified by the program.

Condition Codes: H: Not affected.
                 I: Not affected.
                 N: Set if the most significant bit of the index register is set by the operation; cleared otherwise.
                 Z: Set if all bits of the index register are cleared by the operation; cleared otherwise.
                 V: Cleared.
                 C: Not affected.

Boolean Formulae for Condition Codes:

$$N = RH_7$$
$$Z = (\overline{RH_7} \cdot \overline{RH_6} \cdot \overline{RH_5} \cdot \overline{RH_4} \cdot \overline{RH_3} \cdot \overline{RH_2} \cdot \overline{RH_1} \cdot \overline{RH_0}) \cdot$$
$$(\overline{RL_7} \cdot \overline{RL_6} \cdot \overline{RL_5} \cdot \overline{RL_4} \cdot \overline{RL_3} \cdot \overline{RL_2} \cdot \overline{RL_1} \cdot \overline{RL_0})$$
$$V = 0$$

Addressing Modes, Execution Time, and Machine Code (hexadecimal / octal / decimal):

Addressing Modes	Execution Time (No. of cycles)	Number of bytes of machine code	Coding of First (or only) byte of machine code		
			HEX.	OCT.	DEC.
IMM	3	3	CE	316	206
DIR	4	2	DE	336	222
EXT	5	3	FE	376	254
IND	6	2	EE	356	238

Logical Shift Right   # LSR

Operation:

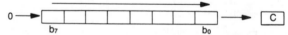

Description: Shifts all bits of ACCX or M one place to the right. Bit 7 is loaded with a zero. The C bit is loaded from the least significant bit of ACCX or M.

Condition Codes: H: Not affected.
                 I: Not affected.
                 N: Cleared.
                 Z: Set if all bits of the result are cleared; cleared otherwise.
                 V: Set if, after the completion of the shift operation, EITHER (N is set and C is cleared) OR (N is cleared and C is set); cleared otherwise.
                 C: Set if, before the operation, the least significant bit of the ACCX or M was set; cleared otherwise.

Boolean Formulae for Condition Codes:

$$N = 0$$
$$Z = \overline{R_7} \cdot \overline{R_6} \cdot \overline{R_5} \cdot \overline{R_4} \cdot \overline{R_3} \cdot \overline{R_2} \cdot \overline{R_1} \cdot \overline{R_0}$$
$$V = N \oplus C = [N \cdot \overline{C}] \odot [\overline{N} \cdot C]$$

           (the foregoing formula assumes values of N and C after the shift operation).

$$C = M_0$$

Addressing Modes, Execution Time, and Machine Code (hexadecimal/octal/decimal):

Addressing Modes	Execution Time (No. of cycles)	Number of bytes of machine code	Coding of First (or only) byte of machine code		
			HEX.	OCT.	DEC.
A	2	1	44	104	068
B	2	1	54	124	084
EXT	6	3	74	164	116
IND	7	2	64	144	100

Negate — NEG

Operation:	ACCX ← − (ACCX) = 00 − (ACCX)
or:	M ← − (M) = 00 − (M)
Description:	Replaces the contents of ACCX or M with its two's complement. Note that 80 is left unchanged.
Condition Codes:	H: Not affected.
	I: Not affected.
	N: Set if most significant bit of the result is set; cleared otherwise.
	Z: Set if all bits of the result are cleared; cleared otherwise.
	V: Set if there would be two's complement overflow as a result of the implied subtraction from zero; this will occur if and only if the contents of ACCX or M is 80.
	C: Set if there would be a borrow in the implied subtraction from zero; the C bit will be set in all cases except when the contents of ACCX or M is 00.

Boolean Formulae for Condition Codes:

$$N = R_7$$
$$Z = \overline{R}_7 \cdot \overline{R}_6 \cdot \overline{R}_5 \cdot \overline{R}_4 \cdot \overline{R}_3 \cdot \overline{R}_2 \cdot \overline{R}_1 \cdot \overline{R}_0$$
$$V = R_7 \cdot \overline{R}_6 \cdot \overline{R}_5 \cdot \overline{R}_4 \cdot \overline{R}_3 \cdot \overline{R}_2 \cdot \overline{R}_1 \cdot \overline{R}_0$$
$$C = R_7 + R_6 + R_5 + R_4 + R_3 + R_2 + R_1 + R_0$$

Addressing Modes, Execution Time, and Machine Code (hexadecimal/octal/decimal):

Addressing Modes	Execution Time (No. of cycles)	Number of bytes of machine code	Coding of First (or only) byte of machine code		
			HEX.	OCT.	DEC.
A	2	1	40	100	064
B	2	1	50	120	080
EXT	6	3	70	160	112
IND	7	2	60	140	096

No Operation — NOP

Description:	This is a single-word instruction which causes only the program counter to be incremented. No other registers are affected.

Condition Codes: Not affected.

Addressing Modes, Execution Time, and Machine Code (hexadecimal/octal/decimal):

Addressing Modes	Execution Time (No. of cycles)	Number of bytes of machine code	Coding of First (or only) byte of machine code		
			HEX.	OCT.	DEC.
INHERENT	2	1	01	001	001

Inclusive OR

ORA

Operation: ACCX ← (ACCX)⊙(M)

Description: Perform logical "OR" between the contents of ACCX and the contents of M and places the result in ACCX. (Each bit of ACCX after the operation will be the logical "OR" of the corresponding bits of M and of ACCX before the operation).

Condition Codes: H: Not affected.
                   I: Not affected.
                  N: Set if most significant bit of the result is set; cleared otherwise.
                  Z: Set if all bits of the result are cleared; cleared otherwise.
                  V: Cleared.
                  C: Not affected.

Boolean Formulae for Condition Codes:

$$N = R_7$$
$$Z = \overline{R_7} \cdot \overline{R_6} \cdot \overline{R_5} \cdot \overline{R_4} \cdot \overline{R_3} \cdot \overline{R_2} \cdot \overline{R_1} \cdot \overline{R_0}$$
$$V = 0$$

Addressing Modes, Execution Time, and Machine Code (hexadecimal/ octal/ decimal):

(DUAL OPERAND)

Addressing Modes	Execution Time (No. of cycles)	Number of bytes of machine code	Coding of First (or only) byte of machine code		
			HEX.	OCT.	DEC.
A IMM	2	2	8A	212	138
A DIR	3	2	9A	232	154
A EXT	4	3	BA	272	186
A IND	5	2	AA	252	170
B IMM	2	2	CA	312	202
B DIR	3	2	DA	332	218
B EXT	4	3	FA	372	250
B IND	5	2	EA	352	234

Push Data Onto Stack

PSH

Operation: ↓ (ACCX)
                  SP ← (SP) − 0001

Description: The contents of ACCX is stored in the stack at the address contained in the stack pointer. The stack pointer is then decremented.

Condition Codes: Not affected.

Addressing Modes, Execution Time, and Machine Code (hexadecimal/ octal/ decimal):

Addressing Modes	Execution Time (No. of cycles)	Number of bytes of machine code	Coding of First (or only) byte of machine code		
			HEX.	OCT.	DEC.
A	4	1	36	066	054
B	4	1	37	067	055

Pull Data from Stack

PUL

Operation: SP ← (SP) + 0001
                  ↑ ACCX

Description: The stack pointer is incremented. The ACCX is then loaded from the stack, from the address which is contained in the stack pointer.

Condition Codes: Not affected.

Addressing Modes, Execution Time, and Machine Code (hexadecimal/octal/decimal):

Addressing Modes	Execution Time (No. of cycles)	Number of bytes of machine code	Coding of First (or only) byte of machine code		
			HEX.	OCT.	DEC.
A	4	1	32	062	050
B	4	1	33	063	051

Rotate Left

ROL

Operation:

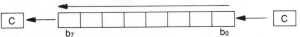

b_7 b_0

Description: Shifts all bits of ACCX or M one place to the left. Bit 0 is loaded from the C bit. The C bit is loaded from the most significant bit of ACCX or M.

Condition Codes: H: Not affected.
                  I: Not affected.
                  N: Set if most significant bit of the result is set; cleared otherwise.
                  Z: Set if all bits of the result are cleared; cleared otherwise.
                  V: Set if, after the completion of the operation, EITHER (N is set and C is cleared) OR (N is cleared and C is set); cleared otherwise.
                  C: Set if, before the operation, the most significant bit of the ACCX or M was set; cleared otherwise.

Boolean Formulae for Condition Codes:

$N = R_7$

$Z = \overline{R}_7 \cdot \overline{R}_6 \cdot \overline{R}_5 \cdot \overline{R}_4 \cdot \overline{R}_3 \cdot \overline{R}_2 \cdot \overline{R}_1 \cdot \overline{R}_0$

$V = N \oplus C = [N \cdot \overline{C}] \odot [\overline{N} \cdot C]$

(the foregoing formula assumes values of N and C after the rotation)

$C = M_7$

Addressing Modes, Execution Time, and Machine Code (hexadecimal/octal/decimal):

Addressing Modes	Execution Time (No. of cycles)	Number of bytes of machine code	Coding of First (or only) byte of machine code		
			HEX.	OCT.	DEC.
A	2	1	49	111	073
B	2	1	59	131	089
EXT	6	3	79	171	121
IND	7	2	69	151	105

Rotate Right

ROR

Operation:

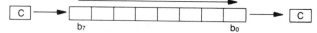

Description: Shifts all bits of ACCX or M one place to the right. Bit 7 is loaded from the C bit. The C bit is loaded from the least significant bit of ACCX or M.

Condition Codes:
- H: Not affected.
- I: Not affected.
- N: Set if most significant bit of the result is set; cleared otherwise.
- Z: Set if all bits of the result are cleared; cleared otherwise.
- V: Set if, after the completion of the operation, EITHER (N is set and C is cleared) OR (N is cleared and C is set); cleared otherwise.
- C: Set if, before the operation, the least significant bit of the ACCX or M was set; cleared otherwise.

Boolean Formulae for Condition Codes:

$N = R_7$

$Z = \bar{R}_7 \cdot \bar{R}_6 \cdot \bar{R}_5 \cdot \bar{R}_4 \cdot \bar{R}_3 \cdot \bar{R}_2 \cdot \bar{R}_1 \cdot \bar{R}_0$

$V = N \oplus C = [N \cdot \bar{C}] \odot [\bar{N} \cdot C]$

(the foregoing formula assumes values of N and C after the rotation)

$C = M_0$

Addressing Modes, Execution Time, and Machine Code (hexadecimal/octal/decimal):

Addressing Modes	Execution Time (No. of cycles)	Number of bytes of machine code	Coding of First (or only) byte of machine code		
			HEX.	OCT.	DEC.
A	2	1	46	106	070
B	2	1	56	126	086
EXT	6	3	76	166	118
IND	7	2	66	146	102

Return from Interrupt

RTI

Operation:
SP ← (SP) + 0001 , ↑CC
SP ← (SP) + 0001 , ↑ACCB
SP ← (SP) + 0001 , ↑ACCA
SP ← (SP) + 0001 , ↑IXH
SP ← (SP) + 0001 , ↑IXL
SP ← (SP) + 0001 , ↑PCH
SP ← (SP) + 0001 , ↑PCL

Description: The condition codes, accumulators B and A, the index register, and the program counter, will be restored to a state pulled from the stack. Note that the interrupt mask bit will be reset if and only if the corresponding bit stored in the stack is zero.

Condition Codes: Restored to the states pulled from the stack.

Addressing Modes, Execution Time, and Machine Code (hexadecimal/octal/decimal):

Addressing Modes	Execution Time (No. of cycles)	Number of bytes of machine code	Coding of First (or only) byte of machine code		
			HEX.	OCT.	DEC.
INHERENT	10	1	3B	073	059

Return from Subroutine **RTS**

Operation: SP ← (SP) + 0001
↑ PCH
SP ← (SP) + 0001
↑ PCL

Description: The stack pointer is incremented (by 1). The contents of the byte of memory, at the address now contained in the stack pointer, are loaded into the 8 bits of highest significance in the program counter. The stack pointer is again incremented (by 1). The contents of the byte of memory, at the address now contained in the stack pointer, are loaded into the 8 bits of lowest significiance in the program counter.

Condition Codes: Not affected.

Addressing Modes, Execution Time, and Machine Code (hexadecimal/ octal/ decimal):

Addressing Modes	Execution Time (No. of cycles)	Number of bytes of machine code	Coding of First (or only) byte of machine code		
			HEX.	OCT.	DEC.
INHERENT	5	1	39	071	057

Subtract Accumulators **SBA**

Operation: ACCA ← (ACCA) − (ACCB)

Description: Subtracts the contents of ACCB from the contents of ACCA and places the result in ACCA. The contents of ACCB are not affected.

Condition Codes: H: Not affected.
I: Not affected.
N: Set if most significant bit of the result is set; cleared otherwise.
Z: Set if all bits of the result are cleared; cleared otherwise.
V: Set if there was two's complement overflow as a result of the operation.
C: Carry is set if the absolute value of accumulator B plus previous carry is larger than the absolute value of accumulator A; reset otherwise.

Boolean Formulae for Condition Codes:
$$N = R_7$$
$$Z = \overline{R_7} \cdot \overline{R_6} \cdot \overline{R_5} \cdot \overline{R_4} \cdot \overline{R_3} \cdot \overline{R_2} \cdot \overline{R_1} \cdot \overline{R_0}$$
$$V = A_7 \cdot \overline{B_7} \cdot \overline{R_7} + \overline{A_7} \cdot B_7 \cdot R_7$$
$$C = \overline{A_7} \cdot B_7 + B_7 \cdot R_7 + R_7 \cdot \overline{A_7}$$

Addressing Modes, Execution Time, and Machine Code (hexadecimal/ octal/ decimal):

Addressing Modes	Execution Time (No. of cycles)	Number of bytes of machine code	Coding of First (or only) byte of machine code		
			HEX.	OCT.	DEC.
INHERENT	2	1	10	020	016

Subtract with Carry **SBC**

Operation: ACCX ← (ACCX) − (M) − (C)

Description: Subtracts the contents of M and C from the contents of ACCX and places the result in ACCX.

Condition Codes: H: Not affected.

I: Not affected.

N: Set if most significant bit of the result is set; cleared otherwise.

Z: Set if all bits of the result are cleared; cleared otherwise.

V: Set if there was two's complement overflow as a result of the operation; cleared otherwise.

C: Carry is set if the absolute value of the contents of memory plus previous carry is larger than the absolute value of the accumulator; reset otherwise.

Boolean Formulae for Condition Codes:

$$N = R_7$$
$$Z = \overline{R}_7 \cdot \overline{R}_6 \cdot \overline{R}_5 \cdot \overline{R}_4 \cdot \overline{R}_3 \cdot \overline{R}_2 \cdot \overline{R}_1 \cdot \overline{R}_0$$
$$V = X_7 \cdot \overline{M}_7 \cdot \overline{R}_7 + \overline{X}_7 \cdot M_7 \cdot R_7$$
$$C = \overline{X}_7 \cdot M_7 + M_7 \cdot R_7 + R_7 \cdot \overline{X}_7$$

Addressing Modes, Execution Time, and Machine Code (hexadecimal / octal / decimal):

(DUAL OPERAND)

Addressing Modes	Execution Time (No. of cycles)	Number of bytes of machine code	Coding of First (or only) byte of machine code		
			HEX.	OCT.	DEC.
A IMM	2	2	82	202	130
A DIR	3	2	92	222	146
A EXT	4	3	B2	262	178
A IND	5	2	A2	242	162
B IMM	2	2	C2	302	194
B DIR	3	2	D2	322	210
B EXT	4	3	F2	362	242
B IND	5	2	E2	342	226

Set Carry

SEC

Operation: C bit ← 1

Description: Sets the carry bit in the processor condition codes register.

Condition Codes: H: Not affected.

I: Not affected.

N: Not affected.

Z: Not affected.

V: Not affected.

C: Set.

Boolean Formulae for Condition Codes:

$$C = 1$$

Addressing Modes, Execution Time, and Machine Code (hexadecimal / octal / decimal):

Addressing Modes	Execution Time (No. of cycles)	Number of bytes of machine code	Coding of First (or only) byte of machine code		
			HEX.	OCT.	DEC.
INHERENT	2	1	0D	015	013

Set Interrupt Mask — SEI

Operation: I bit ← 1

Description: Sets the interrupt mask bit in the processor condition codes register. The microprocessor is inhibited from servicing an interrupt from a peripheral device, and will continue with execution of the instructions of the program, until the interrupt mask bit has been cleared.

Condition Codes:
- H: Not affected.
- I: Set.
- N: Not affected.
- Z: Not affected.
- V: Not affected.
- C: Not affected.

Boolean Formulae for Condition Codes:
I = 1

Addressing Modes, Execution Time, and Machine Code (hexadecimal/octal/decimal):

Addressing Modes	Execution Time (No. of cycles)	Number of bytes of machine code	HEX.	OCT.	DEC.
INHERENT	2	1	0F	017	015

Set Two's Complement Overflow Bit — SEV

Operation: V bit ← 1

Description: Sets the two's complement overflow bit in the processor condition codes register.

Condition Codes:
- H: Not affected.
- I: Not affected.
- N: Not affected.
- Z: Not affected.
- V: Set.
- C: Not affected.

Boolean Formulae for Condition Codes:
V = 1

Addressing Modes, Execution Time, and Machine Code (hexadecimal/octal/decimal):

Addressing Modes	Execution Time (No. of cycles)	Number of bytes of machine code	HEX.	OCT.	DEC.
INHERENT	2	1	0B	013	011

Store Accumulator — STA

Operation: M ← (ACCX)

Description: Stores the contents of ACCX in memory. The contents of ACCX remains unchanged.

Condition Codes: H: Not affected.
                 I: Not affected.
                 N: Set if the most significant bit of the contents of ACCX is set; cleared otherwise.
                 Z: Set if all bits of the contents of ACCX are cleared; cleared otherwise.
                 V: Cleared.
                 C: Not affected.

Boolean Formulae for Condition Codes:

$$N = X_7$$
$$Z = \overline{X_7} \cdot \overline{X_6} \cdot \overline{X_5} \cdot \overline{X_4} \cdot \overline{X_3} \cdot \overline{X_2} \cdot \overline{X_1} \cdot \overline{X_0}$$
$$V = 0$$

Addressing Modes, Execution Time, and Machine Code (hexadecimal/octal/decimal):

Addressing Modes	Execution Time (No. of cycles)	Number of bytes of machine code	Coding of First (or only) byte of machine code		
			HEX.	OCT.	DEC.
A DIR	4	2	97	227	151
A EXT	5	3	B7	267	183
A IND	6	2	A7	247	167
B DIR	4	2	D7	327	215
B EXT	5	3	F7	367	247
B IND	6	2	E7	347	231

Store Stack Pointer   **STS**

Operation: M ← (SPH)
                      M + 1 ← (SPL)

Description: Stores the more significant byte of the stack pointer in memory at the address specified by the program, and stores the less significant byte of the stack pointer at the next location in memory, at one plus the address specified by the program.

Condition Codes: H: Not affected.
                 I: Not affected.
                 N: Set if the most significant bit of the stack pointer is set; cleared otherwise.
                 Z: Set if all bits of the stack pointer are cleared; cleared otherwise.
                 V: Cleared.
                 C: Not affected.

Boolean Formulae for Condition Codes:

$$N = SPH_7$$
$$Z = (\overline{SPH_7} \cdot \overline{SPH_6} \cdot \overline{SPH_5} \cdot \overline{SPH_4} \cdot \overline{SPH_3} \cdot \overline{SPH_2} \cdot \overline{SPH_1} \cdot \overline{SPH_0}) \cdot$$
$$(\overline{SPL_7} \cdot \overline{SPL_6} \cdot \overline{SPL_5} \cdot \overline{SPL_4} \cdot \overline{SPL_3} \cdot \overline{SPL_2} \cdot \overline{SPL_1} \cdot \overline{SPL_0})$$
$$V = 0$$

Addressing Modes, Execution Time, and Machine Code (hexadecimal/octal/decimal):

Addressing Modes	Execution Time (No. of cycles)	Number of bytes of machine code	Coding of First (or only) byte of machine code		
			HEX.	OCT.	DEC.
DIR	5	2	9F	237	159
EXT	6	3	BF	277	191
IND	7	2	AF	257	175

STX

Store Index Register

Operation: $M \leftarrow (IXH)$
$M + 1 \leftarrow (IXL)$

Description: Stores the more significant byte of the index register in memory at the address specified by the program, and stores the less significant byte of the index register at the next location in memory, at one plus the address specified by the program.

Condition Codes:
H: Not affected.
I: Not affected.
N: Set if the most significant bite of the index register is set; cleared otherwise.
Z: Set if all bits of the index register are cleared; cleared otherwise.
V: Cleared.
C: Not affected.

Boolean Formulae for Condition Codes:
$N = IXH_7$
$Z = (\overline{IXH_7} \cdot \overline{IXH_6} \cdot \overline{IXH_5} \cdot \overline{IXH_4} \cdot \overline{IXH_3} \cdot \overline{IXH_2} \cdot \overline{IXH_1} \cdot \overline{IXH_0}) \cdot$
$(\overline{IXL_7} \cdot \overline{IXL_6} \cdot \overline{IXL_5} \cdot \overline{IXL_4} \cdot \overline{IXL_3} \cdot \overline{IXL_2} \cdot \overline{IXL_1} \cdot \overline{IXL_0})$
$V = 0$

Addressing Modes, Execution Time, and Machine Code (hexadecimal/octal/decimal):

Addressing Modes	Execution Time (No. of cycles)	Number of bytes of machine code	Coding of First (or only) byte of machine code		
			HEX.	OCT.	DEC.
DIR	5	2	DF	337	223
EXT	6	3	FF	377	255
IND	7	2	EF	357	239

SUB

Subtract

Operation: $ACCX \leftarrow (ACCX) - (M)$

Description: Subtracts the contents of M from the contents of ACCX and places the result in ACCX.

Condition Codes:
H: Not affected.
I: Not affected.
N: Set if most significant bit of the result is set; cleared otherwise.
Z: Set if all bits of the result are cleared; cleared otherwise.
V: Set if there was two's complement overflow as a result of the operation; cleared otherwise.
C: Set if the absolute value of the contents of memory are larger than the absolute value of the accumulator; reset otherwise.

Boolean Formulae for Condition Codes:
$N = R_7$
$Z = \overline{R_7} \cdot \overline{R_6} \cdot \overline{R_5} \cdot \overline{R_4} \cdot \overline{R_3} \cdot \overline{R_2} \cdot \overline{R_1} \cdot \overline{R_0}$
$V = X_7 \cdot \overline{M_7} \cdot \overline{R_7} \cdot \overline{X_7} \cdot M_7 \cdot R_7$
$C = \overline{X_7} \cdot M_7 + M_7 \cdot R_7 + R_7 \cdot \overline{X_7}$

Addressing Modes, Execution Time, and Machine Code (hexadecimal/octal/decimal):
(DUAL OPERAND)

Addressing Modes	Execution Time (No. of cycles)	Number of bytes of machine code	Coding of First (or only) byte of machine code		
			HEX.	OCT.	DEC.
A IMM	2	2	80	200	128
A DIR	3	2	90	220	144
A EXT	4	3	B0	260	176
A IND	5	2	A0	240	160
B IMM	2	2	C0	300	192
B DIR	3	2	D0	320	208
B EXT	4	3	F0	360	240
B IND	5	2	E0	340	224

Software Interrupt

SWI

Operation:

PC ← (PC) + 0001
↓ (PCL) , SP ← (SP)-0001
↓ (PCH) , SP ← (SP)-0001
↓ (IXL) , SP ← (SP)-0001
↓ (IXH) , SP ← (SP)-0001
↓ (ACCA) , SP ← (SP)-0001
↓ (ACCB) , SP ← (SP)-0001
↓ (CC) , SP ← (SP)-0001
I ← 1
PCH ← (n-0005)
PCL ← (n-0004)

Description:

The program counter is incremented (by 1). The program counter, index register, and accumulator A and B, are pushed into the stack. The condition codes register is then pushed into the stack, with condition codes H, I, N, Z, V, C going respectively into bit positions 5 thru 0, and the top two bits (in bit positions 7 and 6) are set (to the 1 state). The stack pointer is decremented (by 1) after each byte of data is stored in the stack.

The interrupt mask bit is then set. The program counter is then loaded with the address stored in the software interrupt pointer at memory locations (n-5) and (n-4), where n is the address corresponding to a high state on all lines of the address bus.

Condition Codes:

H: Not affected.
I: Set.
N: Not affected.
Z: Not affected.
V: Not affected.
C: Not affected.

Boolean Formula for Condition Codes:

I = 1

Addressing Modes, Execution Time, and Machine Code (hexadecimal/octal/decimal):

Addressing Modes	Execution Time (No. of cycles)	Number of bytes of machine code	Coding of First (or only) byte of machine code		
			HEX.	OCT.	DEC.
INHERENT	12	1	3F	077	063

Transfer from Accumulator A to Accumulator B   **TAB**

Operation: ACCB ← (ACCA)

Description: Moves the contents of ACCA to ACCB. The former contents of ACCB are lost.
                     The contents of ACCA are not affected.

Condition Codes: H: Not affected.
                   I: Not affected.
                   N: Set if the most significant bit of the contents of the accumulator is set; cleared
                       otherwise.
                   Z: Set if all bits of the contents of the accumulator are cleared; cleared other-
                       wise.
                   V: Cleared.
                   C: Not affected.

Boolean Formulae for Condition Codes:
                   $N = R_7$
                   $Z = \overline{R}_7 \cdot \overline{R}_6 \cdot \overline{R}_5 \cdot \overline{R}_4 \cdot \overline{R}_3 \cdot \overline{R}_2 \cdot \overline{R}_1 \cdot \overline{R}_0$
                   $V = 0$

Addressing Modes, Execution Time, and Machine Code (hexadecimal/octal/decimal):

Addressing Modes	Execution Time (No. of cycles)	Number of bytes of machine code	Coding of First (or only) byte of machine code		
			HEX.	OCT.	DEC.
INHERENT	2	1	16	026	022

Transfer from Accumulator A   **TAP**
to Processor Condition Codes Register

Operation: CC ← (ACCA)

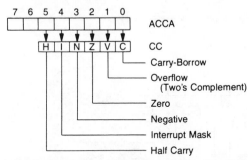

Description: Transfers the contents of bit positions 0 thru 5 of accumulator A to the correspond-
                     ing bit positions of the processor condition codes register. The contents of
                     accumulator A remain unchanged.

Condition Codes: Set or reset according to the contents of the respective bits 0 thru 5 of accumulator
                     A.

Addressing Modes, Execution Time, and Machine Code (hexadecimal/octal/decimal):

Addressing Modes	Execution Time (No. of cycles)	Number of bytes of machine code	Coding of First (or only) byte of machine code		
			HEX.	OCT.	DEC.
INHERENT	2	1	06	006	006

Transfer from Accumulator B to Accumulator A   **TBA**

Operation: ACCA ← (ACCB)

Description: Moves the contents of ACCB to ACCA. The former contents of ACCA are lost. The contents of ACCB are not affected.

Condition Codes: H: Not affected.
                    I: Not affected.
                    N: Set if the most significant accumulator bit is set; cleared otherwise.
                    Z: Set if all accumulator bits are cleared; cleared otherwise.
                    V: Cleared.
                    C: Not affected.

Boolean Formulae for Condition Codes:

$$N = R_7$$
$$Z = \overline{R_7} \cdot \overline{R_6} \cdot \overline{R_5} \cdot \overline{R_4} \cdot \overline{R_3} \cdot \overline{R_2} \cdot \overline{R_1} \cdot \overline{R_0}$$
$$V = 0$$

Addressing Modes, Execution Time, and Machine Code (hexadecimal/octal/decimal):

Addressing Modes	Execution Time (No. of cycles)	Number of bytes of machine code	Coding of First (or only) byte of machine code		
			HEX.	OCT.	DEC.
INHERENT	2	1	17	027	023

Transfer from Processor Condition Codes Register to Accumulator A   **TPA**

Operation: ACCA ← (CC)

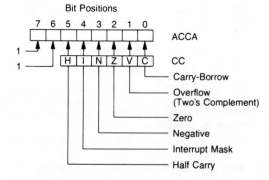

Description: Transfers the contents of the processor condition codes register to corresponding bit positions 0 thru 5 of accumulator A. Bit positions 6 and 7 of accumulator A are set (i.e. go to the "1" state). The processor condition codes register remains unchanged.

Condition Codes: Not affected.

Addressing Modes, Execution Time, and Machine Code (hexadecimal/octal/decimal):

Addressing Modes	Execution Time (No. of cycles)	Number of bytes of machine code	Coding of First (or only) byte of machine code		
			HEX.	OCT.	DEC.
INHERENT	2	1	07	007	007

Test TST

Operation:	(ACCX) − 00
	(M) − 00
Description:	Set condition codes N and Z according to the contents of ACCX or M.
Condition Codes:	H: Not affected.
	I: Not affected.
	N: Set if most significant bit of the contents of ACCX or M is set; cleared otherwise.
	Z: Set if all bits of the contents of ACCX or M are cleared; cleared otherwise.
	V: Cleared.
	C: Cleared.

Boolean Formulae for Condition Codes:

$$N = M_7$$
$$Z = \overline{M_7} \cdot \overline{M_6} \cdot \overline{M_5} \cdot \overline{M_4} \cdot \overline{M_3} \cdot \overline{M_2} \cdot \overline{M_1} \cdot \overline{M_0}$$
$$V = 0$$
$$C = 0$$

Addressing Modes, Execution Time, and Machine Code (hexadecimal/octal/decimal):

Addressing Modes	Execution Time (No. of cycles)	Number of bytes of machine code	Coding of First (or only) byte of machine code		
			HEX.	OCT.	DEC.
A	2	1	4D	115	077
B	2	1	5D	135	093
EXT	6	3	7D	175	125
IND	7	2	6D	155	109

Transfer from Stack Pointer to Index Register TSX

Operation:	IX ← (SP) + 0001
Description:	Loads the index register with one plus the contents of the stack pointer. The contents of the stack pointer remain unchanged.
Condition Codes:	Not affected.

Addressing Modes, Execution Time, and Machine Code (hexadecimal/octal/decimal):

Addressing Modes	Execution Time (No. of cycles)	Number of bytes of machine code	Coding of First (or only) byte of machine code		
			HEX.	OCT.	DEC.
INHERENT	4	1	30	060	048

Transfer From Index Register to Stack Pointer $\qquad$ **TXS**

Operation: $\quad$ SP ← (IX) − 0001

Description: $\quad$ Loads the stack pointer with the contents of the index register, minus one. The contents of the index register remain unchanged.

Condition Codes: Not affected.

Addressing Modes, Execution Time, and Machine Code (hexadecimal/octal/decimal):

Addressing Modes	Execution Time (No. of cycles)	Number of bytes of machine code	Coding of First (or only) byte of machine code		
			HEX.	OCT.	DEC.
INHERENT	4	1	35	.065	053

Wait for Interrupt $\qquad$ **WAI**

Operation: $\quad$ PC ← (PC) + 0001
↓ (PCL) , SP ← (SP)-0001
↓ (PCH) , SP ← (SP)-0001
↓ (IXL) , SP ← (SP)-0001
↓ (IXH) , SP ← (SP)-0001
↓ (ACCA) , SP ← (SP)-0001
↓ (ACCB) , SP ← (SP)-0001
↓ (CC) , SP ← (SP)-0001

Condition Codes: $\quad$ Not affected.

Description: $\quad$ The program counter is incremented (by 1). The program counter, index register, and accumulators A and B, are pushed into the stack. The condition codes register is then pushed into the stack, with condition codes H, I, N, Z, V, C going respectively into bit positions 5 thru 0, and the top two bits (in bit positions 7 and 6) are set (to the 1 state). The stack pointer is decremented (by 1) after each byte of data is stored in the stack.

Execution of the program is then suspended until an interrupt from a peripheral device is signalled, by the interrupt request control input going to a low state.

When an interrupt is signalled on the interrupt request line, and provided the I bit is clear, execution proceeds as follows. The interrupt mask bit is set. The program counter is then loaded with the address stored in the internal interrupt pointer at memory locations (n-7) and (n-6), where n is the address corresponding to a high state on all lines of the address bus.

Condition Codes: $\quad$ H: $\quad$ Not affected.
I: $\quad$ Not affected until an interrupt request signal is detected on the interrupt request control line. When the interrupt request is received the I bit is set and further execution takes place, provided the I bit was initially clear.
N: $\quad$ Not affected.
Z: $\quad$ Not affected.
V: $\quad$ Not affected.
C: $\quad$ Not affected.

Addressing Modes, Execution Time, and Machine Code (hexadecimal/octal/decimal):

Addressing Modes	Execution Time (No. of cycles)	Number of bytes of machine code	Coding of First (or only) byte of machine code		
			HEX.	OCT.	DEC.
INHERENT	9	1	3E	076	062

Hexadecimal Values of Machine Codes

Code	Op	Acc	Mode	Code	Op	Acc	Mode	Code	Op	Acc	Mode	Code	Op	Acc	Mode
00	*			40	NEG	A		80	SUB	A	IMM	C0	SUB	B	IMM
01	NOP			41	*			81	CMP	A	IMM	C1	CMP	B	IMM
02	*			42	*			82	SBC	A	IMM	C2	SBC	B	IMM
03	*			43	COM	A		83	*			C3	*		
04	*			44	LSR	A		84	AND	A	IMM	C4	AND	B	IMM
05	*			45	*			85	BIT	A	IMM	C5	BIT	B	IMM
06	TAP			46	ROR	A		86	LDA	A	IMM	C6	LDA	B	IMM
07	TPA			47	ASR	A		87	*			C7	*		
08	INX			48	ASL	A		88	EOR	A	IMM	C8	EOR	B	IMM
09	DEX			49	ROL	A		89	ADC	A	IMM	C9	ADC	B	IMM
0A	CLV			4A	DEC	A		8A	ORA	A	IMM	CA	ORA	B	IMM
0B	SEV			4B	*			8B	ADD	A	IMM	CB	ADD	B	IMM
0C	CLC			4C	INC	A		8C	CPX	A	IMM	CC	*		
0D	SEC			4D	TST	A		8D	BSR		REL	CD	*		
0E	CLI			4E	*			8E	LDS		IMM	CE	LDX		IMM
0F	SEI			4F	CLR	A		8F	*			CF	*		
10	SBA			50	NEG	B		90	SUB	A	DIR	D0	SUB	B	DIR
11	CBA			51	*			91	CMP	A	DIR	D1	CMP	B	DIR
12	*			52	*			92	SBC	A	DIR	D2	SBC	B	DIR
13	*			53	COM	B		93	*			D3	*		
14	*			54	LSR	B		94	AND	A	DIR	D4	AND	B	DIR
15	*			55	*			95	BIT	A	DIR	D5	BIT	B	DIR
16	TAB			56	ROR	B		96	LDA	A	DIR	D6	LDA	B	DIR
17	TBA			57	ASR	B		97	STA	A	DIR	D7	STA	B	DIR
18	*			58	ASL	B		98	EOR	A	DIR	D8	EOR	B	DIR
19	DAA			59	ROL	B		99	ADC	A	DIR	D9	ADC	B	DIR
1A	*			5A	DEC	B		9A	ORA	A	DIR	DA	ORA	B	DIR
1B	ABA			5B	*			9B	ADD	A	DIR	DB	ADD	B	DIR
1C	*			5C	INC	B		9C	CPX		DIR	DC	*		
1D	*			5D	TST	B		9D	*			DD	*		
1E	*			5E	*			9E	LDS		DIR	DE	LDX		DIR
1F	*			5F	CLR	B		9F	STS		DIR	DF	STX		DIR
20	BRA		REL	60	NEG		IND	A0	SUB	A	IND	E0	SUB	B	IND
21	*			61	*			A1	CMP	A	IND	E1	CMP	B	IND
22	BHI		REL	62	*			A2	SBC	A	IND	E2	SBC	B	IND
23	BLS		REL	63	COM		IND	A3	*			E3	*		
24	BCC		REL	64	LSR		IND	A4	AND	A	IND	E4	AND	B	IND
25	BCS		REL	65	*			A5	BIT	A	IND	E5	BIT	B	IND
26	BNE		REL	66	ROR		IND	A6	LDA	A	IND	E6	LDA	B	IND
27	BEQ		REL	67	ASR		IND	A7	STA	A	IND	E7	STA	B	IND
28	BVC		REL	68	ASL		IND	A8	EOR	A	IND	E8	EOR	B	IND
29	BVS		REL	69	ROL		IND	A9	ADC	A	IND	E9	ADC	B	IND
2A	BPL		REL	6A	DEC		IND	AA	ORA	A	IND	EA	ORA	B	IND
2B	BMI		REL	6B	*			AB	ADD	A	IND	EB	ADD	B	IND
2C	BGE		REL	6C	INC		IND	AC	CPX		IND	EC	*		
2D	BLT		REL	6D	TST		IND	AD	JSR		IND	ED	*		
2E	BGT		REL	6E	JMP		IND	AE	LDS		IND	EE	LDX		IND
2F	BLE		REL	6F	CLR		IND	AF	STS		IND	EF	STX		IND
30	TSX			70	NEG		EXT	B0	SUB	A	EXT	F0	SUB	B	EXT
31	INS			71	*			B1	CMP	A	EXT	F1	CMP	B	EXT
32	PUL	A		72	*			B2	SBC	A	EXT	F2	SBC	B	EXT
33	PUL	B		73	COM		EXT	B3	*			F3	*		
34	DES			74	LSR		EXT	B4	AND	A	EXT	F4	AND	B	EXT
35	TXS			75	*			B5	BIT	A	EXT	F5	BIT	B	EXT
36	PSH	A		76	ROR		EXT	B6	LDA	A	EXT	F6	LDA	B	EXT
37	PSH	B		77	ASR		EXT	B7	STA	A	EXT	F7	STA	B	EXT
38	*			78	ASL		EXT	B8	EOR	A	EXT	F8	EOR	B	EXT
39	RTS			79	ROL		EXT	B9	ADC	A	EXT	F9	ADC	B	EXT
3A	*			7A	DEC		EXT	BA	ORA	A	EXT	FA	ORA	B	EXT
3B	RTI			7B	*			BB	ADD	A	EXT	FB	ADD	B	EXT
3C	*			7C	INC		EXT	BC	CPX		EXT	FC	*		
3D	*			7D	TST		EXT	BD	JSR		EXT	FD	*		
3E	WAI			7E	JMP		EXT	BE	LDS		EXT	FE	LDX		EXT
3F	SWI			7F	CLR		EXT	BF	STS		EXT	FF	STX		EXT

Notes: 1. Addressing Modes:
A = Accumulator A IMM = Immediate
B = Accumulator B DIR = Direct
REL = Relative
IND = Indexed

2. Unassigned code indicated by "*"

Some instructions have immediate addressing as well as others, and where this occurs the addressing mode always describes that which accompanies the op-code as memory (M). In these cases, the interpretation is as follows: substitute immediate in lieu of memory; consider the instruction ADC (add with carry).

$$\text{ACCX} \leftarrow (\text{ACCX}) + (\text{M}) + (\text{C})$$

means that: the contents of the specified accumulator being replaced by ($\leftarrow$): the contents of the specified accumulator (ACCX) + contents of the specified memory location (M) + the carry flag (if set). This is equivalent to the contents of the specified accumulator being replaced by: the contents of the specified accumulator + the immediate data (second byte of the instruction) + the carry flag (if set).

Example (ADC).   Add the contents of memory location 0137 to the contents of Acc-A, + carry flag (if set):

Instruction B9 0137

Add the number 4F to the contents of Acc-B + carry flag if set:

Instruction C9 4F

The status of the V bit, or overflow flag, and C bit, or carry flag, is determined by the status of bit 7 of the binary numbers used in the program as data for the minuend and subtrahend, in the case of subtraction, and the addend and augend, in the case of addition, as well as the status of bit 7 of the result. For example, from the Boolean formula for the condition codes for subtraction (SUB) it can be seen that the V flag will be set if bit 7 of the minuend, subtrahend, and result have the following states:

	Bit 7			*Bit 7*	
Original minuend data	0	$(\overline{X_7})$		1	(X_7)
Original subtrahend data	1	(M_7)	or	0	$(\overline{M_7})$
Result	1	(R_7)		0	$(\overline{R_7})$

$$\therefore \quad V = \overline{X_7} \cdot M_7 \cdot R_7 + X_7 \cdot \overline{M_7} \cdot \overline{R_7}$$

Note: The result is obtained, in the case of subtraction, by taking the 2's complement of the subtrahend and adding it to the minuend as is demonstrated next.

Hexadecimal Number
Used as Data in
the Program   *Binary Equivalent*

$$X_7 = 0 \quad \therefore \quad \overline{X}_7 = 1$$

7F $\boxed{0}\,1\ 1\ 1\quad 1\ 1\ 1\ 1$ Minuend

81 $^{-}$ $\boxed{1}\,0\ 0\ 0\quad 0\ 0\ 0\ 1$ $^{-}$ Subtrahend

$$M_7$$

Form the 2's complement of 81:

$$81 = 0111\ 1111$$

$$\therefore \quad \begin{array}{r} 0111\ 1111 \\ 0111\ 1111 \end{array} \ + \quad \begin{array}{l} \text{Original number} \\ \text{2's complement of 81} \end{array}$$

$$\boxed{1}\,111\ 1110 \quad \leftarrow R_0$$

$$\begin{array}{cc} \nearrow & \uparrow \\ R_7 & R_1 \end{array}$$

Q: Is the V flag set as a result of the subtraction?

A: Yes; since $X_7 = 0$, therefore, $\overline{X}_7 = 1$, $M_7 = 1$ and $R_7 = 1$

$$\therefore \quad V = \overline{X}_7 \cdot M_7 \cdot R_7$$

Thus $V = 1 \cdot 1 \cdot 1 = 1$ (the V flag is set).
Similarly, with the C flag:

$$C = \overline{X}_7 \cdot M_7 + M_7 \cdot R_7 + \overline{X}_7 \cdot R_7$$

From the example just performed,

$$\overline{X}_7 = 1, \qquad M_7 = 1, \qquad R_7 = 1$$

Therefore, the C flag is set.

The Z and N flags are similarly determined from the Boolean formulas for the condition codes for subtraction:

$$N = R_7$$

$$Z = \overline{R}_7 \cdot \overline{R}_6 \cdot \overline{R}_5 \cdot \overline{R}_4 \cdot \overline{R}_3 \cdot \overline{R}_2 \cdot \overline{R}_1 \cdot \overline{R}_0$$

By inspection of the result,

$$R_7 = 1, \quad R_6 = 1, \quad R_5 = 1, \quad R_4 = 1,$$
$$R_3 = 1, \quad R_2 = 1, \quad R_1 = 1, \quad R_0 = 0$$

Thus N = 1 and Z = 0.

The use of the Boolean formulas for the condition codes is very important in determining which flag is set and will be verified after a program operation has been performed. The processor makes use of the flags of its condition codes register in its decision-making process. Two's complement numbers are used extensively in certain branch operations, that is, BGE, BGT, BLE, and BLT. This means that the two binary numbers that form the data for a subtract operation or compare operation, and so on, are represented as 2's complement binary numbers. Details of this number system are to be found in Section 1.8.

Similarly, branch operations BHI and BLS use unsigned binary numbers as the number representation system. An example comparing the use of the two number systems to perform the same operation will now be offered.

Subtract 7F from the contents of Acc-A, previously loaded with 80. The result is 01. When will a branch occur as a result of this operation?

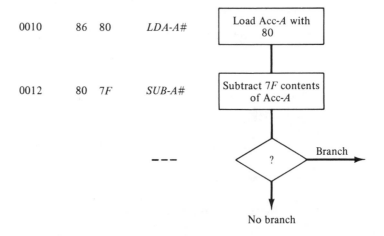

This example will be solved initially by determining which of the flags in the condition codes register are set as a result of the subtraction.

$$V = X_7 \cdot \overline{M_7} \cdot \overline{R_7} = 1$$
$$N = R_7 \qquad\qquad = 0$$
$$Z = \overline{R_7} \cdot \overline{R_6} \cdot \overline{R_5} \cdot \overline{R_4} \cdot \overline{R_3} \cdot \overline{R_2} \cdot \overline{R_1} \cdot \overline{R_0} = 1 \cdot 1 \cdot 1 \cdot 1 \cdot 1 \cdot 1 \cdot 1 \cdot 0 = 0$$

Hexadecimal
Representation   *Binary Representation*

X_7

$$
\begin{array}{ccc}
80 & \;\rightarrow\; \boxed{1}\;0\;0\;0\;\;0\;0\;0\;0 \\
7F & \;\;\;\;\;\;\;\; \boxed{0}\;1\;1\;1\;\;1\;1\;1\;1
\end{array}
$$

M_7

$$
\begin{array}{cl}
\begin{array}{c}
1\;0\;0\;0\;\;0\;0\;0\;0 \\
1\;0\;0\;0\;\;0\;0\;0\;1 \\ \hline
\boxed{0}\;0\;0\;0\;\;0\;0\;0\;1
\end{array}^{+} & \;\;\begin{array}{l} \text{2's complement} \\ \text{of 7F} \\ \\ \text{Result} \end{array}
\end{array}
$$

R_7

$$C = \overline{X_7} \cdot M_7 + M_7 \cdot R_7 + R_7 \cdot \overline{X_7} = 0 \cdot 0 + 0 \cdot 0 + 0 \cdot 0 = 0$$

$\therefore$ V is set

N is reset (clear)

Z is reset (clear)

C is reset (clear)

Consider the branch instruction BHI. A branch will occur if the C flag is clear *and* the Z flag is clear. From the flags given previously, a branch will occur if BHI (op-code 22) is used at decision point.

Q: Would a branch occur if BGE (op-code 2C) has been used in lieu of BHI?

A: No.

Q: Why would the branch *not* occur, as clearly 80 *appears* to be larger than 7F?

A: 80 is larger than 7F when BHI is used, as its number system is unsigned binary numbers, whereas 80 is smaller than 7F when BGE is used, as its number representation is a 2's complement number system.

This example illustrates the importance of understanding the number systems used by the processor and the method of flag status determination. Each instruction in the instruction set gives the relevant Boolean formula, where applicable.

Appendix B

Programmable Controller

Network Controls

These controls are discussed as follows; pushbuttons indicated with an asterisk (*) will not function unless Memory Protect is OFF:

START NEXT - This pushbutton causes a new network to be created immediately after the network on the screen. All networks following the current network will have their step numbers increased by one. If no logic is on the screen, the new network will created at network number one (before all existing logic). All existing networks will have their step numbers increased by one, providing space for the new logic at network one.

SUPERVISORY - This pushbutton allows the operator to enter the most powerful level of programming the 484 controller. This mode should be used only with great care, since major changes such as clear all logic can occur once in the Supervisory mode. Depressing the SUPERVISOR pushbutton clears the screen and displays seven options available to the operator as follows:

```
 0 = Exit Supervisory State
*1 = Stop Controller Sweep
*2 = Start Controller Sweep
*3 = Clear Controller Memory
 4 = Load Memory through ASCII Port
*5 = Dump Memory from ASCII Port
 6 = Verify Memory against ASCII Port
```

Entering the proper numerical digit from the keyboard will cause that function to occur. Exit returns CRT to normal functions; Stop/Start control scanning of controller with all outputs OFF when sweep stopped. Clear removes all stored logic from controller memory. Dump Memory causes the entire 484 Controller memory (logic and coil/register storage) to be outputted via an ASCII port built into the P180 Programmer. The ASCII device connected to the P180 port should be a simple ASCII tape loader capable of operating at 600 baud. The scanning of the controller is NOT halted when a dump is made. The Load Memory allows a previous ASCII Dump to be placed into the 484 controller memory; scanning must be stopped prior to a load, and requires either a start command or AC power to be cycled on the controller to restart the scanning following a successful load. Verify will compare controller's memory against tape record made by the ASCII device.

To prevent accidental changes to the control system, whenever a 1 (Stop) or a 3 (Clear) is selected, the CRT will display the option "7 — Confirm." The operation is performed only after the digit 7 is selected as the second step. To cancel an operation before it is executed, select any other option (0-6). The controller must be stopped (option 1) prior to selecting a Clear (option 3) or Load (option 4).

*DELETE - When depressed, this pushbutton will cause the element of the ladder diagram where the cursor is positioned to be deleted. Relay elements can be deleted only from the bottom of a column or from the right of the top rung when only one element remains in the column. Elements can always be replaced by horizontal and vertical opens. To delete a non-relay function, the cursor is placed at the top of the function; all numerical elements of this function will be simultaneously deleted. If the shift has been previously selected, the DELETE pushbutton will cause the entire network that the cursor is on to be removed. All existing networks that follow the deleted network will have their step numbers decreased by one. Deletions occur simultaneously both from the CRT screen as well as the memory of the controller.

*DISABLE - Both logic coils and discrete inputs can be disconnected from their normal control when this pushbutton is depressed. The normal control still exists within the controller, but is temporarily bypassed; disables are retentive upon power failure. Coils (0XXX) are disabled by placing the cursor on the coil in the logic area and depressing the DISABLE pushbutton. Inputs (1XXX) are similarly disabled by placing the cursor on the input in the reference area. Once disabled, these references are no longer under control of the controller until manually changed. Disabled references "freeze" their state (ON/OFF); as many references as desired can be simultaneously disabled. All contacts/outputs controlled by this reference wherever they are in the ladder diagram, will reflect the ON/OFF state of the disabled reference. Depressing this pushbutton a second time for a disabled reference will return it to the normal (enabled) condition.

*Force - This pushbutton can be used to alter the ON/OFF state of any disabled reference. The cursor is placed on a previously disabled reference in the ladder diagram; only logic coils (0XXX) and discrete inputs (1XXX) can be disabled. When the pushbuttom is depressed, the state of the reference will be altered (e.g., OFF to ON or ON to OFF). Successive depressing of this pushbutton will cause the reference to be toggled OFF to ON to OFF to ON, etc.

Relay Contact Controls

When programming relay contacts into the assembly area, the ten pushbuttons (plus two spares) will control which contact type is selected. Changes are easily made by entering the corrected contact type anytime prior to entry into the network, or construction of a new element if the existing logic element of a network is to be altered. The relay element can be either a normally open contact (⊣ ⊢), or a normally closed contact (⊣⧸⊢), or a horizontal shunt (—), or a horizontal open (• •). In addition, any relay element can have a vertical connection to the next lower rung (|), or no vertical connection (:). Vertical connections are possible only to the right of the element; vertical connections are not possible with coils or the bottom (seventh) rung of a network. If enhanced capabilities are available, transitional contacts can be used (⊣↑⊢ or ⊣↓⊢); otherwise an error message will appear when these contacts are selected. Transitional contacts pass power for exactly one scan, when their referenced coil or input goes from OFF to ON or ON to OFF. Coils of any network can be normal (—( )—) or latched (—(L)—); all logic coils are latchable. Normal coils will be de-energized if a power failure occurs; latched coils are restored following a power failure to the state (ON or OFF) that they held prior to the power failure.

Numerical Entry

This set of controls is used basically to enter numerical values into the assembly area. These values can be discrete references to control relay contacts, fixed values or register references for numerical elements, or value for storage in a holding register. All four digits must be entered, with existing digits moved one position to the left with each new digit entered at the units position. Numerical values must first be entered into the assembly area, prior to use as part of the controller's logic.

All numerical keys, except for the digit five, have dual functions controlled by the shift key. In addition to the entry of numerical values, these keys specify non-relay functions to the assembly area. The upper element of the non-relay function is specified by these keys. For example, if the SHIFT is depressed and then the digit zero, the assembly area will be prepared to construct the preset of a counter logic. When this element is placed into the ladder diagram, the next lower element will be coded to accept a holding register reference (4XXX), since the counter is a two element function. If a SHIFT, then the seven key is depressed, the assembly area is prepared to construct the upper element of an ADD function. When this element is placed into the ladder diagram, the next two elements will be coded for arithmetic references, since the add is a three element function. The non-relay functions are listed under the SHIFT operation discussed below.

SHIFT This pushbutton can be depressed prior to another key to alter its function. A letter S inside a small rectangle is displayed in the message space next to the assembly area, after the SHIFT is depressed. The shift will be removed after the depressing of the key a second time, or after any of the twelve dual function keys are depressed. The SHIFT key operates similar to the upper case shift control on a typewriter. A complete discussion of shifted controls is provided as part of the discussion of basic keys functions. The following is a list of dual function keys

Basic Key Label	Shifted (Upper Case) Function	Upper Label
DELETE Element	DELETE Network	Netwrk
T → R	Register-to-Table	R → T
Commence SEARCH	Continue Search	Cont
0 (zero)	Counter	Ctr
1 (one)	Timer 1/100 Sec	T 01
2 (two)	Timer 1/10 Sec	T0.1
3 (three)	Timer Seconds	T1.0
4 (four)	Subtract	—
6 (six)	Divide	÷
7 (seven)	Add	+
8 (eight)	Convert	CONV
9 (nine)	Multiply	X
CLEAR Assembly Area	CLEAR Entire Screen	ALL

CLEAR - When depressed, this pushbutton clears the assembly area. If a SHIFT has been commenced prior to the CLEAR, the entire CRT screen will be cleared including assembly area, error codes, and step (network) number.

Entry Controls

These controls relate to the entry of the logic elements into the ladder diagram from the assembly area. Changes are made to the controller's memory and these elements checked for validity only when their entry into the ladder diagram is attempted. The element is moved from the assembly area to the ladder diagram where the cursor is positioned. If logic exists at the cursor position, it will be replaced by the element in the assembly area; any portion of the assembly area not specified (i.e., left blank - indicated by white area), will not be altered when new logic is entered. All portions of the assembly area must be specified if new logic is to be entered into vacant spaces of the ladder diagram.

Any combination of relay contacts, vertical connections, and non-relay functions (timers, counters, arithmetics, etc.), are possible as long as there is space within the 10 x 7 network format. Logic coils can be entered at the end of any rung. The logic format requires the top rung to be complete for any column that will store logic. Each column that stores any logic must be programmed from the top down to the last element desired by the user. Where necessary, elements must be programmed with horizontal shunts or horizontal opens to complete the format.

Non-relay elements can be entered into any column with sufficient space, as long as the existing logic is blank. Non-relay elements cannot be replaced with relay elements directly; the non-relay functions must first be deleted, and then the relay functions entered. Programming starts at the top of the left most column, and can then procede along the top rung towards the right or down the first column.

There are four pushbuttons marked with arrows ($\leftarrow$, $\uparrow$, $\rightarrow$, or $\downarrow$) that control the position of the cursor. When one of these controls is depressed, the cursor is moved one position in the direction indicated, unless the cursor is at the boundary of the screen. At the right and left boundary the cursor will "wrap around" when forced beyond either side to the opposite side. At the top or bottom of the screen, here is no "wrap around" and the cursor will not move beyond these boundaries. Other controls are as follows.

ERROR RESET - If an error is detected in the operation of the P180 Programmer, a message will be flashed and the ENTER and Network controls will be locked out. When any other pushbutton (including ERROR RESET) is depressed, the message is erased, assuming the error condition does not continue to exist, and then the keyboard is completely functional.

ENTER - If Memory Protect is OFF, the ENTER pushbutton will cause the assembly area to be copied into the location selected by the cursor; the assembly area is not altered and the cursor remains at its previous location. If this location is in the logic area, the entire assembly area will be moved as a logic element, after passing appropriate error checking. If the cursor is in the reference status area and on a holding register reference, the numerical content of the assembly area only will be moved into the register.

SEARCH - This pushbutton initiates a search of all logic entered into the controller's data base. The search will be started at the first (upper left) contact of network one, and continue through all networks column by column, until either the desired element is located or end of logic is detected. Seaches are performed based upon data in the assembly area; portions of the element left blank (undefined), will not be considered during the search. For example, if all references to input 1029 are desired (normally open or normally closed contacts with or without vertical connector), only the reference value 1029 is entered into the assembly area. When this pushbutton is depressed, the search begins and the first network using this reference is displayed in its entirety. If additional networks are desired, the shift key is depressed prior to closing this pushbutton a second time; the shifted search will continue the search from where the previous match was found (not at start of logic).

GET - This pushbutton is used to load the reference status area. The desired reference is entered into the assembly area, and the cursor positioned in the reference (right side) status area where the operator desires to display its status. The status of logic coils (0XXX), discrete inputs (1XXX), or registers (3XXX or 4XXX) can be displayed; sequencer references (2YXX) *cannot* be displayed. Up to six references of any type can be displayed simultaneously, with the reference number on top and the status immediately below it. Discrete status will be provided as ON or OFF; a D prior to the input status indicates a disabled reference. Register statuses are provided as a three digit value indicating the content of the register.

Successive depressions of the GET pushbutton with the cursor on a reference in the status area will cause that reference to increase by one for each depression. The assembly area will also copy this reference regardless of its previous content. To remove a reference from the status area, a DELETE is selected while the cursor is on that reference area.

GET NEXT - When depressed, the network following the one currently on the screen will be displayed. For example, if the network on the screen is step number 23, this pushbutton will cause the network step 24 to be displayed. If no network is on the screen, network one will be displayed.

GET PREV - This pushbutton operates similar to the GET NEXT, except that when depressed the previous network is obtained. Using previous example, with network 23 on screen,this pushbutton causes network 22 to be displayed.

Starting a Network

When power is first applied to the P180 Programmer, the screen logic area will apear blank after a short warmup time period. To enter a new network, the START NEXT pushbutton is depressed. This will cause the left power rail to appear, and the cursor to move to the top of the screen -The first element of the network is entered into the Assembly area. If the first element is to be a normally open contact, referenced to 1025, and with a vertical connection, the following steps can be used (in any order):

1 Depress normally open contact (⊣ ⊢).

2 Depress vertical shunt

3 Enter value : 0, 2, 5 from numerical keypad.

The ENTER pushbutton is depressed after all portions of the element are selected, this will cause the element to be moved into the cursor location   The assembly area is not cleared, and the cursor is not moved. The memory of the controller now contains this contact

Entering Timer/Counter Elements

To develop a Timer/Counter in a network, the SHIFT key is depressed followed by the appropriate digit on the numerical keyboard. The digit 2 is depressed for timers in tenths of a second (upper label on digit 2 is TO.1). If a preset of 14.7 seconds is desired, the preset is entered from the keyboard as 0, 1, 4, 7.   The cursor is moved to the location where the timer is to begin (preset displayed), and the ENTER depressed   Note that the value 4000 is placed by the CRT into the current time storage address. This reference must be replaced by a valid holding register reference (4XXX), before the timer will operate. If the timer is to store its value into register 4049, that reference is entered into the assembly area from the keyboard as 4, 0, 4, 9. The cursor is moved down, and the ENTER pushbutton depressed; this causes the value 4049 to be placed into the current time reference location

Entering Arithmetic Operations

If an arithmetic operation, such as subtraction, is desired immediately to the right of the previously discussed timer, the cursor is moved to the top of the next column (up one and to the right one position). The SHIFT and then the digit 4 are depressed; into the assembly area is placed the upper element of a subtract operation. The value (fixed or register reference) is then placed into the assembly area. If the reference is to register 4005, the digits 4, 0, 0, 5 are entered from the keypad. When the ENTER pushbutton is depressed, the preset is moved to the cursor location, and the subtract format is displayed   If the middle and lower elements are to be 0251 and 4099 respectively, the cursor is moved down one rung at a time, their values are entered into the assembly area (0, 2, 5, 1 and 4, 0, 9, 9), and then moved into the proper position as indicated by the cursor position with the ENTER pushbutton

Altering an Element

Any element of a network can be altered at any time, as long as Memory Protect is in the OFF position. In general, the following steps are used:

1. The network is placed on the CRT screen.
2. The cursor is placed on the element to be altered.
3. The changed portion only is placed in the assembly area.
4. The ENTER pushbutton is depressed, causing both the controller's memory and the CRT display to be altered.

For example, if a relay contact is to be altered to change type (NO to NC), reference number, or vertical connection, the CLEAR pushbutton is depressed. This clears the assembly area (all white assembly area). Only the altered position, such as a normally closed contact, new reference number, or vertical connection, is placed in the assembly area; the portion of the relay element to remain unaltered, remains white in the assembly area. The ENTER pushbutton is depressed to change the portion of the element as indicated in the assembly area. Changes to relay elements can be as follows:

1. Change contact type (normally open to normally closed, or normally closed to normally open).
2. Change reference number to any legal discrete reference (0XXX, 1XXX, or 2XXX).
3. Add or delete vertical connection.
4. Any combination of the above three including all three.

Non-relay elements can be changed from fixed values to register references, or from register references to fixed values (when legal) at any time. The cursor is placed on the element to be changed, the new value (0XXX, 30XX, or 4XXX) is placed in the assembly area, and then the ENTER pushbutton is depressed to perform the change. Relay elements can be changed to numerical elements whenever there is sufficient space within the network. Timer/counter functions can start at any rung except the last (7th) rung; arithmetic functions similarly can be placed at any rung except the last two (6th and 7th) rungs. To change relay elements to numerical references, delete relay elements (starting at bottom of column) and then enter numerical references as discussed under enter above. Non-relay functions must be entered starting with their upper element; no relay contacts can be below this element where numerical references are required for the function selected.

Non-relay functions, since they are inter-dependent, cannot be independently changed directly to relay contacts. To convert non-relay elements to relay contacts, the non-relay function must be first deleted. The function must be the last elements of a column of logic. The cursor is placed at the top of the function and then the DELETE pushbutton depressed. All elements of the non-relay functions are removed from the network, and any or all can be now entered as relay contacts.

Monitoring Status

Up to six references, in any order selected by this operator, can be placed at the bottom right of the CRT screen.   If these references are discrete logic coils or inputs (0XXX and 1XXX) their status will be shown as ON or OFF; the letter D procedes the status of the input if it is Disabled. Sequencer references (2YXX) CANNOT be monitored. If the reference is a register (30XX or 4XXX, including sequencer holding registers 4051-4058) the three digit value stored in the registers will be displayed immediately below the reference.   The status of these six references, in any order chosen by the operator, will be updated as often as the refresh rate of the CRT at 9600 baud will allow. To place a reference in the display area, the cursor is positioned in any of the six areas allocated to status display. The desired reference is entered into the assembly area, and the GET pushbutton depressed. The status is always shown directly below the reference.

To disable an input reference, it is first placed in the status area. With the cursor placed on the reference to be disabled, the DISABLE pushbutton is depressed. This reference and all contacts referred to it will hold their current (ON or OFF) state. The only way to alter its state (ON to OFF or OFF to ON) is by depressing the FORCE pushbutton; every time this pushbutton is depressed, the disabled reference under the cursor will be changed. Successive activation of the FORCE pushbutton will cause the disabled reference to cycle ON to OFF to ON to OFF etc. Any network can be called up or deleted, reference status entered or replaced, power turned ON or OFF, etc. without changing the disable status of any input reference. As many inputs as desired can be disabled, each either ON or OFF. To restore the normal operation of any disabled reference, the reference is placed into the status area as discussed above, with the cursor on it. When the DISABLE pushbutton is depressed a second time, the reference is restored to its normal (enabled) condition.

Search

The logic in the controller can be searched for specific elements. Networks containing the desired element will be placed on the CRT screen, one at a time. The desired element is first built into the assembly area; only those portions of a relay contact that is pertinent to the search need be entered. For example, if all networks using input 1009 are desired, the assembly area is cleared with the CLEAR pushbutton. The reference 1009 is entered from the keypad as 1, 0, 0, 9; all other portions of the assembly area are left blank (white).   When the SEARCH pushbutton is depressed, the first network to use this input (1009) on any contact (NO or NC), will be placed on the CRT screen with the cursor on that contact.

To search for other networks using this element, the SHIFT pushbutton is depressed and then the SEARCH. Successive networks will be placed on the CRT screen until the end of the stored logic is detected. At the end of the logic a message will appear to inform the operator that all logic has been searched. If normally closed contacts (with or without vertical connections) referenced to 0265 are to be located, the assembly area is cleared, and then the desired element entered (0, 2, 6, 5); the vertical connector is left unaltered.   Searching is then begun with the SEARCH pushbutton as previously discussed. Searches can similarly be done on coils (latched or normal), or register references (30XX or 4XXX).

Inserting/Deleting Networks

To insert a new network between existing networks, the lower network step number is placed on the CRT screen. For example, to insert a network between steps 7 and 8, network 7 is called up. With the cursor on this network, the START NEXT pushbutton is depressed. This causes all networks following step number 7 to be increased by one; for example, step 8 becomes 9, 9 becomes 10, 10 becomes 11, etc. to the end of the stored logic. A new network is started on the CRT screen and any old networks removed. The step number becomes the value 8, and the logic for this network can now be entered. To delete a network, it is placed on the CRT screen with the cursor on it. The SHIFT pushbutton is depressed and then the DELETE pushbutton. This action causes the entire network to be deleted both from the CRT screen and the controller's memory; all subsequent networks will have their step numbers decreased by one. For example, if network 8 is to be deleted, network 9 becomes network 8, 10 becomes 9, 11 becomes 10, etc.

ASCII Code

Hex	ASCII	Hex	ASCII	Hex	ASCII	Hex	ASCII
00	nul	20	sp	40	@	60	
01	soh	21	!	41	A	61	a
02	stx	22	"	42	B	62	b
03	etx	23	#	43	C.	63	c
04	eot	24	$	44	D	64	d
05	enq	25	%	45	E	65	e
06	ack	26	&	46	F	66	f
07	bel	27	'	47	G	67	g
08	bs	28	(	48	H	68	h
09	ht	29	)	49	I	69	i
0A	nl	2A	*	4A	J	6A	j
0B	vt	2B	+	4B	K	6B	k
0C	ff	2C	,	4C	L	6C	l
0D	cr	2D	-	4D	M	6D	m
0E	so	2E	.	4E	N	6E	n
0F	si	2F	/	4F	O	6F	o
10	dle	30	0	50	P	70	p
11	dc1	31	1	51	Q	71	q
12	dc2	32	2	52	R	72	r
13	dc3	33	3	53	S	73	s
14	dc4	34	4	54	T	74	t
15	nak	35	5	55	U	75	u
16	syn	36	6	56	V	76	v
17	etb	37	7	57	W	77	w
18	can	38	8	58	X	78	x
19	em	39	9	59	Y	79	y
1A	sub	3A	:	5A	Z	7A	z
1B	esc	3B	;	5B	[	7B	{
1C	fs	3C	<	5C	\	7C	¦
1D	gs	3D	=	5D	]	7D	}
1E	rs	3E	>	5E	Λ	7E	~
1F	us	3F	?	5F	—	7F	del

6805 Instruction Set and Summary

Nomenclature

The following nomenclature is used in the executable instructions that follow this paragraph.

(a) Operators:

( )	indirection, i.e., (SP) means the value pointed to by SP
←	is loaded with (read: "gets")
·	Boolean AND
v	Boolean (inclusive) OR
⊕	Boolean Exclusive OR
~	Boolean NOT
−	negation (2's complement)

(b) Registers in the MPU:

ACCA	Accumulator
CC	Condition Code Register
X	Index Register
PC	Program Counter
PCH	Program Counter High Byte
PCL	Program Counter Low Byte
SP	Stack Pointer

(c) Memory and Addressing:

M   Contents of any memory location (one byte)

Rel   Relative address (i.e., the 2's complement number stored in the second byte of machine code in a branch instruction)

(d) Bits in the Condition Code Register:

C   Carry/Borrow, Bit 0

Z   Zero Indicator, Bit 1

N   Negative Indicator, Bit 2

I   Interrupt Mask, Bit 3

H   Half-Carry Indicator, Bit 4

(e) Status of Individual Bits *before* Execution of an Instruction

An   Bit n of ACCA (n = 7, 6, 5, 4, 3, 2, 1, 0)

Xn   Bit n of X (n = 7, 6, 5, 4, 3, 2, 1, 0)

Mn   Bit n of M (n = 7, 6, 5, 4, 3, 2, 1, 0). In read/modify/write instructions, Mn is used to represent bit n of M, A or X.

(f) Status of Individual Bits *after* Execution of an Instruction:

Rn   Bit n of the result (n = 7, 6, 5, 4, 3, 2, 1, 0)

(g) Source Forms:

P   Operands with IMMediate, DIRect, EXTended and INDexed (0, 1, 2 byte offset) addressing modes

Q   Operands with DIRect, INDexed (0 and 1 byte offset) addressing modes

dd   Relative operands

DR   Operands with DIRect addressing mode only.

(h) iff

abbreviation for if-and-only-if.

ADD   Add   ADD

Operation: ACCA ← ACCA + M

Description: Adds the contents of ACCA and the contents of M and places the result in ACCA.

Condition Codes:

H: Set if there was a carry from bit 3; cleared otherwise.

I: Not affected.

N: Set if the most significant bit of the result is set; cleared otherwise.

Z: Set if all bits of the result are cleared; cleared otherwise.

C: Set if there was a carry from the most significant bit of the result; cleared otherwise.

Boolean Formulae for Condition Codes:

$$H = A3 \cdot M3 v M3 \cdot R3 v R3 \cdot A3$$
$$N = R7$$
$$Z = \overline{R7} \cdot \overline{R6} \cdot \overline{R5} \cdot \overline{R4} \cdot \overline{R3} \cdot \overline{R2} \cdot \overline{R1} \cdot \overline{R0}$$
$$C = A7 \cdot M7 v M7 \cdot \overline{R7} v \overline{R7} \cdot A7$$

Addressing Mode	Cycles HMOS	CMOS	Bytes	Opcode
Immediate	2	2	2	AB
Direct	4	3	2	BB
Extended	5	4	3	CB
Indexed 0 Offset	4	3	1	FB
Indexed 1-Byte	5	4	2	EB
Indexed 2-Byte	6	5	3	DB

AND   Logical AND   AND

Operation: ACCA — ACCA M

Description: Performs logical AND between the contents of ACCA and the contents of M and places the result in ACCA. Each bit of ACCA after the operation will be the logical AND result of the corresponding bits of M and of ACCA before the operation.

Condition Codes:

H: Not affected.
I: Not affected.
N: Set if the most significant bit of the result is set; cleared otherwise.
Z: Set if all bits of the result are cleared; cleared otherwise.
C: Not affected.

Boolean Formulae for Condition Codes:

$$N = R7$$
$$Z = \overline{R7} \cdot \overline{R6} \cdot \overline{R5} \cdot \overline{R4} \cdot \overline{R3} \cdot \overline{R2} \cdot \overline{R1} \cdot \overline{R0}$$

Addressing Mode	Cycles HMOS	CMOS	Bytes	Opcode
Immediate	2	2	2	A4
Direct	4	3	2	B4
Extended	5	4	3	C4
Indexed 0 Offset	4	3	1	F4
Indexed 1-Byte	5	4	2	E4
Indexed 2-Byte	6	5	3	D4

ASL
Arithmetic Shift
ASL

Operation:

$$C \leftarrow \boxed{b7 \quad | \quad | \quad | \quad | \quad | \quad b0} \leftarrow 0$$

Description: Shifts all bits of ACCA, X or M one place to the left. Bit 0 is loaded with a zero. The C bit is loaded from the most significant bit of ACCA, X or M.

Condition Codes:

H: Not affected.
I: Not affected.
N: Set if the most significant bit of the result is set; cleared otherwise.
Z: Set if all bits of the result are cleared; cleared otherwise.
C: Set if, before the operation, the most significant bit of ACCA, X or M was set; cleared otherwise.

Boolean Formulae for Condition Codes:

$N = R7$
$Z = \overline{R7} \cdot \overline{R6} \cdot \overline{R5} \cdot \overline{R4} \cdot \overline{R3} \cdot \overline{R2} \cdot \overline{R1} \cdot \overline{R0}$
$C = M7$

Comments: Same opcode as LSL

Addressing Mode	Cycles		Bytes	Opcode
	HMOS	CMOS		
Accumulator	4	3	1	48
Index Register	4	3	1	58
Direct	6	5	2	38
Indexed 0 Offset	6	5	1	78
Indexed 1-Byte	7	6	2	68
Indexed 2-Byte				

ASR
Arithmetic Shift Right
ASR

Operation:

$$\boxed{b7 \quad | \quad | \quad | \quad | \quad | \quad b0} \rightarrow C$$

Description: Shifts all bits of ACCA, X or M one place to the right. Bit 7 is held constant. Bit 0 is loaded into the C bit.

Condition Codes:

H: Not affected.
I: Not affected.
N: Set if the most significant bit of the result is set; cleared otherwise.
Z: Set if all bits of the result are cleared; cleared otherwise.
C: Set if, before the operation, the least significant bit of ACCA, X or M was set; cleared otherwise.

Boolean Formulae for Condition Codes:

$N = R7$
$Z = \overline{R7} \cdot \overline{R6} \cdot \overline{R5} \cdot \overline{R4} \cdot \overline{R3} \cdot \overline{R2} \cdot \overline{R1} \cdot \overline{R0}$
$C = M0$

Addressing Mode	Cycles		Bytes	Opcode
	HMOS	CMOS		
Accumulator	4	3	1	47
Index Register	4	3	1	57
Direct	6	5	2	37
Indexed 0 Offset	6	5	1	77
Indexed 1-Byte	7	6	2	67
Indexed 2-Byte				

BCC   Branch if Carry Clear   BCC

Operation: $PC \leftarrow PC + 0002 + Rel$ iff $C = 0$

Description: Tests the state of the C bit and causes a branch iff C is clear. See BRA instruction for further details of the execution of the branch.

Condition Codes: Not affected.

Comments: Same opcode as BHS

Addressing Mode	Cycles		Bytes	Opcode
	HMOS	CMOS		
Relative	4	3	2	24

BCLR n   Bit Clear Bit n   BCLR n

Operation: $Mn \leftarrow 0$

Description: Clear bit n (n = 0, 7) in location M. All other bits in M are unaffected.

Condition Codes: Not affected.

Addressing Mode	Cycles		Bytes	Opcode
	HMOS	CMOS		
Direct	7	5	2	$11 + 2 \cdot n$

BHCC

Branch if Half Carry Clear

BHCC

Operation: PC ← PC + 0002 + Rel iff H = 0

Description: Tests the state of the H bit and causes a branch iff H is clear. See BRA instruction for further details of the execution of the branch.

**Condition
Codes:**   Not affected.

Addressing Mode	Cycles		Bytes	Opcode
	HMOS	CMOS		
Relative	4	3	2	28

BHCS

Branch if Half Carry Set

BHCS

Operation: PC ← PC + 0002 + Rel iff H = 1

Description: Tests the state of the H bit and causes a branch iff H is set. See BRA instruction for further details of the execution of the branch.

**Condition
Codes:**   Not affected.

Addressing Mode	Cycles		Bytes	Opcode
	HMOS	CMOS		
Relative	4	3	2	29

BHI

Branch if Higher

BHI

Operation: PC ← PC + 0002 + Rel iff (C v Z) = 0
                   i.e., if ACCA > M (unsigned binary numbers)

Description: Causes a branch iff both C and Z are zero. If the BHI instruction is executed immediately after execution of either of the CMP or SUB instructions, the branch will occur if and only if the unsigned binary number represented by the minuend (i.e., ACCA) was greater than the unsigned binary number represented by the subtrahend (i.e., M). See BRA instruction for further details of the execution of the branch.

**Condition
Codes:**   Not affected.

Addressing Mode	Cycles		Bytes	Opcode
	HMOS	CMOS		
Relative	4	3	2	22

BHS

Branch iff Higher or Same

BHS

Operation: PC — PC + 0002 + Rel iff C = 0

Description: Following an unsigned compare or subtract, BHS will cause a branch iff the register was higher than or the same as the location in memory. See BRA instruction for further details of the execution of the branch.

Condition Codes: Not affected.

Comments: Same opcode as BCC

Addressing Mode	Cycles		Bytes	Opcode
	HMOS	CMOS		
Relative	4	3	2	24

BIH

Branch iff Interrupt Line is High

BIH

Operation: PC — PC + 0002 + Rel iff INT = 1

Description: Tests the state of the external interrupt pin and branches iff it is high. See BRA instruction for further details of the execution of the branch.

Condition Codes: Not affected.

Comments: In systems not using interrupts, this instruction and BIL can be used to create an extra I/O input bit. This instruction does NOT test the state of the interrupt mask bit nor does it indicate whether an interrupt is pending. All it does is indicate whether the INT line is high.

Addressing Mode	Cycles		Bytes	Opcode
	HMOS	CMOS		
Relative	4	3	2	2F

BIL

Branch if Interrupt Line is Low

BIL

Operation: PC — PC + 0002 + Rel iff INT = 0

Description: Tests the state of the external interrupt pin and branches iff it is low. See BRA instruction for further details of the execution of the branch.

Condition Codes: Not affected.

Comments: In systems not using interrupts, this instruction and BIH can be used to create an extra I/O input bit. This instruction does NOT test the state of the interrupt mask bit nor does it indicate whether an interrupt is pending. All it does is indicate whether the INT line is Low.

| Addressing Mode | Cycles | | Bytes | Opcode |
	HMOS	CMOS		
Relative	4	3	2	2E

BIT

Bit Test Memory with Accumulator

BIT

Operation: ACCA · M

Description: Performs the logical AND comparison of the contents of ACCA and the contents of M and modifies the condition codes accordingly. The contents of ACCA and M are unchanged.

Condition Codes:

H: Not affected.
I: Not affected.
N: Set if the most significant bit of the result of the AND is set; cleared otherwise.
Z: Set if all bits of the result of the AND are cleared; cleared otherwise.
C: Not affected.

Boolean Formulae for Condition Codes:

$N = R7$

$Z = \overline{R7} \cdot \overline{R6} \cdot \overline{R5} \cdot \overline{R4} \cdot \overline{R3} \cdot \overline{R2} \cdot \overline{R1} \cdot \overline{R0}$

| Address' · Mode | Cycles | | Bytes | Opcode |
	HMOS	CMOS		
Immediate	2	2	2	A5
Direct	4	3	2	B5
Extended	5	4	3	C5
Indexed 0 Offset	4	3	1	F5
Indexed 1-Byte	5	4	2	E5
Indexed 2-Byte	6	5	3	D5

BLO

Branch if Lower

BLO

BLO

Operation: PC — PC + 0002 + Rel iff C = 1

Description: Following a compare, BLO will branch iff the register was lower than the memory location. See BRA instruction for further details of the execution of the branch.

Condition Codes: Not affected.

Comments: Same opcode as BCS

Addressing Mode	Cycles HMOS	Cycles CMOS	Bytes	Opcode
Relative	4	3	2	25

BLS

Branch iff Lower or Same

BLS

Operation: PC — PC + 0002 + Rel iff (C v Z) = 1
i.e., if ACCA — M (unsigned binary numbers)

Description: Causes a branch if (C is set) OR (Z is set). If the BLS instruction is executed immediately after execution of either of the instructions CMP or SUB, the branch will occur if and only if the unsigned binary number represented by the minuend (i.e., ACCA) was less than or equal to the unsigned binary number represented by the subtrahend (i.e., M). See BRA instruction for further details of the execution of the branch.

Condition Codes: Not affected.

Addressing Mode	Cycles HMOS	Cycles CMOS	Bytes	Opcode
Relative	4	3	2	23

BMC

Branch if Interrupt Mask is Clear

BMC

Operation: PC — PC + 0002 + Rel iff I = 0

Description: Tests the state of the I bit and causes a branch iff I is clear. See BRA instruction for further details of the execution of the branch.

Condition Codes: Not affected.

Comments: This instruction does NOT branch on the condition of the external interrupt line. The test is performed only on the interrupt mask bit.

Addressing Mode	Cycles HMOS	Cycles CMOS	Bytes	Opcode
Relative	4	3	2	2C

BMI

Branch if Minus

BMI

Operation: PC — PC + 0002 + Rel iff N = 1

Description: Tests the state of the N bit and causes a branch iff N is set. See BRA instruction for further details of the execution of the branch.

Condition Codes: Not affected.

Addressing Mode	Cycles		Bytes	Opcode
	HMOS	CMOS		
Relative	4	3	2	2B

BMS

Branch if Interrupt Mask Bit is Set

BMS

Operation: PC — PC + 0002 + Rel iff I = 1

Description: Tests the state of the I bit and causes a branch iff I is set. See BRA instruction for further details of the execution of the branch.

Condition Codes: Not affected.

Comments: This instruction does NOT branch on the condition of the external interrupt line. The test is performed only on the interrupt mask bit.

Addressing Mode	Cycles		Bytes	Opcode
	HMOS	CMOS		
Relative	4	3	2	2D

BNE

Branch if Not Equal

BNE

Operation: PC — PC + 0002 + Rel iff Z = 0

Description: Tests the state of the Z bit and causes a branch iff Z is clear. Following a compare or subtract instruction BNE will cause a branch if the arguments were different. See BRA instruction for further details of the execution of the branch.

Condition Codes: Not affected.

Addressing Mode	Cycles		Bytes	Opcode
	HMOS	CMOS		
Relative	4	3	2	26

BPL

Branch if Plus

BPL

Operation: PC — PC + 0002 + Rel iff N = 0

Description: Tests the state of the N bit and causes a branch iff N is clear. See BRA instruction for further details of the execution of the branch.

**Condition
Codes:** Not affected.

Addressing Mode	Cycles HMOS	Cycles CMOS	Bytes	Opcode
Relative	4	3	2	2A

BRA

Branch Always

BRA

Operation: PC — PC + 0002 + Rel

Description: Unconditional branch to the address given by the foregoing formula, in which Rel is the relative address stored as a two's complement number in the second byte of machine code corresponding to the branch instruction.

NOTE: The source program specifies the destination of any branch instruction by its absolute address, either as a numerical value or as a symbol or expression which can be evaluated by the assembler. The assembler obtains the relative address Rel from the absolute address and the current value of the program counter.

**Condition
Codes:** Not affected.

Addressing Mode	Cycles HMOS	Cycles CMOS	Bytes	Opcode
Relative	4	3	2	20

BRCLR n

Branch if Bit n is Clear

BRCLR n

Operation: PC — PC + 0003 + Rel iff bit n of M is zero

Description: Tests bit n (n = 0, 7) of location M and branches iff the bit is clear.

**Condition
Codes:**
H: Not affected.
I: Not affected.
N: Not affected.
Z: Not affected.
C: Set if Mn = 1; cleared otherwise.

Boolean Formulae for Condition Codes:
C = Mn

Comments: The C bit is set to the state of the bit tested. Used with an appropriate rotate instruction, this instruction is an easy way to do serial to parallel conversions.

Addressing Mode	Cycles HMOS	Cycles CMOS	Bytes	Opcode
Relative	10	5	3	$01 + 2 \cdot n$

BRN   Branch Never   BRN

Description: Never branches. Branch never is a 2 byte 4 cycle NOP.

Condition Codes: Not affected.

Comments: BRN is included here to demonstrate the nature of branches on the M6805 Family. Each branch is matched with an inverse that varies only in the least significant bit of the opcode. BRN is the inverse of BRA. This instruction may have some use during program debugging.

Addressing Mode	Cycles HMOS	Cycles CMOS	Bytes	Opcode
Relative	4	3	2	21

BRSET   Branch if Bit n is Set   BRSET

Operation: PC — PC + 0003 + Rel iff Bit n of M is not zero

Description: Tests bit n (n = 0, 7) of location M and branches iff the bit is set.

Condition Codes:
H: Not affected.
I: Not affected.
N: Not affected.
Z: Not affected.
C: Set if Mn = 1; cleared otherwise.

Boolean Formulae for Condition Codes:
C = Mn

Comments: The C bit is set to the state of the bit tested. Used with an appropriate rotate instruction, this instruction is an easy way to provide serial to parallel conversions.

Addressing Mode	Cycles HMOS	Cycles CMOS	Bytes	Opcode
Relative	10	5	3	$2 \cdot n$

BSET n

Set Bit in Memory

BSET n

Operation: Mn — 1

Description: Set bit n (n = 0, 7) in location M. All other bits in M are unaffected.

Condition Codes: Not affected.

Addressing Modes	Cycles HMOS	CMOS	Bytes	Opcode
Direct	7	5	2	10 + 2•n

BSR

Branch to Subroutine

BSR

Operation:
PC — PC + 0002
(SP) — PCL; SP — SP − 0001
(SP) — PCH; SP — SP − 0001
PC — PC + Rel

Description: The program counter is incremented by 2. The least (low) significant byte of the program counter contents is pushed onto the stack. The stack pointer is then decremented (by one). The most (high) signficant byte of the program counter contents is then pushed onto the stack. Unused bits in the Program Counter high byte are stored as 1's on the stack. The stack pointer is again decremented (by one). A branch then occurs to the location specified by the relative offset. See the BRA instruction for details of the branch execution.

Condition Codes: Not affected.

Addressing Mode	Cycles HMOS	CMOS	Bytes	Opcode
Relative	8	6	2	AD

CLC
Clear Carry Bit
CLC

Operation: C bit — 0

Description: Clears the carry bit in the processor condition code register.

**Condition
Codes:**

H: Not affected.
I: Not affected.
N: Not affected.
Z: Not affected.
C: Cleared.

Boolean Formulae for Condition Codes:
C = 0

Addressing Mode	Cycles		Bytes	Opcode
	HMOS	CMOS		
Inherent	2	2	1	98

CLI
Clear Interrupt Mask Bit
CLI

Operation: I bit — 0

Description: Clears the interrupt mask bit in the processor condition code register. This enables the microprocessor to service interrupts. Interrupts that were pending while the I bit was set will now begin to have effect.

**Condition
Codes:**

H: Not affected.
I: Cleared
N: Not affected.
Z: Not affected.
C: Not affected.

Boolean Formulae for Condition Codes:
I = 0

Addressing Mode	Cycles		Bytes	Opcode
	HMOS	CMOS		
Inherent	2	2	1	9A

CLR

Clear

CLR

Operation: X — 00 or,
ACCA — 00 or,
M — 00

Description: The contents of ACCA, X or M are replaced with zeroes.

**Condition
Codes:**
H: Not affected.
I: Not affected.
N: Cleared.
Z: Set.
C: Not affected.

Boolean Formulae for Condition Codes:
N = 0
Z = 1

Addressing Mode	Cycles		Bytes	Opcode
	HMOS	CMOS		
Accumulator	4	3	1	4F
Index Register	4	3	1	5F
Direct	6	5	2	3F
Indexed 0 Offset	6	5	1	7F
Indexed 1-Byte	7	6	2	6F

CMP

Compare Accumulator with Memory

CMP

Operation: ACCA — M

Description: Compares the contents of ACCA and the contents of M and sets the condition codes, which may then be used for controlling the conditional branches. Both operands are unaffected.

**Condition
Codes:**
H: Not affected.
I: Not affected.
N: Set if the most significant bit of the result of the subtraction is set; cleared otherwise.
Z: Set if all bits of the result of the subtraction are cleared; cleared otherwise.
C: Set if the absolute value of the contents of memory is larger than the absolute value of the accumulator; cleared otherwise.

Boolean Formulae for Condition Codes:

$$N = R7$$
$$Z = \overline{R7} \cdot \overline{R6} \cdot \overline{R5} \cdot \overline{R4} \cdot \overline{R3} \cdot \overline{R2} \cdot \overline{R1} \cdot \overline{R0}$$
$$C = A7 \cdot M7 v M7 \cdot \overline{R7} v \overline{R7} \cdot A7$$

Addressing Mode	Cycles HMOS	Cycles CMOS	Bytes	Opcode
Immediate	2	2	2	A1
Direct	4	3	2	B1
Extended	5	4	3	C1
Indexed 0 Offset	4	3	1	F1
Indexed 1-Byte	5	4	2	E1
Indexed 2-Byte	6	5	3	D1

COM Complement COM

Operation:
$$X \leftarrow \sim X = \$FF - X \text{ or,}$$
$$ACCA \leftarrow \sim ACCA = \$FF - ACCA \text{ or,}$$
$$M \leftarrow \sim M = \$FF - M$$

Description: Replaces the contents of ACCA, X or M with the one's complement. Each bit of the operand is replaced with the complement of that bit.

Condition Codes:

H: Not affected.
I: Not affected.
N: Set if the most significant bit of the result is set; cleared otherwise.
Z: Set if all bits of the result are cleared; cleared otherwise.
C: Set.

Boolean Formulae for Condition Codes:

$$N = R7$$
$$Z = \overline{R7} \cdot \overline{R6} \cdot \overline{R5} \cdot \overline{R4} \cdot \overline{R3} \cdot \overline{R2} \cdot \overline{R1} \cdot \overline{R0}$$
$$C = 1$$

Addressing Mode	Cycles HMOS	Cycles CMOS	Bytes	Opcode
Accumulator	4	3	1	43
Index Register	4	3	1	53
Direct	6	5	2	33
Indexed 0 Offset	6	5	1	73
Indexed 1-Byte	7	6	2	63

CPX

Compare Index Register with Memory

CPX

Operation: X – M

Description: Compares the contents of X to the contents of M and sets the condition codes, which may then be used for controlling the conditional branches. Both operands are unaffected.

Condition Codes:

H: Not affected.

I: Not affected.

N: Set if the most significant bit of the result of the subtraction is set; cleared otherwise.

Z: Set if all bits of the result of the subtraction are cleared; cleared otherwise.

C: Set if the absolute value of the contents of memory is larger than the absolute value of the index register; cleared otherwise.

Boolean Formulae for Condition Codes:

$N = R7$

$Z = \overline{R7} \cdot \overline{R6} \cdot \overline{R5} \cdot \overline{R4} \cdot \overline{R3} \cdot \overline{R2} \cdot \overline{R1} \cdot \overline{R0}$

$C = X7 \cdot M7 v M7 \cdot \overline{R7} v \overline{R7} \cdot X7$

Addressing Mode	Cycles HMOS	Cycles CMOS	Bytes	Opcode
Immediate	2	2	2	A3
Direct	4	3	2	B3
Extended	5	4	3	C3
Indexed 0 Offset	4	3	1	F3
Indexed 1-Byte	5	4	2	E3
Indexed 2-Byte	6	5	3	D3

DEC

Decrement

DEC

Operation: X – X-01 or,
ACCA – ACCA-01 or,
M – M-01

Description: Subtract one from the contents of ACCA, X or M. The N and Z bits are set or reset according to the result of this operation. The C bit is not affected by this operation.

Condition Codes:

H: Not affected.

I: Not affected.

N: Set if the most significant bit of the result is set; cleared otherwise.

Z: Set if all bits of the result are cleared; cleared otherwise.

C: Not affected.

Boolean Formulae for Condition Codes:

$$N = R7$$
$$Z = \overline{R7} \cdot \overline{R6} \cdot \overline{R5} \cdot \overline{R4} \cdot \overline{R3} \cdot \overline{R2} \cdot \overline{R1} \cdot \overline{R0}$$

Addressing Mode	Cycles HMOS	Cycles CMOS	Bytes	Opcode
Accumulator	4	3	1	4A
Index Register	4	3	1	5A
Direct	6	5	2	3A
Indexed 0 Offset	6	5	1	7A
Indexed 1-Byte	7	6	2	6A

EOR   Exclusive Or Memory with Accumulator   EOR

Operation: ACCA — ACCA ⊕ M

Description: Performs the logical EXCLUSIVE OR between the contents of ACCA and the contents of M, and places the result in ACCA. Each bit of ACCA after the operation will be the logical EXCUSIVE OR of the corresponding bit of M and ACCA before the operation.

Condition Codes:

H: Not affected.
I: Not affected.
N: Set if the most significant bit of the result is set; cleared otherwise.
Z: Set if all bits of the result are cleared; cleared otherwise.
C: Not affected.

Boolean Formulae for Condition Codes:

$$N = R7$$
$$Z = \overline{R7} \cdot \overline{R6} \cdot \overline{R5} \cdot \overline{R4} \cdot \overline{R3} \cdot \overline{R2} \cdot \overline{R1} \cdot \overline{R0}$$

Addressing Mode	Cycles HMOS	Cycles CMOS	Bytes	Opcode
Immediate	2	2	2	A8
Direct	4	3	2	B8
Extended	5	4	3	C8
Indexed 0 Offset	4	3	1	F8
Indexed 1-Byte	5	4	2	E8
Indexed 2-Byte	6	5	3	D8

INC

Increment

INC

Operation: $X \leftarrow X + 01$ or,
ACCA $\leftarrow$ ACCA + 01 or,
$M \leftarrow M + 01$

Description: Add one to the contents of ACCA, X or M. The N and Z bits are set or reset according to the result of this operation. The C bit is not affected by this operation.

Condition Codes:
    H: Not affected.
    I: Not affected.
    N: Set if the most significant bit of the result is set; cleared otherwise.
    Z: Set if all bits of the result are cleared; cleared otherwise.
    C: Not affected.

Boolean Formulae for Condition Codes:
$$N = R7$$
$$Z = \overline{R7} \cdot \overline{R6} \cdot \overline{R5} \cdot \overline{R4} \cdot \overline{R3} \cdot \overline{R2} \cdot \overline{R1} \cdot \overline{R0}$$

Addressing Mode	Cycles		Bytes	Opcode
	HMOS	CMOS		
Accumulator	4	3	1	4C
Index Register	4	3	1	5C
Direct	6	5	2	3C
Indexed 0 Offset	6	5	1	7C
Indexed 1-Byte	7	6	2	6C

JMP

Jump

JMP

Operation: PC $\leftarrow$ effective address

Description: A jump occurs to the instruction stored at the effective address. The effective address is obtained according to the rules for EXTended, DIRect or INDexed addressing.

Condition Codes: Not affected.

Addressing Mode	Cycles HMOS	Cycles CMOS	Bytes	Opcode
Direct	3	2	2	BC
Extended	4	3	3	CC
Indexed 0 Offset	3	2	1	FC
Indexed 1-Byte	4	3	2	EC
Indexed 2-Byte	5	4	3	DC

JSR   Jump to Subroutine   JSR

Operation:
PC — PC + N
(SP) — PCL; SP — SP − 0001
(SP) — PCH ; SP — SP − 0001
PC — effective address

Description: The program counter is incremented by N (N = 1, 2 or 3 depending on the addressing mode), and is then pushed onto the stack (least significant byte first). Unused bits in the Program Counter high byte are stored as 1's on the stack. The stack pointer points to the next empty location on the stack. A jump occurs to the instruction stored at the effective address. The effective address is obtained according to the rules for EXTended, DIRect, or INDexed addressing.

Condition Codes: Not affected.

Addressing Mode	Cycles HMOS	Cycles CMOS	Bytes	Opcode
Direct	7	5	2	BD
Extended	8	6	3	CD
Indexed 0 Offset	7	5	1	FD
Indexed 1-Byte	8	6	2	ED
Indexed 2-Byte	9	7	3	DD

LDA   Load Accumulator from Memory   LDA

Operation: ACCA — M

Description: Loads the contents of memory into the accumulator. The condition codes are set according to the data.

Condition
Codes:
 H: Not affected.
 I: Not affected.
 N: Set if the most significant bit of the accumulator is set; cleared otherwise.
 Z: Set if all bits of the accumulator are cleared; cleared otherwise.
 C: Not affected.

Boolean Formulae for Condition Codes:

$$N = R7$$
$$Z = \overline{R7} \cdot \overline{R5} \cdot \overline{R4} \cdot \overline{R3} \cdot \overline{R2} \cdot \overline{R1} \cdot \overline{R0}$$

Addressing Mode	Cycles		Bytes	Opcode
	HMOS	CMOS		
Immediate	2	2	2	A6
Direct	4	3	2	B6
Extended	5	4	3	C6
Indexed 0 Offset	4	3	1	F6
Indexed 1-Byte	5	4	2	E6
Indexed 2-Byte	6	5	3	D6

LDX   Load Index Register from Memory   LDX

Operation: X – M

Description: Loads the contents of memory into the index register. The condition codes are set according to the data.

Condition
Codes:
 H: Not affected.
 I: Not affected.
 N: Set if the most significant bit of the index register is set; cleared otherwise.
 Z: Set if all bits of the index register are cleared; cleared otherwise.
 C: Not affected.

Boolean Formulae for Condition Codes:

$$N = R7$$
$$Z = \overline{R7} \cdot \overline{R6} \cdot \overline{R5} \cdot \overline{R4} \cdot \overline{R3} \cdot \overline{R2} \cdot \overline{R1} \cdot \overline{R0}$$

Addressing Mode	Cycles		Bytes	Opcode
	HMOS	CMOS		
Immediate	2	2	2	AE
Direct	4	3	2	BE
Extended	5	4	3	CE
Indexed 0 Offset	4	3	1	FE
Indexed 1-Byte	5	4	2	EE
Indexed 2-Byte	6	5	3	DE

LSL

Logical Shift Left

LSL

Operation: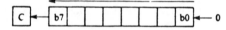

Description: Shifts all bits of the ACCA, X or M one place to the left. Bit 0 is loaded with a zero. The C bit is loaded from the most signficant bit of ACCA, X or M.

Condition Codes:

H: Not affected.
I: Not affected.
N: Set if the most significant bit of the result is set; cleared otherwise.
Z: Set if all bits of the result are cleared; cleared otherwise.
C: Set if, before the operation, the most significant bit of ACCA, X or M was set; cleared otherwise.

Boolean Formulae for Condition Codes:

$N = R7$
$Z = \overline{R7} \cdot \overline{R6} \cdot \overline{R5} \cdot \overline{R4} \cdot \overline{R3} \cdot \overline{R2} \cdot \overline{R1} \cdot \overline{R0}$
$C = M7$

Comments: Same as ASL

Addressing Mode	Cycles		Bytes	Opcode
	HMOS	CMOS		
Accumulator	4	3	1	48
Index Register	4	3	1	58
Direct	6	5	2	38
Indexed 0 Offset	6	5	1	78
Indexed 1-Byte	7	6	2	68

LSR

Logical Shift Right

LSR

Operation: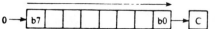

Description: Shifts all bits of ACCA, X or M one place to the right. Bit 7 is loaded with a zero. Bit 0 is loaded into the C bit.

Condition Codes:

H: Not affected.
I: Not affected.
N: Cleared.
Z: Set if all bits of the result are cleared; cleared otherwise.
C: Set if, before the operation, the least significant bit of ACCA, X or M was set; cleared otherwise.

Boolean Formulae for Condition Codes:

$N = 0$
$Z = \overline{R7} \cdot \overline{R6} \cdot \overline{R5} \cdot \overline{R4} \cdot \overline{R3} \cdot \overline{R2} \cdot \overline{R1} \cdot \overline{R0}$
$C = M0$

| Addressing Mode | Cycles | | Bytes | Opcode |
	HMOS	CMOS		
Accumulator	4	3	1	44
Index Register	4	3	1	54
Direct	6	5	2	34
Indexed 0 Offset	6	5	1	74
Indexed 1-Byte	7	6	2	64

NEG
Negate
NEG

Operation: $-X - X = 00 - X$ or,
$-ACCA - ACCA = 00 - ACCA$ or,
$-M - M = 00 - M$

Description: Replaces the contents of ACCA, X or M with its two's complement. Note that $80 is left unchanged.

Condition Codes:

H: Not affected.

I: Not affected.

N: Set if the most significant bit of the result is set; cleared otherwise.

Z: Set if all bits of the result are cleared; cleared otherwise.

C: Set if there would be a borrow in the implied subtraction from zero; the C bit will be set in all cases except when the contents of ACCA, X or M before the NEG is 00.

Boolean Formulae for Condition Codes:

$N = R7$

$Z = \overline{R7} \cdot \overline{R6} \cdot \overline{R5} \cdot \overline{R4} \cdot \overline{R3} \cdot \overline{R2} \cdot \overline{R1} \cdot \overline{R0}$

$C = R7vR6vR5vR4vR3vR2vR1vR0$

| Addressing Mode | Cycles | | Bytes | Opcode |
	HMOS	CMOS		
Accumulator	4	3	1	40
Index Register	4	3	1	50
Direct	6	5	2	30
Indexed 0 Offset	6	5	1	70
Indexed 1-Byte	7	6	2	60

NOP

No Operation

NOP

Description: This is a single-byte instruction which causes only the program counter to be incremented. No other registers are changed.

Condition Codes: Not affected.

Addressing Mode	Cycles HMOS	CMOS	Bytes	Opcode
Inherent	2	2	1	9D

ORA

Inclusive OR

ORA

Operation: ACCA — ACCA V M

Description: Performs logical OR between the contents of ACCA and the contents of M and place the result in ACCA. Each bit of ACCA after the operation will be the logical (inclusive) OR result of the corresponding bits of M and ACCA before the operation.

Condition Codes:

H: Not affected.
I: Not affected.
N: Set if the most significant bit of the result is set; cleared otherwise.
Z: Set if all bits of the result are cleared; cleared otherwise.
C: Not affected.

Boolean Formulae for Condition Codes:

$N = R7$
$Z = \overline{R7} \cdot \overline{R6} \cdot \overline{R5} \cdot \overline{R4} \cdot \overline{R3} \cdot \overline{R2} \cdot \overline{R1} \cdot \overline{R0}$

Addressing Mode	Cycles HMOS	CMOS	Bytes	Opcode
Immediate	2	2	2	AA
Direct	4	3	2	BA
Extended	5	4	3	CA
Indexed 0 Offset	4	3	1	FA
Indexed 1-Byte	5	4	2	EA
Indexed 2-Byte	6	5	3	DA

ROL   Rotate Left thru Carry   ROL

Operation:

Description: Shifts all bits of the ACCA, X or M one place to the left. Bit 0 is loaded from the C bit. The C bit is loaded from the most significant bit of ACCA, X or M.

Condition Codes:

H: Not affected.
I: Not affected.
N: Set if the most significant bit of the result is set; cleared otherwise.
Z: Set if all bits of the result are cleared; cleared otherwise.
C: Set if, before the operation, the most significant bit of ACCA, X or M was set; cleared otherwise.

Boolean Formulae for Condition Codes:

$N = R7$
$Z = \overline{R7} \cdot \overline{R6} \cdot \overline{R5} \cdot \overline{R4} \cdot \overline{R3} \cdot \overline{R2} \cdot \overline{R1} \cdot \overline{R0}$
$C = M7$

Addressing Mode	Cycles HMOS	Cycles CMOS	Bytes	Opcode
Accumulator	4	3	1	49
Index Register	4	3	1	59
Direct	6	5	2	39
Indexed 0 Offset	6	5	1	79
Indexed 1-Byte	7	6	2	69

ROR   Rotate Right Thru Carry   ROR

Operation:

Description: Shifts all bits of ACCA, X or M one place to the right. Bit 7 is loaded from the C bit. Bit 0 is loaded into the C bit.

Condition Codes:

H: Not affected.
I: Not affected.
N: Set if the most significant bit of the result is set; cleared otherwise.
Z: Set if all bits of the result are cleared; cleared otherwise.
C: Set if, before the operation, the least significant bit of ACCA, X or M was set; cleared otherwise.

Boolean Formulae for Condition Codes:

$N = R7$
$Z = \overline{R7} \cdot \overline{R6} \cdot \overline{R5} \cdot \overline{R4} \cdot \overline{R3} \cdot \overline{R2} \cdot \overline{R1} \cdot \overline{R0}$
$C = M0$

Addressing Mode	Cycles HMOS	CMOS	Bytes	Opcode
Accumulator	4	3	1	46
Index Register	4	3	1	56
Direct	6	5	2	36
Indexed 0 Offset	6	5	1	76
Indexed 1-Byte	7	6	2	66

RSP

Reset Stack Pointer

RSP

Operation: SP — $7F

Description: Resets the stack pointer to the top of the stack.

Condition Codes: Not affected.

Addressing Mode	Cycles HMOS	CMOS	Bytes	Opcode
Inherent	2	2	1	9C

RTI

Return from Interrupt

RTI

Operation:
SP — SP + 0001 ; CC — (SP)
SP — SP + 0001 ; ACCA — (SP)
SP — SP + 0001 ; X — (SP)
SP — SP + 0001 ; PCH — (SP)
SP — SP + 0001 ; PCL — (SP)

Description: The Condition Codes, Accumulator, Index Register and the Program Counter are restored according to the state previously saved on the stack. Note that the interrupt mask bit (I bit) will be reset if and only if the corresponding bit stored on the stack is zero.

Condition Codes: Set or cleared according to the first byte pulled from the stack.

Addressing Mode	Cycles HMOS	CMOS	Bytes	Opcode
Inherent	9	9	1	80

RTS

Return from Subroutine

RTS

Operation: $\quad$ SP — SP + 0001 ; PCH — (SP)

$\qquad\qquad$ SP — SP + 0001 ; PCL — (SP)

Description: $\quad$ The stack pointer is incremented (by one). The contents of the byte of memory, pointed to by the stack pointer, are loaded into the high byte of the program counter. The stack pointer is again incremented (by one). The byte pointed to by the stack pointer is loaded into the low byte of the program counter.

Condition Codes: $\quad$ Not affected.

Addressing Mode	Cycles		Bytes	Opcode
	HMOS	CMOS		
Inherent	6	6	1	81

SBC

Subtract with Carry

SBC

Operation: $\quad$ ACCA — ACCA − M − C

Description: $\quad$ Subtracts the contents of M and C from the contents of ACCA, and places the result in ACCA.

Condition Codes:

H: $\quad$ Not affected.

I: $\quad$ Not affected.

N: $\quad$ Set if the most significant bit of the result is set; cleared otherwise.

Z: $\quad$ Set if all bits of the result are cleared; cleared otherwise.

C: $\quad$ Set if the absolute value of the contents of memory plus the previous carry is larger than the absolute value of the accumulator; cleared otherwise.

Boolean Formulae for Condition Codes:

$N = R7$

$Z = \overline{R7} \cdot \overline{R6} \cdot \overline{R5} \cdot \overline{R4} \cdot \overline{R3} \cdot \overline{R2} \cdot \overline{R1} \cdot \overline{R0}$

$C = A7 \cdot M7 \vee M7 \cdot \overline{R7} \vee \overline{R7} \cdot A7$

Addressing Mode	Cycles		Bytes	Opcode
	HMOS	CMOS		
Immediate	2	2	2	A2
Direct	4	3	2	B2
Extended	5	4	3	C2
Indexed 0 Offset	4	3	1	F2
Indexed 1-Byte	5	4	2	E2
Indexed 2-Byte	6	5	3	D2

SEC

Set Carry Bit

SEC

Operation: C bit — 1

Description: Sets the carry bit in the processor condition code register.

**Condition
Codes:**

H: Not affected.
I: Not affected.
N: Not affected.
Z: Not affected.
C: Set.

Boolean Formulae for Condition Codes:
C = 1

| Addressing Mode | Cycles | | Bytes | Opcode |
	HMOS	CMOS		
Inherent	2	2	1	99

SEI

Set Interrupt Mask Bit

SEI

Operation: I bit — 1

Description: Sets the interrupt mask bit in the processor condition code register. The microprocessor is inhibited from servicing interrupts, and will continue with execution of the instructions of the program until the interrupt mask bit is cleared.

**Condition
Codes:**

H: Not affected.
I: Set
N: Not Affected.
Z: Not affected.
C: Not affected.

Boolean Formulae for Condition Codes:
I = 1

| Addressing Mode | Cycles | | Bytes | Opcode |
	HMOS	CMOS		
Inherent	2	2	1	9B

STA

Store Accumulator in Memory

STA

Operation: M — ACCA

Description: Stores the contents of ACCA in memory. The contents of ACCA remain the same.

Condition Codes:

H: Not affected.
I: Not affected.
N: Set if the most significant bit of the accumulator is set; cleared otherwise.
Z: Set if all bits of the accumulator are clear; cleared otherwise.
C: Not Affected.

Boolean Formulae for Condition Codes:

$N = A7$

$Z = \overline{A7} \cdot \overline{A6} \cdot \overline{A5} \cdot \overline{A4} \cdot \overline{A3} \cdot \overline{A2} \cdot \overline{A1} \cdot \overline{A0}$

Addressing Mode	Cycles		Bytes	Opcode
	HMOS	CMOS		
Direct	5	4	2	B7
Extended	6	5	3	C7
Indexed 0 Offset	5	4	1	F7
Indexed 1-Byte	6	5	2	E7
Indexed 2-Byte	7	6	3	D7

STOP

Enable IRQ, Stop Oscillator

STOP

Description: Reduces power consumption by eliminating all dynamic power dissipation. Results in: (1) timer prescaler to clear; (2) disabling of timer interrupts (3) timer interrupt flag bit to clear; (4) external interrupt request enabling; and (5) inhibiting of oscillator.

When $\overline{RESET}$ or $\overline{IRQ}$ input goes low: (1) oscillator is enabled, (2) a delay of 1920 instruction cycles allows oscillator to stabilize, (3) the interrupt request vector is fetched, and (4) service routine is executed.

External interrupts are enabled following the RTI command.

Condition Codes:

H: Not Affected.
I: Cleared.
N: Not Affected.
Z: Not Affected.
C: Not Affected.

Addressing Mode	Cycles		Bytes	Opcode
	HMOS	CMOS		
Inherent	—	2	1	8E

STX

Store Index Register in Memory

STX

Operation: M ← X

Description: Stores the contents of X in memory. The contents of X remain the same.

Condition Codes:

H: Not Affected.
I: Not affected.
N: Set if the most significant bit of the index register is set; cleared otherwise.
Z: Set if all bits of the index register are clear; cleared otherwise.
C: Not affected.

Boolean Formulae for Condition Codes:

$N = X7$

$Z = \overline{X7} \cdot \overline{X6} \cdot \overline{X5} \cdot \overline{X4} \cdot \overline{X3} \cdot \overline{X2} \cdot \overline{X1} \cdot \overline{X0}$

Addressing Mode	Cycles		Bytes	Opcode
	HMOS	CMOS		
Direct	5	4	2	BF
Extended	6	5	3	CF
Indexed 0 Offset	5	4	1	FF
Indexed 1-Byte	6	5	2	EF
Indexed 2-Byte	7	6	3	DF

SUB

Subtract

SUB

Operation: ACCA ← ACCA − M

Description: Subtracts the contents of M from the contents of ACCA and places the result in ACCA.

Condition Codes:

H: Not affected.
I: Not affected.
N: Set if the most significant bit of the result is set; cleared otherwise.
Z: Set if all bits of the results are cleared; cleared otherwise.
C: Set if the absolute value of the contents of memory are larger than the absolute value of the accumulator; cleared otherwise.

Boolean Formulae for Condition Codes:

$N = R7$

$Z = \overline{R7} \cdot \overline{R6} \cdot \overline{R5} \cdot \overline{R4} \cdot \overline{R3} \cdot \overline{R2} \cdot \overline{R1} \cdot \overline{R0}$

$C = A7 \cdot M7 \vee M7 \cdot \overline{R7} \vee \overline{R7} \cdot A7$

Addressing Mode	Cycles HMOS	Cycles CMOS	Bytes	Opcode
Immediate	2	2	2	A0
Direct	4	3	2	B0
Extended	5	4	3	C0
Indexed 0 Offset	4	3	1	F0
Indexed 1-Byte	5	4	2	E0
Indexed 2-Byte	6	5	3	D0

SWI
Software Interrupt.
SWI

Operation:
PC — PC + 0001
(SP) — PCL ; SP — SP − 0001
(SP) — PCH ; SP — SP − 0001
(SP) — X ; SP — SP − 0001
(SP) — ACCA ; SP — SP − 0001
(SP) — CC ; SP — SP − 0001
1 bit — 1
PCH — n − 0003
PCL — n − 0002

Description: The program counter is incremented (by one). The Program Counter, Index Register and Accumulator are pushed onto the stack. The Condition Code register bits are then pushed onto the stack with bits H, I, N, Z and C going into bit positions 4 through 0 with the top three bits (7, 6 and 5) containing ones. The stack pointer is decremented by one after each byte is stored on the stack.

The interrupt mask bit is then set. The program counter is then loaded with the address stored in the software interrupt vector located at memory locations n − 0002 and n − 0003, where n is the address corresponding to a high state on all lines of the address bus.

Condition Codes:
H: Not affected.
I: Set.
N: Not affected.
Z: Not affected.
C: Not affected.

Boolean Formulae for Condition Codes:
1 = 1

Caution: This instruction is used by Motorola in some of its software products and may be unavailable for general use.

Addressing Mode	Cycles HMOS	Cycles CMOS	Bytes	Opcode
Inherent	11	10	1	83

TAX

Transfer Accumulator to Index Register

TAX

Operation: X − ACCA

Description: Loads the index register with the contents of the accumulator. The contents of the accumulator are unchanged.

Condition Codes: Not affected.

Addressing Mode	Cycles		Bytes	Opcode
	HMOS	CMOS		
Inherent	2	2	1	97

TST

Test for Negative or Zero

TST

Operation: X − 00 or,
ACCA − 00 or,
M − 0

Description: Sets the condition codes N and Z according to the contents of ACCA, X or M.

Condition Codes:

H: Not affected.
I: Not affected.
N: Set if the most significant bit of the contents of ACCA, X or M is set; cleared otherwise.
Z: Set if all bits of ACCA, X or M are clear; cleared otherwise.
C: Not affected.

Boolean Formulae for Condition Codes:

$$N = M7$$
$$Z = \overline{M7} \cdot \overline{M6} \cdot \overline{M5} \cdot \overline{M4} \cdot \overline{M3} \cdot \overline{M2} \cdot \overline{M1} \cdot \overline{M0}$$

Addressing Mode	Cycles		Bytes	Opcode
	HMOS	CMOS		
Accumulator	4	3	1	4D
Index Register	4	3	1	5D
Direct	6	4	2	3D
Indexed 0 Offset	6	4	1	7D
Indexed 1-Byte	7	5	2	6D

TXA

Transfer Index Register to Accumulator

TXA

Operation: ACCA — X

Description: Loads the accumulator with the contents of the index register. The contents of the index register are unchanged.

Condition Codes: Not affected.

Addressing Mode	Cycles		Bytes	Opcode
	HMOS	CMOS		
Inherent	2	2	1	9F

WAIT

Enable Interrupt, Stop Processor

WAIT

Description: Reduces power consumption by eliminating dynamic power dissipation in all circuits except the timer and timer prescaler. Causes enabling of external interrupts and stops clocking or processor circuits.

Timer interrupts may be enabled or disabled by programmer prior to execution of WAIT.

When RESET or IRQ input goes low, or timer counter reaches zero with counter interrupt enabled: (1) processor clocks are enabled, and (2) interrupt request, reset, and timer interrupt vectors are fetched.

Interrupts are enabled following the RTI command.

Condition Codes:
H: Not affected.
I: Cleared.
N: Not affected.
Z: Not affected.
C: Not affected.

Addressing Mode	Cycles		Bytes	Opcode
	HMOS	CMOS		
Inherent	—	2	1	8F

Appendix E

FCC Rules and Regulations

On April 9, 1980, the FCC released its final report (FCC 80-148/27114) amending Part 15 of its rules governing emission limitations for all computing devices. (originally released as report FCC 79-555/14686 on October 11, 1979 as docket 20780).

This report, adopted March 27, 1980, defines the marketing rules, compliance dates, exemptions, definitions, limits, classes of equipment, and labeling requirements.

As stated in Appendix B of the report the FCC defines a computing device as:

"Any electronic device or system that generates and uses timing signals or pulses at a rate in excess of 10,000 pulses (cycles) per second and uses digital techniques; inclusive of telephone equipment that utilizes digital techniques or any device or system that generates and utilizes radio frequency energy for the purpose of performing data processing functions, such as electronic computations, operations, transformations, recording, filing, sorting, storage, retrival, or transfer. Radio transmitters, receivers, industrial, scientific and medical equipment and any other radio frequency device which are specifically subject to an emanation requirement elsewhere in this Chapter are excluded from this definition."

Computing devices covered by this definition were subdivided into two classes of equipment. They are:

Class A: A computing device that is marketed for use in a commercial, industrial or business environment: exclusive of a device which is marketed for use by the general public, or which is intended to be used in the home.

Class B: A computing device that is marketed for use in a residential environment notwithstanding use in commercial, business and industrial environment. Examples of such devices include, but are not limited to, electronic games, personal computers, calculators, and similar electronic devices that are marketed for use by the general public.

Class A: (1) All other computing devices first manufactured after *October 1, 1981* shall be verified for compliance by the manufacturer.

(2) All other computing devices in production before *October 1, 1981* shall comply by *October 1, 1983.*

In the interim between *January 1, 1981* and compliance with technical requirements, the FCC is requiring labelling that will inform users of the interference potential of equipment.

The device must have permanently attached in a conspicuous location for the user to observe, a label with the following statements:

"This equipment has not been tested to show compliance with new FCC Rules (47 CFR Part 15) designed to limit interference to radio communication requiring the operator to take whatever steps are necessary to correct the interference."

Additionally, the instruction manual for the device shall contain the following information:

Warning: This equipment generates, uses, and can radiate radio frequency energy and if not installed and used in accordance with the instructions manual, may cause interference to radio communications. As temporarily permitted by regulation it has not been tested for compliance with the limits for Class A computing devices pursuant to Subpart J of Part 15 of FCC Rules, which are designed to provide reasonable protection against such interference. Operation of this equipment in a residential area is likely to cause interference in which case the user at his own expense will be required to take whatever measures may be required to correct the interference.

There are other labelling requirements stated in Appendix B of the FCC report for peripherals that are found to comply when connected to a typical system and/or tested alone, and for Class A and Class B equipment that do comply with the specified requirements. Therefore, it is advisable that *each manufacturer obtain and read the requirements* and determine, on an individual basis, their applicability.

Temporarily exempt from this ruling are:

1. Electronic Test Equipment. (oscillator, scope, counters, etc.)
2. Home Appliances (solid state devices that control dishwashers, etc.)
3. Automotive Electronics (electronic fuel injection, etc.)
4. Industrial Control Systems (computer controlled mills, etc.).

Microwave ovens, medical equipment, ultra sonic cleaners, etc. are governed by a separate FCC document.

Components and subassemblies (not to be confused with self-contained peripherals) such as power supplies, floppy disk drives, tape cartridge, etc. that are not considered as stand alone systems are exempt from this ruling at the OEM Level. The burden of compliance falls on the shoulders of the manufacturer that incorporates these devices to form a complete system or peripheral.

The dates for complying have been rescheduled from the original *July 1, 1980* to the following:

Class B: Electronic games and personal computers, including peripherals that attach to personal computers, manufactured after January 1, 1981 shall be certificated by the Commission.

For further information about this you may contact Mr. Art Wall, office of Science and Technology, FCC, Washington, DC 20554, phone (202) 653-8128.

Although the FCC test methods follow those now being used by the VDE and CISPR the requirements differ.

The VDE and DISPR requirements begin at 10 KHz while the FCC requirements begin at 450 KHz. For this reason, it is safe to assume that if a system complies with the applicable VDE Class requirement, it will also comply with the same FCC requirement, but not the other way around.

At present, when a system is evaluated in one of Cornell-Dubilier's EMI test facilities, tests are performed to both VDE and FCC. That is to say we begin our tests at 10 KHz. This provides the necessary data in the event that the manufacturer chooses to market his product in Europe as well as in the U.S.A.

Hexadecimal and Decimal Conversion

From hex: locate each hex digit in its corresponding column position and note the decimal equivalents. Add these to obtain the decimal value.

From decimal: (1) locate the largest decimal value in the table that will fit into the decimal number to be converted, and (2) note its hex equivalent and hex column position. (3) Find the decimal remainder. Repeat the process on this and subsequent remainders.

HEXADECIMAL COLUMNS											
6		**5**		**4**		**3**		**2**		**1**	
HEX	DEC	HEX	DEC	HEX	DEC	HEX	DEC	HEX	DEC	HEX	DEC
0	0	0	0	0	0	0	0	0	0	0	0
1	1,048,576	1	65,536	1	4,096	1	256	1	16	1	1
2	2,097,152	2	131,072	2	8,192	2	512	2	32	2	2
3	3,145,728	3	196,608	3	12,288	3	768	3	48	3	3
4	4,194,304	4	262,144	4	16,384	4	1,024	4	64	4	4
5	5,242,880	5	327,680	5	20,480	5	1,280	5	80	5	5
6	6,291,456	6	393,216	6	24,576	6	1,536	6	96	6	6
7	7,340,032	7	458,752	7	28,672	7	1,792	7	112	7	7
8	8,388,608	8	524,288	8	32,768	8	2,048	8	128	8	8
9	9,437,184	9	589,824	9	36,864	9	2,304	9	144	9	9
A	10,485,760	A	655,360	A	40,960	A	2,560	A	160	A	10
B	11,534,336	B	720,896	B	45,056	B	2,816	B	176	B	11
C	12,582,912	C	786,432	C	49,152	C	3,072	C	192	C	12
D	13,631,488	D	851,968	D	53,248	D	3,328	D	208	D	13
E	14,680,064	E	917,504	E	54,344	E	3,584	E	224	E	14
F	15,728,640	F	983,040	F	61,440	F	3,840	F	240	F	15
0 1 2 3		4 5 6 7		0 1 2 3		4 5 6 7		0 1 2 3		4 5 6 7	
BYTE				BYTE				BYTE			

POWERS OF 2

2^n	n
256	8
512	9
1 024	10
2 048	11
4 096	12
8 192	13
16 384	14
32 768	15
65 536	16
131 072	17
262 144	18
524 288	19
1 048 576	20
2 097 152	21
4 194 304	22
8 388 608	23
16 777 216	24

$2^0 = 16^0$
$2^4 = 16^1$
$2^8 = 16^2$
$2^{12} = 16^3$
$2^{16} = 16^4$
$2^{20} = 16^5$
$2^{24} = 16^6$
$2^{28} = 16^7$
$2^{32} = 16^8$
$2^{36} = 16^9$
$2^{40} = 16^{10}$
$2^{44} = 16^{11}$
$2^{48} = 16^{12}$
$2^{52} = 16^{13}$
$2^{56} = 16^{14}$
$2^{60} = 16^{15}$

POWERS OF 16

16^n	n
1	0
16	1
256	2
4 096	3
65 536	4
1 048 576	5
16 777 216	6
268 435 456	7
4 294 967 296	8
68 719 476 736	9
1 099 511 627 776	10
17 592 186 044 416	11
281 474 976 710 656	12
4 503 599 627 370 496	13
72 057 594 037 927 936	14
1 152 921 504 606 846 976	15

Signed Binary Numbers

Much of the literature on binary arithmetic is written in terms of signed binary numbers, and this is one of the operating number systems of 6800, 6805, and other microprocessors.

Signed binary number notation uses the most significant digit bit 7 as its sign bit; the remaining digits are weighted as straight binary numbers, as the accompanying table illustrates.

Binary weighting	$\pm$	64	32	16	8	4	2	1	
Bit number	b_7	b_6	b_5	b_4	b_3	b_2	b_1	b_0	
	1	1	1	1	1	1	1	1	-127
	1	0	0	0	0	0	0	1	-1
	0	0	0	0	0	0	0	0	0
	0	0	0	0	0	0	0	1	$+1$
	0	1	1	1	1	1	1	1	$+127$

The ALU of the microprocessor *always* performs its addition operations with the number in 2's complement format. Therefore, if numbers are entered into the microprocessor as signed numbers, they *must* be converted to 2's complement representation by an internal program. This is quite easily accomplished if a comparison is made of the number systems; for example

2's Complement	Signed Binary	
1000 0001	1111 1111	− 127
1111 1111	1000 0001	− 1
0000 0000	0000 0000	0
0000 0001	0000 0001	+ 1
0111 1111	0111 1111	+ 127

Positive numbers are the same in both systems, negative number are reversed. That is, − 127 in 2's complement representation is − 1 in signed binary number representation. The accompanying simple program illustrates a conversion program from the signed binary number system to the 2's complement number system.

Assume the number to be converted is already loaded in Acc-A.

There is no such number as − 128_{10} in signed number binary arithmetic; there is in 2's complement representation. Therefore, the numbers that can be converted are in the range from + 127_{10} to − 127_{10}, as is illustrated.

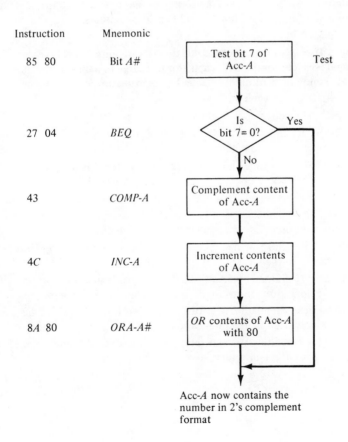

Instruction	Mnemonic
85 80	Bit A#
27 04	BEQ
43	COMP-A
4C	INC-A
8A 80	ORA-A#

Acc-A now contains the number in 2's complement format

Glossary

Accumulator. A register that is the source of one operand and the destination of the result for most arithmetic and logical operations.

ACIA. See UART.

Acquisition Time. For a sample hold, the time required, after the sample command is given, for the hold capacitor to charge to a full-scale voltage change and then remain within a specified error band around final value.

Active High. The active state is the 1 state.

Active Low. The active state is the 0 state.

Address. The identification code that distinguishes one memory location or input/output port from another and that can be used to select a specific one.

Addressing Mode. Specifies how the selected register(s) is (are) to be used when locating the source operand and/or when locating the destination operand.

Analog. Continuous signal or representation of a quantity that can take any value.

Anode. An electrode that functions with a positive charge.

Architecture. Structure of a system. Microprocessor architecture often refers specifically to the CPU.

Arithmetic-Logic Unit (ALU). A device that can perform any of a variety of arithmetic or logical functions under the control of function inputs.

ASCII. American Standard Code for Information Interchange, a 7-bit character code widely used in communications.

Asynchronous. Operating without reference to an overall timing source, that is, operating at irregular intervals.

Asynchronous Shift Register.   A shift register that does not require a clock. Register segments are loaded and shifted only at data entry.

Base (Base Number).   The radix of a number system; 10 is the radix for the decimal system; 2 is the radix for the binary system (base 2).

Baud Rate.   A type of measurement of data flow in which the number of signal elements per second is based on the duration of the shortest element. When each element carries 1 bit, the baud rate is numerically equal to bits per second (bps).

BCD (Binary-Coded Decimal).   A type of positional value code in which each decimal digit is binary coded into 4-bit "words."

Bidirectional.   Generally refers to interface ports or bus lines that can be used to transfer data in either direction, for example, to or from the microprocessor.

Bidirectional Bus Driver.   Circuitry that provides for both electrical isolation and increased current load capability or drive in both signal-flow directions. When arrangement provides for multiple line handling, it becomes a bus driver.

Binary.   Number system with base 2; having two distinct levels.

Bipolar Mode.   For a data converter, when the analog signal range includes both positive and negative values.

Bit.   A binary digit, possible values of 0 or 1.

Bit Rate.   The rate at which binary digits, or pulses representing them, pass a given point in a communication line.

Block Diagram.   A simplified schematic drawing.

Boolean Algebra.   Shorthand notation for expressing logic functions.

Boolean Equation.   Expression of relations between logic functions.

Branch.   To depart from the normal sequence of executing instruction in a microprocessor; synonymous with jump.

Branching.   A method of selecting, on the basis of the previous result, the next operation to execute while a program is in progress.

Breakpoint.   A location specified by the user at which program execution is to end temporarily. Used as an aid in program debugging.

Buffer.   A device designed to be inserted between devices or program elements to match impedances or peripheral equipment speeds, to prevent mixed interactions, to supply additional drive or relay capability, or simply to delay the rate of information flow; classified as inverting or noninverting.

Bug.   A program defect or error; also refers to any circuit fault due to improper design or construction; a mistake or malfunction.

Bus.   One or more conductors used as a path over which information is transmitted.

Bus Driver.   A device that amplifies outputs sufficiently so that they can be recognized by the devices on a bus.

Byte.   The basic grouping of bits that the microprocessor handles as a unit; most often 8 bits in length.

Carry Bit. A status bit that is 1 if the last operation generated a carry from the most significant bit.

Cathode. An electrode that functions with a negative charge.

Central Processing Unit (CPU). The control section of a microprocessor. It contains the arithmetic unit, registers, instruction-decoding mechanism, and timing and control circuitry.

Character. One symbol of a set of elementary symbols, such as a letter of the alphabet or a decimal numeral.

Check Sum. A logical sum of data that is included as a guard against recording or transmission errors.

Chip. A substrate containing a single integrated circuit.

Clear. Set state to 0; an input to a device that sets the state to 0.

Clock. A pulse generator that synchronizes the timing of various logic circuits and memory in the processor.

Clock Rate. The speed (frequency) at which the processor operates, as determined by the rate at which words or bits are transferred through internal logic sequences.

CMOS. Complementary metal-oxide semiconductor, a logic family that uses complementary N-channel and P-channel MOS field-effect transistors to provide high noise immunity and low power consumption.

Common-Mode Rejection Ratio. For an amplifier, the ratio of differential voltage gain to common-mode voltage gain, generally expressed in decibels.

$$CMRR = 20 \log_{10} \frac{A_D}{A_{CM}}$$

where A_D is differential voltage gain and A_{CM} is common-mode voltage gain.

Condition Code (or Flag). A single bit that indicates a condition within the microprocessor; often used to choose between alternate instruction sequences.

Condition Code Register. A register that contains one or more condition codes.

Contact Symbology Diagram. Commonly referred to as a ladder diagram; it expresses the user-programmed logic of the controller in relay-equivalent symbology.

Control Bus. A group of lines originating either at the CPU or the peripheral equipment that are bidirectional in nature and generally used to control transfer or reception of signals to or from the CPU.

Control Register. Stores the current instruction governing the operation of the microprocessor for a cycle; also called instruction register.

Control Unit. The section that directs the sequence of operations, interrupts coded instructions, and sends the proper signals to other circuits to carry out instructions.

Control Word (Data). One or more items of data whose 0 and 1 arrangement determines the mode of operation, direction, or selection of a particular device, port, program flow, and so on.

Conversion Rate. The number of repetitive A/D or D/A conversions per second for a full-scale change to specified resolution and linearity.

Conversion Time. The time required for an A/D converter to complete a single conversion to specified resolution and linearity for a full-scale analog input charge.

Current-Loop Interface (or Teletype Interface). An interface that allows connections between digital logic and a device that uses current-loop signals; that is, typically the presence of 20 mA in the loop is a logic 1 and the absence of that current is a logic 0.

Cursor. Various position indicators frequently employed in a display on a video terminal to indicate a character to be corrected or a position in which data are to be entered.

Daisy Chain. Bus lines that are interconnected with units in such a manner that the signal passes from one unit to the next in serial fashion.

Data. A general term used to denote any or all facts, numbers, letters, symbols, and so on, that can be processed or produced by a computer.

Data Bus. Usually eight bidirectional lines capable of transferring data to and from the CPU, storage, and peripheral devices.

Debouncing. Eliminating unwanted pulse variations caused by mechanically generated pulses when contacts repeatedly make and break in a bouncing manner.

Debug. An instruction, program, or action designed in microprocessor software to search for, correct, and/or eliminate sources of errors in programming routines.

Differential Linearity Error. The maximum deviation of any quantum (LSB change) in the transfer function of a data converter from its ideal size of $FSR/2n$.

Digital. Having discrete levels, quantized into a series of distinct levels.

Direct Memory Access (DMA). An input/output method whereby an external controller directly transfers data between the memory and input/output sections without processor intervention.

Dual-Slope A/D Converter. An indirect method of A/D conversion whereby an analog voltage is converted into a time period by an integrator and reference and then measured by a clock and counter. The method is relatively slow but capable of high accuracy.

Edge Triggered. Circuit action is initiated at the rising and falling edge of the control pulse.

Enabled. A state of the central processing unit that allows the occurrence of certain types of interruptions.

EPROM (or EROM). Erasable PROM, a PROM that can be completely erased by exposure to ultraviolet light.

Fan-in. The number of inputs connected to a gate.

Fan-out. The maximum number of outputs of the same family that can be connected to a gate without causing current overload.

Fetch. The particular portion of a microprocessor cycle during which the location of the next instruction is determined. The instruction is taken from memory, modified if necessary, and then entered into the instruction register.

Firmware. Microprograms, usually implemented in read-only memories.

Flag. A bit (or bits) used to store 1 bit of information; has two stable states and is the software analogy of a flip-flop.

Flip-flop. A digital electronic device with two stable states that can be made to switch from one state to the other.

Floating-Point Arithmetic. Arithmetic used in a microprocessor where the microprocessor keeps track of the decimal point (contrasted with fixed-point arithmetic).

Flow Chart. A programmer's tool for determining a sequence of operations as charted using sets of symbols, directional marks, and other representations to indicate stepped procedures of a microprocessor's operation; a chart containing all the logical steps in a particular microprocessor program; also called flow diagram.

Full Duplex. A mode of data transmission that is the equivalent of two paths, one in each direction simultaneously.

Full-Scale Range (FSR). The difference between maximum and minimum analog values for an A/D converter input or D/A converter output.

Gain Error. The difference in slope between the actual and ideal transfer functions for a data converter or other circuit. It is expressed as a percentage of analog magnitude.

Gate. A circuit having one output and several inputs, the output remaining unenergized until certain input conditions have been met.

Half-Carry (or Auxiliary Carry) Bit. A status bit that is 1 if the last operation produced a carry from bit 3 of an 8-bit word. Used on 8-bit microprocessors to make the correction between binary and decimal (BCD) arithmetic.

Half-Duplex. Permits electrical communications in one direction between stations.

Handshaking. A descriptive term indicating that electrical provision has been made for verification that a proper data transfer has occurred.

Hardware. Physical equipment forming a microprocessor system.

Hexadecimal. Base 16 number system.

High-Level Language. A language in which each instruction or statement corresponds to several machine-code instructions.

High Z. A condition of high-impedance characteristics causing a low current load effect.

Indexed Addressing. An addressing method in which the address included in the instruction is modified by the contents of an index register in order to find the actual address of the data.

Index Register. A register that can be used to modify memory addresses.

Indirect Addressing. An addressing method in which the address of the data, rather than the data themselves, is in the memory location specified by the instruction.

Input/Output (Section). The section of the microprocessor that handles communications with external devices.

Instruction. A group of bits that defines a microprocessor operation and is part of the instruction set.

Instruction Cycle. The process of fetching, decoding, and executing an instruction.

Integral Linearity Error. The maximum deviation of a data converter transfer function from the ideal straight line with offset and gain errors zeroed. It is generally expressed in LSBs or in percentage of FSR.

Interfacing. The process of developing an electrical circuit that enables a device to yield information to and/or acquire it from another device.

Interference. Any undesired electrical signal induced into a conductor by electrostatic or electromagnetic means.

Interlock. To arrange the control of machines or devices so that their operation is interdependent in order to assure their proper coordination.

Interrupt. A microprocessor input that temporarily suspends the normal sequence of operations and transfers control to a special routine.

Interrupt-driven System. A system that depends on interrupts to handle input and output or that idles until it receives an interrupt.

Interrupt Mask (Interrupt Enable). A mechanism that allows the program to specify whether interrupts will be accepted.

Interrupt Service Routine. A program that performs the actions required to respond to an interrupt.

Jumper. A short length of conductor used to make a connection between terminals, around a break in a circuit, or around an instrument.

Jump Instruction. An instruction that places a new value in the program counter, thus departing from the normal one-step incrementing. Jump instructions may be conditional; that is, the new value may only be placed in the program counter if certain conditions are met.

Ladder Diagram. An industry standard for representing control logic relay systems (see Contact Symbology Diagram).

Large-Scale Integration (LSI). The accumulation and design of a large number of circuits (1000 or more) on a single chip of semiconductor.

Latch. A circuit that may be locked into a particular condition and will remain stable until changed; also, to hold a particular condition of output.

Leading Edge. The rising or falling edge of a pulse that appears first in time.

Lease Significant Bit (LSB). The rightmost bit in a data converter code. The analog size of the LSB can be found from the converter resolution:

$$\text{LSB size} = \frac{\text{FSR}}{2^n}$$

where FSR is full-scale range and n is the resolution in bits.

Least Significant Digit. The digit that represents the smallest value.

Limit Switch. An input device, consisting of a switch and an arm assembly; operated by external mechanical action.

Line. In communications, describes cables, telephone lines, and the like over which data are transmitted to and received from the terminal.

Line Driver. An integrated circuit specifically designed to transmit digital information over long lines, that is, extended distances.

Look-up Table. A procedure for obtaining the function value corresponding to an argument from a table of function values.

Loop. A self-contained series of instructions in which the last instruction can modify and repeat itself until a terminal condition is reached.

Machine Cycle. The shortest complete process or action that is repeated in order; the minimum length of time in which the foregoing can be performed.

Machine Language. The basic binary code used by all microprocessors; it may be written in either hexadecimal or octal.

Major Transition. In a data converter, the change from a code of 1000 . . . 000 to 0111 . . . 1111, or vice versa. This transition is the most difficult one to make from a linearity standpoint since the MSB weight must ideally be precisely one LSB larger than the sum of all other bit weights.

Maskable Interrupt. An interrupt that the system can disable.

Masking. A technique for sensing specific binary conditions and ignoring others; typically accomplished by placing 0s in bit positions of no interest and 1s in bit positions to be sensed.

Medium-Scale Integration (MSI). An integrated circuit with a complexity of between 10 and 100 gates.

Memory (MEM). Stores information for future use; accepts and holds binary numbers or images.

Memory-Mapped I/O. The process of connecting memory address lines to I/O decoding systems to enable I/O devices to be handled and treated as memory locations.

Microprocessor. An electronic computer processor section implemented in relatively few IC chips (typically LSI), which contain arithmetic, logic, register, control, and memory functions. The microprocessor is characterized by having instructions that reference micro-operations. Functional equivalence of minicomputer instructions accomplished by programming series of microinstructions.

Microsecond (μs). One millionth of a second: 1×10^{-6} or 0.000001 second.

Mil. Unit of measure equal to $\frac{1}{1000}$ of an inch.

Missing Code. In an A/D converter, the characteristic whereby not all output codes are present in the transfer function of the converter. This is caused by a nonmonotonic D/A converter inside the A/D.

Mnemonics. Symbolic names or abbreviations for instructions, registers, memory locations, and so on, that suggest their actual functions or purposes.

Modem. Modulator/demodulator; a device that adds or removes a carrier frequency, thereby allowing data to be transmitted on a high-frequency channel or received from such a channel.

Monotonicity. For a D/A converter, the characteristic of the transfer function whereby an increasing input code produces a continuously increasing analog output. Nonmonotonicity may occur if the converter differential linearity error exceeds +1 LSB.

Most Significant Bit (MSB). The leftmost bit in a data converter code. It has the largest weight, equal to one-half of full-scale range.

Multiplexing. The time-shared scanning of a number of data lines into a single channel. Only one data line is enabled at any instant.

Multiplying D/A Converter. A type of digital-to-analog converter in which the reference voltage can be varied over a wide range to produce an analog output that is the product of the input code and input reference voltage. Multiplication can be accomplished in one, two, or four algebraic quadrants.

Natural Binary. A number system to the base (radix 2) in which the ones and zeros have weighted value in accordance with their relative position in the binary word. Carries may affect many digits. (Contrasted with Gray Code, which permits only one digit to change state.)

Nesting. Constructing subroutines or interrupt service routines so that one transfers control to another and so on. The nesting level is the number of transfers required to reach a particular routine without returning.

Nibble. A sequence of 4 bits operated on as a unit.

Noise. Extraneous signals; any disturbance that causes interference with the desired signal of operation (see Interference).

Noise Immunity. The ability to reject unwanted noise signals.

Noise Spike. Voltage or current surge produced in the industrial operating environment.

Nonvolatile. A memory type that holds data even if power has been disconnected.

Octal. Number system with base 8; for example, the decimal number 324 would be written in octal notation as 504_8. Only the digits 0 through 7 are used.

Offset Drift. The change with temperature of analog zero for a data converter operating in the bipolar mode. It is generally expressed in ppm/°C of FSR.

Offset Error. The error at analog zero for a data converter operating in the bipolar mode.

On-Line System. A microprocessor system in which information reflecting current activity is introduced as soon as it occurs.

Operation Code (Op-Code). The part of an instruction that specifies the operation to be performed during the next cycle.

Optoisolator. Semiconductor device consisting of an LED and a photodiode or phototransistor in close proximity. Current through the LED causes internal light emission that forces current to flow in the phototransistor. Voltage differences have no effect because the devices are electrically separated.

Overflow Bit. A status bit that is 1 if the last operation produced a 2's complement overflow.

Page. A subdivision of the memory section.

Page Zero. The first page of memory; the most significant address bits (or page number) are 00.

Parallel Operation. Type of information transfer whereby all digits of a word are handled simultaneously.

Parallel Output. Simultaneous availability of two or more bits, channels, or digits.

Parity. A method of verifying the accuracy of recorded data.

Parity Bit. An additional bit added to a memory word to make the sum of the number of 1s in a word always "even parity" or "odd parity" (always even or odd).

Parity Check. A check that tests whether the number of 1s in an array of binary digits is odd or even.

Peripheral Equipment. Units that work in conjunction with a microprocessor but are not part of it, for example, a tape reader, analog-to-digital converter, typewriter, peripheral interface adapter (PIA).

Pointer. Register or memory location that contains an address rather than data.

Polling. Determining the state of peripherals or other devices by examining each one in succession.

Pop (or Pull). Remove an operand from a stack.

Port. The basic addressable unit of the computer input/output section.

Power-On Reset. A circuit that automatically causes a RESET signal when the power is turned on, thus starting the system in a known state.

Priority Interrupt System. An interrupt system in which some interrupts have precedence over others, that is, will be serviced first or can interrupt the other's service routines.

Program. A sequence of instructions properly ordered to perform a particular task.

Program Counter. A register that specifies the address of the next instruction to be fetched from program memory.

Programmable Controller. A solid-state control system that has a user programmable memory for storage of instructions to implement specific functions such as I/O control logic, timing, counting, arithmetic, and data manipulation. A PC consists of central processor, input/output interface, memory, and programming device, which typically uses relay-equivalent symbols. PC is purposely designed as an industrial control system that can perform functions equivalent to a relay panel or a wired solid-state logic control system.

PROM. Programmable read-only memory, generally any type that is not recorded during its fabrication but that requires a physical operation to program it; a semiconductor diode array that is programmed by fusing or burning out diode junctions.

Protocol. A defined means of establishing criteria for receiving and transmitting data through communication channels.

Pulse. A brief voltage or current surge of measurable duration.

Queue (or FIFO).   A set of registers or memory locations that are accessed in a first-in, first-out manner. That is, the first data entered into the queue will be the first data read.

RAM.   Random-access (read/write) memory; a memory that can be both read and altered (written) in normal operation.

Random Access.   All internal storage locations can be accessed in the same amount of time.

Ratiometric A/D Converter.   An analog-to-digital converter that uses a variable reference to measure the ratio of the input voltage to the reference.

Read.   The process of taking in data from an external device or system; to sense information contained in some source and transmit this information to an internal storage.

Register.   A memory device capable of containing one or more microprocessor bits or words.

Relative Accuracy.   The worst-case input to output error of a data converter, as a percentage of full scale, referred to the converter reference. The error consists of offset, gain, and linearity components.

Resolution.   The smallest change that can be distinguished by an A/D converter or produced by a D/A converter. Resolution may be stated in percentage of full scale, but is commonly expressed as the number of bits n, where the converter has $2n$ possible states.

Bits	Steps $2n$	LSB Size (% of Full Scale)	LSB Size (10-V Full Scale, mV)
8	256	0.3906	39.06
10	1,024	0.0976	9.76
12	4,096	0.0244	2.44
14	16,384	0.0061	0.61
16	65,536	0.0015	0.15

ROM.   Read-only memory; programmed by a mask pattern as part of the final manufacturing stage. Information is stored permanently or semipermanently and is read out but not altered in operation.

Root-Mean-Square Current.   The alternating value that corresponds to the direct-current value that will produce the same heating effect. Abbreviated rms.

Rung.   A grouping of PC instructions that controls one output. This is represented as one section of a logic ladder diagram. (See Contact Symbology Diagram.)

Scan Time.   The time necessary to completely execute the entire PC program one time.

Schematic.   A diagram of a circuit in which symbols illustrate circuit components.

Scratch-Pad Memory.   Memory locations or registers that are used to store temporary or intermediate results.

Settling Time.   The time elapsed from the application of a full-scale step input to a circuit to the time when the output has entered and remained within a specified error band around its final value.

Shield. Any barrier to the passage of interference-causing electrostatic or electromagnetic fields. An electrostatic shield is formed by a conductive layer (usually foil) surrounding a cable core. An electromagnetic shield is a ferrous metal cabinet or wireway.

Shielding. The practice of confining the electrical field around a conductor to the primary insulation of the cable by putting a conducting layer over and/or under the cable insulation. (External shielding is a conducting layer on the outside of the cable insulation. Strand or internal shielding is a conducting layer over the wire insulation.)

Significant Digit. A digit that contributes to the precision of a number. The number of significant digits is counted beginning with the digit contributing the most value, called the most significant digit, and ending with the one contributing the least value, called the least significant digit.

Solid-State Devices (Semiconductors). Electronic components that control electron flow through solid materials such as crystals (e.g., transistors, diodes, integrated circuits).

Stack. A sequence of registers or memory locations that are used in a last-in, first-out manner; that is, the last data entered is the first to be removed, and vice versa.

Stack Pointer. A register or memory location that is used to address a stack.

Subroutine. Part of a master program or routine that may be jumped or branched to an independent program in itself but usually of smaller size or importance; also a series of computer instructions to perform a specific task for many other routines.

Successive Approximation A/D Converter. An A/D conversion method that compares in sequence a series of binary weighted values with the analog input to produce an output digital word in just n steps, where n is the resolution in bits. The process is efficient and is analogous to weighing an unknown quantity on a balance scale using a set of binary standard weights.

Surge. A transient variation in the current and/or potential at a point in the circuit.

Synchronous Shift Register. Shift register that uses a clock for timing of a system operation and where only one state change per clock pulse occurs.

Top-down Design. A design method whereby the overall structure is designed first and parts of the structure are subsequently defined in greater detail.

Transducer. A device used to convert physical parameters, such as temperature, pressure, and weight, into electrical signals.

Triac. A solid-state component capable of switching alternating current.

Tristate (or Three State). Logic outputs with three possible states: high, low, and an inactive (high impedance or open circuit) state that can be combined with other similar outputs in a busing structure.

Tristate Enable (or Select). An input that, if not active, forces the outputs of a tristate device into the inactive or open-circuit state.

Truth Table. A matrix that describes a logic function by listing all possible combinations of inputs and by indicating the outputs for each combination.

TTL (Transistor-Transistor Logic). The most widely used bipolar technology for digital integrated circuits. Popular variants include high-speed Schottky TTL and low-power Schottky (or LS) TTL.

TTL-Compatible. Uses voltage levels that are within the range of TTL devices and can be used with TTL devices without level shifting, although buffering may be necessary.

TTY. An abbreviation of teletype.

Twisted Pair. Any two individual insulated conductors that are twisted together.

Two's Complement. A binary number that, when added to the original number in a binary adder, produces a zero result. The 2's complement is the 1's complement plus 1.

Two's Complement Overflow. A situation in which a signed arithmetic operation produces a result that cannot be represented correctly; that is, the magnitude overflows into the sign bit.

UART. Universal asynchronous receiver/transmitter; in a UART, the transmitter converts parallel data bits into serial form for transmission; the receive section does the reverse operation.

Ultraviolet. Electromagnetic radiation at frequencies higher than those of visible light and with wavelengths of about 200 to 4000 angstrom units.

USART. Universal synchronous/asynchronous transmitter/receiver chip.

USRT. Universal synchronous receiver/transmitter.

UV Erasable PROM. An ultraviolet erasable PROM is a programmable read-only memory that can be cleared (set to 0) by exposure to intense ultraviolet light. After being cleared, it may be reprogrammed. (See EPROM.)

Vector. A software routine's entry address; also the address that points to the beginning of a service routine as it applies to interrupting devices.

Vectored Interrupt. Term indicating an automatic branch operation to a predetermined start point when an interrupt occurs.

Volatile Memory. A memory that loses its information if the power is removed from it.

Weighted Value. The numerical value assigned to any single bit as a function of its position in the code word.

Wired-OR. Connecting outputs together without gates to form a busing structure.

Wire Wrap. An alternative to soldering; consists basically of winding a number of turns of wire around a metal post that has at least two sharp edges.

Word. A group of characters occupying one storage location in a computer; treated by the computer circuits as an entity, by the control unit as an instruction, and by the arithmetic unit as a quantity.

Write. The process of sending data to an external device or system; to record information in a register, location, or other storage device or medium.

Zero Page Addressing. In some systems the zero page instructions allow for shorter code and execution times by fetching only the second byte of the instruction and assuming a zero high address byte.

Index